VLSI Circuit Layout:
Theory and Design

Monolithic Microwave Integrated Circuits, *Edited by Robert A. Pucel*
Kalman Filtering: Theory and Application, *Edited by H. W. Sorenson*
Spectrum Management and Engineering, *Edited by F. Matos*
Digital VLSI Systems, *Edited by M. T. Elmasry*
Introduction to Magnetic Recording, *Edited by R. M. White*
Insights Into Personal Computers, *Edited by A. Gupta and H. D. Toong*
Television Technology Today, *Edited by T. S. Rzeszewski*
The Space Station: An Idea Whose Time Has Come, *By T. R. Simpson*
Marketing Technical Ideas and Products Successfully! *Edited by L. K. Moore and D. L. Plung*
The Making of a Profession: A Century of Electrical Engineering in America, *By A. M. McMahon*
Power Transistors: Device Design and Applications, *Edited by B. J. Baliga and D. Y. Chen*
VLSI: Technology and Design, *Edited by O. G. Folberth and W. D. Grobman*
General and Industrial Management, *By H. Fayol; revised by I. Gray*
A Century of Honors, *An IEEE Centennial Directory*
MOS Switched-Capacitor Filters: Analysis and Design, *Edited by G. S. Moschytz*
Distributed Computing: Concepts and Implementations, *Edited by P. L. McEntire, J. G. O'Reilly, and R. E. Larson*
Engineers and Electrons, *By J. D. Ryder and D. G. Fink*
Land-Mobile Communications Engineering, *Edited by D. Bodson, G. F. McClure, and S. R. McConoughey*
Frequency Stability: Fundamentals and Measurement, *Edited by V. F. Kroupa*
Electronic Displays, *Edited by H. I. Refioglu*
Spread-Spectrum Communications, *Edited by C. E. Cook, F. W. Ellersick, L. B. Milstein, and D. L. Schilling*
Color Television, *Edited by T. Rzeszewski*
Advanced Microprocessors, *Edited by A. Gupta and H. D. Toong*
Biological Effects of Electromagnetic Radiation, *Edited by J. M. Osepchuk*
Engineering Contributions to Biophysical Electrocardiography, *Edited by T. C. Pilkington and R. Plonsey*
The World of Large Scale Systems, *Edited by J. D. Palmer and R. Saeks*
Electronic Switching: Digital Central Systems of the World, *Edited by A. E. Joel, Jr.*
A Guide for Writing Better Technical Papers, *Edited by C. Harkins and D. L. Plung*
Low-Noise Microwave Transistors and Amplifiers, *Edited by H. Fukui*
Digital MOS Integrated Circuits, *Edited by M. I. Elmasry*
Geometric Theory of Diffraction, *Edited by R. C. Hansen*
Modern Active Filter Design, *Edited by R. Schaumann, M. A. Soderstrand, and K. B. Laker*
Adjustable Speed AC Drive Systems, *Edited by B. K. Bose*
Optical Fiber Technology, II, *Edited by C. K. Kao*
Protective Relaying for Power Systems, *Edited by S. H. Horowitz*
Analog MOS Integrated Circuits, *Edited by P. R. Gray, D. A. Hodges, and R. W. Broderson*
Interference Analysis of Communication Systems, *Edited by P. Stavroulakis*
Integrated Injection Logic, *Edited by J. E. Smith*
Sensory Aids for the Hearing Impaired, *Edited by H. Levitt, J. M. Pickett, and R. A. Houde*
Data Conversion Integrated Circuits, *Edited by D. J. Dooley*
Semiconductor Injection Lasers, *Edited by J. K. Butler*
Satellite Communications, *Edited by H. L. Van Trees*
Frequency-Response Methods in Control Systems, *Edited by A. G. J. MacFarlane*
Programs for Digital Signal Processing, *Edited by the Digital Signal Processing Committee, IEEE*
Automatic Speech & Speaker Recognition, *Edited by N. R. Dixon and T. B. Martin*
Speech Analysis, *Edited by R. W. Schafer and J. D. Markel*
The Engineer in Transition to Management, *By I. Gray*
Multidimensional Systems: Theory & Applications, *Edited by N. K. Bose*
Analog Integrated Circuits, *Edited by A. B. Grebene*
Integrated-Circuit Operational Amplifiers, *Edited by R. G. Meyer*
Modern Spectrum Analysis, *Edited by D. G. Childers*
Digital Image Processing for Remote Sensing, *Edited by R. Bernstein*
Reflector Antennas, *Edited by W. Love*
Phase-Locked Loops & Their Application, *Edited by W. C. Lindsey and M. K. Simon*
Digital Signal Computers and Processors, *Edited by A. C. Salazar*
Systems Engineering: Methodology and Applications, *Edited by A. P. Sage*
Modern Crystal and Mechanical Filters, *Edited by D. F. Sheahan and R. A. Johnson*
Electrical Noise: Fundamentals and Sources, *Edited by M. S. Gupta*
Computer Methods in Image Analysis, *Edited by J. K. Aggarwal, R. O. Duda, and A. Rosenfeld*

VLSI Circuit Layout:
Theory and Design

Edited by
T. C. Hu

Department of
  Electrical Engineering and Computer Science
University of California, San Diego

Ernest S. Kuh

Department of
  Electrical Engineering and Computer Science
University of California, Berkeley

A volume in the IEEE PRESS Selected Reprint
Series, prepared under the sponsorship of the
IEEE Circuits and Systems Society.

**IEEE
PRESS**

The Institute of Electrical and Electronics Engineers, Inc., New York

Copyright © 1985 by
THE INSTITUTE OF ELECTRICAL AND ELECTRONICS ENGINEERS, INC.
345 East 47th Street, New York, NY 10017-2394
All rights reserved.

PRINTED IN THE UNITED STATES OF AMERICA

IEEE Order Number: PC01875

Library of Congress Cataloging-in-Publication Data
Main entry under title:

VLSI circuit layout.

(IEEE Press selected reprint series)
Includes indexes.
1. Integrated circuits—Very large scale integration—Design and
construction. I. Hu, T. C. (Te Chiang), 1930-   II. Kuh, Ernest
S.   III. Institute of Electrical and Electronics Engineers.
TK7874.V5573   1985   621.395   85-14313

ISBN 0-87942-193-2

Contents

Part I
Overview

Theory and Concepts of Circuit Layout

T. C. HU AND ERNEST S. KUH, FELLOW, IEEE

Abstract—Circuit layout is an important part in the design process of VLSI chips. Circuit layout has induced a new discipline, a new geometry where lines have width and points have dimensions. In this new geometry, configurations are not drawn on one plane but on two or more planes (layers). The multilayer model is a very accurate model for circuit layout. Most problems in circuit layout can be formulated mathematically and, in principle, solved by CAD (computer-aided-design) tools. This is why we choose circuit layout as the topic of this book.

In Section I we start with an informal introduction. In Section II we describe the chip architecture based on the gate-array approach. In Section III, we state and formulate the problems of placement, routing, etc. In Section IV, we review past and current literature in general and give pointers to the 160 references and 25 papers in the book. In Section V we discuss refinements of the mathematical models and open questions.

Papers and references are from various journals: computer science and electrical engineering journals, mathematics journals, as well as operations research journals. Because the contributors are from different fields, there is no standard terminology. Hopefully, this book will help to unify the terminology and bring into focus some of the outstanding unsolved problems.

I. INTRODUCTION

FOR simplicity, we shall divide the VLSI design process into eight parts, namely:

1) System Specification (Architectural Design I)
2) Functional Design (Architectural Design II)
3) Logic Design
4) Circuit Design
5) Circuit Layout
6) Design Verification
7) Test and Debugging
8) Prototype Test and Manufacture.

We shall briefly describe all the parts and then concentrate on part 5), the circuit layout, in Section II.

In the system specifications, we define the goals and constraints of the system: what the system will do, the criteria of optimization, the speed requirements, the space or area requirements, the power requirements, and so forth. If we compare the design of a chip (or printed circuit board) with the creation of a city, then the system specification is the highest-level planning of a city. We decide the future population of the city, the area of the city, the connecting highways to the city, etc. The city has to be partitioned into different areas such as residential areas, industrial areas, business sections, etc. Likewise, the system has to be partitioned into subunits or modules. In the functional design, we decide the functional

Manuscript received August 14, 1984; revised January 11, 1985.

T. C. Hu is with the Department of Electrical Engineering and Computer Science, University of California, San Diego, La Jolla, CA 92093.

E. S. Kuh is with the Department of Electrical Engineering and Computer Science, University of California, Berkeley, CA 94720.

relationships among the subunits. Sometimes we combine system design and functional design and call it architectural design.

In the logic design, we want to realize a set of Boolean expressions or a representation of a finite state machine; we may decide to use RAM (random access memory), ROM (read-only-memory), or PLA (programmable logic array), for example.

The logic networks have to be converted into electronic circuits. In the circuit design, we specify the speed and power requirements of the circuit, as well as its size or shape. For our purpose, the final product of the circuit design is a set of functional blocks (or modules) which are rectangles in shape, or shapes composed of two or three rectangles pasted together.

There are three kinds of chips: custom designed, standard cell, and gate-array.

In the custom designed chips, these modules (or functional blocks) are of arbitrary shape. In the standard cell chips, these modules are rectangles of the same height but of different widths. In the gate-array chips, these modules are rectangles which are multiples of connected squares.

The modules have pins (terminals) fixed on their perimeters. In circuit layout, the modules are first placed on different parts of the chip and then certain subsets of pins must be connected by mutually noninterfering wires according to a given wiring list. The problems of placement of modules and the routing of wires will be discussed in later sections.

In the design verification, we verify and check if all design rules are satisfied. Then we test the chip and debug it if there are errors. Finally, we may build a prototype and test it before mass production.

One thing should be emphasized—although we have broken down the design process into eight parts, the design process is *not* a clear-cut sequential process. Rather, the results of one part influence the subsequential parts and feed back to its earlier parts. The iterations of top-down and bottom-up approaches are very important. This is not only true in the total design of a system but also true in circuit layout itself. Usually, the circuit layout is divided into

1) Chip Planning, Wirability Analysis, Partitioning, and Placement
2) Global and Detailed Routing.

Before we go into the details, let us describe the architecture of a gate-array chip.

II. THE GATE-ARRAY CHIP ARCHITECTURE

The architecture of a chip is very similar to a map of a city, where the streets run horizontally and the avenues run vertically. Between two streets, a block is divided into equal-size lots, one lot for each house. In a VLSI chip, each house corresponds to a "gate" which serves some special function. The streets are called channels which are just spaces reserved for wiring. Each gate has several terminals facing the channels, just as houses have doors facing the street. Say a gate has ten terminals, five terminals like the front doors and five terminals like the back doors. This is shown in Fig. 1, where there are 24 gates, two channels, and one avenue.

In Fig. 1 we draw eight terminals on gate A, four terminals on gate B, and four terminals on gate C. The solid and dotted lines represent wires used to connect the terminals among the gates. We can think of the chip as a wooden board where lines are drawn on both sides of the board. In practice, all horizontal wires are printed on one side, called the H-layer, the vertical wires are printed on the other side, the V-layer, and the terminals (also called pins) are nailed through the wooden board. If we want to connect two pins with the same y-coordinates, we use a horizontal wire. If we want to connect two pins at $(0, 0)$ and $(8, 4)$ positions, we run a horizontal wire from $(0, 0)$ to $(8, 0)$ and drill a hole through the board, and then run a vertical wire from $(8, 0)$ to $(8, 4)$ position. In a chip, the holes are called "vias" which are really metal contacts. We use solid lines to represent horizontal wires, dotted lines to represent vertical wires, and a small square to represent the "vias." In the H-layer (or the V-layer), two parallel lines (wires) cannot be too close to each other since the currents in both wires would interfere with each other and also the lithographic limitation. Thus we must print the lines at some distance apart in order to satisfy design rules [9]. This is like pasting two strips of scotch tape on the wooden board without the two tapes overlapping each other. Therefore, it is convenient to think of a wire as a strip of width μ which runs horizontally and vertically. If we draw grids of size $\mu \times \mu$, a wire is represented by a sequence of connected small squares of size $\mu \times \mu$.

In a very simplified model, we can call the large wooden board the mother board. There are many small boards (called daughter boards), each small board consists of several gates or a single gate. The daughter boards are designed to serve a special purpose and can be thought of as a small rectangle with pins fixed on it. (The daughter boards are also called functional blocks or modules.) The circuit layout problem is to place the daughter boards on the mother board and connect pins between various daughter boards; e.g., a pin on daughter board A to a pin on daughter board B. A subset of pins to be connected by a wire is called a net. Most nets consist of two or three pins, but some nets may have fifty or more pins; such as the power, ground and clock lines. The specification of all nets is called the wiring list or net-list L. So the circuit layout problem is to put small rectangles (with fixed pin positions) on the large rectangle and connect all the nets on the wiring list L.

The daughter boards are like a single-family house (a single gate) or multifamily condominium (several connected gates) and the pins on the daughter boards are the door bells on the doors of the house. The wiring list then specifies which subsets of door bells are to be connected.

To fix the idea, let us consider a chip used at Hughes Aircraft Company [163]. The chip is of size 300 by 300 mil.

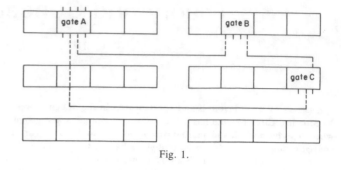
Fig. 1.

(A mil is 0.001 in.). There are 18 rows, each row has 52 positions for gates. These positions are called "slots." There are 16 channels between the rows of gates. In each channel we can have at most 14 parallel horizontal lines. (We say that there are 14 tracks in each channel.) There are no avenues in the chip. The area occupied by the "gates" is called the "active area" and the gates use up the metal on the H-layer. Thus, the horizontal wires cannot run through the active area and must be restricted in the channels. The vertical wires can run anywhere on the chip. Although there are $18 \times 52 = 936$ possible slots for gates, not all of these slots are occupied by "gates." Sometimes, two gates are assigned into one-slot, or half of the slot is unoccupied. In a typical chip, we may have $2000 \sim 5000$ nets. The unoccupied space is used for vertical connections and, sometimes, it is referred to as "feed-through."

The unique feature of the gate-array design approach is that chips with arrays of replicated transistors are prefabricated to a point just prior to two (or more) levels of interconnection. Thus, it is especially attractive to "systems" companies in which various kinds of functional chips are manufactured.

The circuit layout problem is to assign gates into slots and connect pins between the gates. In short, the layout problem consists of two principal phases: a) placement and b) routing.

In the placement phase, the designer selects prefabricated gates of standard size and assigns them into the slots (gate-assignments). The connections between the pins of the gates are specified in advance. Thus, it is desirable to assign the gates such that the pins to be connected are near each other. After all the gates are assigned into slots, we have to connect the pins which now have fixed (x, y)-coordinates.

Let us specify two pins to be connected at

$(3, 6)$ and $(6, 2)$

and another three pins to be connected at

$(5, 1)$, $(7, 8)$ and $(2, 4)$ positions.

A possible connection is shown in Fig. 2. There are two nets in Fig. 2: a two-pin net and a three-pin net. The exact specifications of connecting the nets is called routing.

In the routing phase, we try to connect all the nets on the wiring list. Since a wire has width and the pins have dimensions, it is not obvious that all nets can be connected on the chip. So the first objective is to connect all the nets, and the second objective is to pick among all feasible ways of realizations of the nets, a way which is optimum in a defined

4

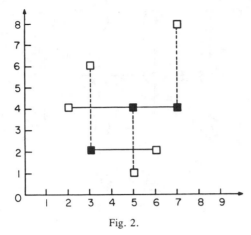

Fig. 2.

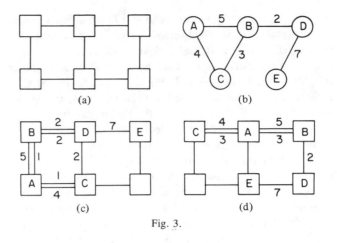

Fig. 3.

criteria such as minimization of total length of all wires used to connect the nets, minimize the longest wire length of connecting two pins, minimize the total number of vias used, etc.

III. MATHEMATICAL FORMULATION

A graph $G = (V, E)$ consists of a set of vertices V and a set of undirected edges E. An edge connecting vertices v_i and v_j is denoted by e_{ij}. Two vertices are called neighbors if there is an edge connecting the two vertices. To allow multiple edges connecting the same pair of vertices, we associate a positive capacity c_{ij} with every edge. The capacity $c_{ij} = k$ means that there are k edges connecting v_i and v_j. Sometimes we also refer to a graph as a network, call the vertices nodes, and call the edges arcs. In network flow theory, the arc capacity c_{ij} is the maximum amount of flow that can pass through the arc from v_i and v_j (see [23], [27], and [28]).

Given two graphs $G_1 = (V_1, E_1)$ and $G_2 = (V_2, E_2)$, we say that the two graphs are isomorphic to each other if we can find one-to-one correspondence between vertices in V_1 and V_2, such that a pair of vertices are neighbors if and only if the corresponding pair of vertices are neighbors in the other graph and the c_{ij} are equal.

A graph $G_j = (V_j, E_j)$ is a subgraph of another graph $G_k = (V_k, E_k)$ if $V_j \subset V_k$ and $E_j \subset E_k$. In graph theory, we often ask: Is a graph G_i isomorphic to a subgraph G_j of G_k? If yes, we say that G_i can be embedded in G_k. If there are many graphs G_i and we want to embed all these G_i in G_k, we say that we want to pack all the graphs G_i in G_k. (Just like we pack odd shaped items into boxes.)

At the end of part 4) of the design process, i.e., circuit design, we have a set of rectangles (modules) and a list of specifications of how these modules should be connected. We want to estimate if all these modules can be put on a single chip and wired according to the specifications. If we decide to put all the modules on one chip, then we may partition the area of the chip into two or more parts and tentatively assign modules into these parts. The partitioning of modules into parts is really rough placement. Then we decide how to place modules into slots. For simplicity, we shall assume that all modules are of the same size and each module consists of a single gate. Now we can represent modules by nodes and indicate the number of wires connecting two modules by the capacity of the edge

connecting the two modules. So we have a network which fully describes the structural relationships between modules to be placed on a chip.

Placing the modules into the slots is called the gate-assignment. The slots are regularly-placed open spaces on the chip, we can represent them as a grid-graph as shown in Fig. 3(a); where we draw six slots. Assume that each module consists of a single gate, and the specified connections between the five gates are shown in Fig. 3(b). The five gates are tentatively assigned into slots as shown in Fig. 3(c). Note that there is no direct arc connecting the slots occupied by B and C. Thus, the three wires connecting B and C have to be split into two groups: one wire goes through A and two wires go through D. Also, if the space between D and E can only allow 6 wires, then one of the seven wires has to be rerouted.

Normally, the exact routing of the wires is decided later in the routing stage. If we have already placed the modules, we can estimate two quantities:

1) The total wire length of all wires connecting all the modules.
2) The number of wires that have to cross each vertical (or horizontal) line which cuts the area of the chip into two parts.

To estimate 1), we can use the square of the Euclidean distances between two modules $[(x_i - x_j)^2 + (y_i - y_j)^2]$ or the Manhattan distances between two modules $[|x_i - x_j| + |y_i - y_j|]$. The distance is multiplied by the number of wires connecting the two modules and summed up between all pairs of modules.

Usually, we want a placement which gives the minimum sum.

To estimate quantity 2), we can draw many vertical lines (or horizontal lines), each cuts the area of a chip into two parts and counts the total number of wires that have to cross that line. If the total number of crossing wires exceeds the total number of tracks available, then we certainly cannot route the wires later. To make the routing easier, we want to distribute the modules in such a way that the maximum number of wires crossing a given line (vertical or horizontal) is a minimum.

One way to handle this problem is to find the minimum number of edges needed to disconnect the network. For the network in Fig. 3(b), this would be the two edges connecting B

5

and D. If the two edges between B and D are deleted, we have three modules A, B, C on one side and D and E on the other side. Then we put A, B, and C on one side of a line, and put D and E on the other side. This line cuts the area of the chip roughly in the ratio of 3 to 2. For the network on one side of the cut, we use the same method. For example, in Fig. 3(d), we first put A, B, C at the top of the horizontal line and D and E in the bottom row. Then we decide that C is on one side of a vertical line, while A and B are on the other side.

To find the minimum number of edges to disconnect a given network, we can use the multiterminal flow algorithm of Gomory and Hu [26] which takes $O(n^4)$ time, where n is the number of modules. When the network is sparse, special algorithms are available to reduce the computation time to $O(n)$ and $O(n^2)$ (see [31], [32]). Assume that we find the minimum cut that will disconnect the network into two parts, say m modules in one part and $n - m$ modules in the other part, there may not exist a line that will cut the area of the chip into exactly that ratio. Kernighan and Lin (see page 76 of this book) study the problem of partitioning the network into two equal parts with the minimum number of edges connecting the two parts.

After the modules are placed, we have to connect the pins among different modules according to the wiring list L. We can represent the total area of the chip as a grid-graph, the vertices of the graph are the potential positions of pins or "vias." In Fig. 4, we show a grid-graph of 16 vertices. There are two nets, the net A consists of three pins and the net B consists of two pins.

In general, there are 10^5 vertices and edges in such a grid-graph and thousands of nets to be connected. Again, we divide the routing into two stages:

1) Global routing (or loose routing)
2) Detailed routing (or embedding).

In the global routing phase, the grid-graph G is partitioned into many subgraphs. For example, the sixteen-node grid-graph shown in Fig. 4 is first partitioned into four sub-grid-graphs, each with four vertices. Then we obtain a simplified grid-graph G' by condensing each sub-grid-graph into a node (called a global cell). The boundary between two global cells is represented by an arc connecting the two nodes in the new grid-graph G'. The capacity of an arc in G' equals the number of wires that can be placed across the boundary of the two adjacent global cells. Thus, the arc capacity in G' is equal to the sum of the corresponding arc capacities in G. In Fig. 5, we show the grid-graph G' where each arc has capacity 2.

The relative positions of the pins within a global cell are ignored. The global routing problem is to embed the nets (trees) in the new grid-graph G'. If we decide to connect the two nets in Fig. 5 using the global routing as shown in Fig. 6, we will have the final routing as shown in Fig. 7.

Note that the grid-graph G' is obtained from the grid-graph G by condensing nodes within a single global cell. This is equivalent to increasing all arc capacities within the global cell to infinite. Thus, the success of global routing is a necessary but not sufficient condition for the success of final routing.

The global routing technique is a hierarchical way of design

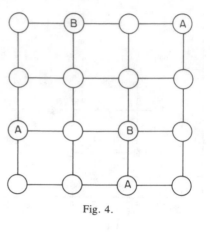

Fig. 4.

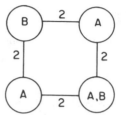

Fig. 5.

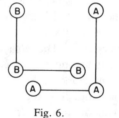

Fig. 6.

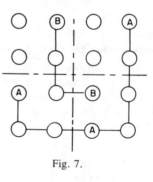

Fig. 7.

and is quite successful in practice. In practice, the number of nets separated by two neighboring cuts do not vary greatly. Thus, the cuts in G' are representatives of the cut in G. The choice of global cells is very important. In the Hughes gate-array chip mentioned in Section II, all the gates occupy the H-layer in regularly placed rows. In doing the global routing, we assign a wire to a channel between two rows of gates, but do not specify which track in the channel. Thus, global routing is sometimes called channel assignment.

Since pins in a net are to be connected by a spanning tree, the problem of global routing is to embed (or pack) many rectilinear Steiner trees in the grid-graph G'.

Neglecting the size of the problem, we can formulate the circuit routing problem as a very large linear program, very much like a multicommodity network flow problem [22], [27].

The problem of routing is very much like a traffic problem. The pins are the origins and destinations of traffic. The wires connecting the pins are the traffic, and the channels are the streets. If there are more wires than the number of tracks in a given channel, some of the wires have to be rerouted just like the rerouting of traffic.

For the real traffic problem, every driver wants to go to his destination in the quickest way, and he may try a different route every day. Finally every driver selects the best route possible for him and the traffic pattern is stabilized. Intuitively, we can do the same for the routing problem. For example, let us associate the cost of routing a wire in an arc. The cost is a function of the number of existing wires in the arc. The cost of routing along a path is the sum of costs of arcs in the path. Every net is to be connected in the cheapest way. We can try various ways of connecting a net until every net is connected in the cheapest way.

Immediately, we can raise two questions for this approach. First, there are so many ways of connecting a net, and there are so many nets, how can we select the best way for connecting each net and make all the connections compatible? Second, how should we assign the cost function defined on an arc? Intuitively, we can let the cost be proportional to the number of wires in the arc, the difference between the number of wires and the number of tracks, or the ratio of wires to the tracks available, etc. Many heuristic cost functions have been suggested. For the first question, the usual approach is to try various configurations based on the heuristic cost function until one satisfactory configuration is found.

Many ingenious heuristic algorithms have been proposed and implemented quite successfully. However, the fundamental question is: Is there an algorithm which can be proved mathematically? (Here, we discount the algorithms, such as backtrack or branch and bound, which can be classified as implicit enumerations.)

The amazing answer is "yes," the algorithm was invented in 1949 by G. B. Dantzig [19] and extended by many others in the early sixties. The way to select various connections for various nets is called the "column-generating" technique in the mathematical programming community. The *correct* cost function of an arc is called the "shadow price." The "shadow price" is obtained as a by-product of the simplex algorithm of linear programming. The reason that all heuristic cost functions are wrong is because they are based on parameters associated with the arc itself. The correct cost depends not only on the arc itself but also on the adjacent arcs, in fact, on the adjacent arcs of adjacent arcs.

We do not claim that the circuit routing problem is solved by the linear programming formulation, since there are many particular features of the circuit routing problem that have to be addressed. However, the concepts of linear programming formulation should be studied very carefully and we shall introduce them very briefly in the following paragraphs. A serious reader should consult books on linear programming.

First, there are many ways to connect a net, each way is a spanning tree connecting the given pins in the net. We associate a variable y_j to each tree which connects a net. The variable y_j is equal to 1 if that particular tree is used and y_j is set to 0 if that particular tree is not used. For example, if there are three ways to connect the first net and five ways to connect the second net, we will set

$$y_1 + y_2 + y_3 \qquad\qquad = 1$$
$$y_4 + y_5 + y_6 + y_7 + y_8 = 1. \qquad (1)$$

Note that there is one equation for each net since only one tree is needed to connect a given net.

In practice, we cannot enumerate all possible ways of connecting all the nets. For the moment, let us assume that there are p nets and a total of n possible ways of connecting all the nets. We shall denote the set of y_j's which correspond to various ways of connecting the kth net by N_k. Each way of connecting a net corresponds to a column of a $(0, 1)$ matrix $[a_{ij}]$.

The matrix $[a_{ij}]$ has m rows, each row corresponds to an arc of the grid-graph G. The (i, j)th entry of the matrix is 1 if the ith arc is used in the jth tree to connect a net and the (i, j)th entry is 0 otherwise. The fact that two trees connecting two different nets must be arc wise disjoint is then expressed as

$$\sum a_{ij} y_j \leq 1. \qquad (2)$$

In general, we can write

$$\sum a_{ij} y_j \leq c_i \qquad (3)$$

where c_i is the capacity of the ith arc.

If we write down all the constraints (1) and (3), we have the typical structure of a linear integer program suitable for decomposition algorithms. Thus the linear program is

$$\max \sum b_j y_j \qquad (j = 1, 2, \cdots, n) \qquad (4)$$

$$\text{subject to} \sum_{y_j \in N_k} y_j = 1 \qquad (k = 1, 2, \cdots, p)$$

$$\sum a_{ij} y_j \leq c_i \qquad (i = 1, 2, \cdots, m)$$

$$0 \leq y_j \leq 1, \qquad \text{integers.}$$

Note that the number of equality constraints in (4) is equal to the number of nets, and the number of inequality constraints of the type $a_{ij} y_j \leq c_i$ is equal to the number of arcs in the grid-graph G'.

The constant b_j is the benefit of connecting a net using the jth tree. For example, we can simply set all b_j to 1 if our objective is to connect all nets; and we can set a large b_j for the nets with many pins situated far apart if we prefer to connect those nets first.

We can view each net j as a company and the b_j is the profit of the company if the net j is connected. The tracks in the arcs are resources available to all companies. Since the resources are limited and all companies want to make a profit, the companies compete and bid for the resources. The so-called "shadow price" is then the price of the resources. The shadow price of an arc depends on the companies currently selected to

7

make the profits (i.e., nets which are currently connected). Intuitively, we want to select a company which will make maximum profit and uses the minimum amount of money (money in terms of shadow prices). The difference between the profit b_j and the sum of shadow prices on arcs is denoted by $\bar{b}_j$. The $\bar{b}_j$ is in a sense the net profit of the company.

Since every arc has a shadow price and each row of the matrix a_{ij} corresponds to an arc, we use π_i to denote the shadow price of the ith arc (or the ith row). The shadow price changes as the basis of the linear program changes.

Let π_i be the shadow price of the ith row under the current basis of the linear program (4), and let $\bar{b}_j$ be

$$\bar{b}_j = b_j - \pi_i a_{ij}. \tag{5}$$

To solve the linear program (4), we will select a column j to enter the basis if $\bar{b}_j$ is positive.

Since the b_j's are the same for all spanning trees of any given net, the best way to connect the net corresponds to the column which maximizes $\bar{b}_j$; and among all possible ways of connecting the net, maximum $\bar{b}_j$ corresponds to minimum $\pi_i a_{ij}$. In other words, we want to find a minimum Steiner tree in a grid-graph where the arc lengths are defined by the shadow prices.

Once a new column enters the basis, and a pivot operation is performed, a new set of shadow prices is used for selecting the best way to connect the next net. This process is iterated until no more nets can be connected. (Note that y_j are bounded, see [20].)

Although there are enormously many columns in the matrix a_{ij} in (4), it is not necessary to write down all the coefficients of these columns. We can generate the best column for each net by finding a minimum Steiner tree connecting the pins in the net. Unfortunately, the number of rows in the matrix a_{ij} is also too large.

To handle the size of the problem, Hu and Shing (see page 144 of this book) developed a cut-and-paste technique.

IV. CURRENT LITERATURE IN CIRCUIT LAYOUT

There are several books on VLSI (see the reference list). All the books treat all aspects of VLSI, not just layout. The first important book on VLSI is Mead and Conway's book [9]. Their book covers devices, fabrication, and many other aspects of VLSI which are needed for the total understanding of the subject. Muroga's book [10] concentrates more on the design aspect of VLSI and includes a chapter on computer-aided design. Ullman's book [16] emphasizes theory as well as computational algorithms and includes substantial material not covered by this tutorial. The book also contains a chapter on layout algorithms, including the layout of trees and planar graphs. The book by Leighton [7] on complexity issues in VLSI and the book by Leiserson [8] on area-efficient VLSI computation should be of special interest to mathematicians. Ayres' book [1] emphasizes the software tools needed for CAD, it covers other aspects of the design process. It is especially useful for architectural design and high-level decisions.

Among the general surveys, we highly recommend Soukup (see page 21 of this book) which deals with circuit layout and

has the same nature as this tutorial. It reviews many papers before 1981. The [2] paper gives a bird's eye view of VLSI. The [4] paper surveys the methods used in industry, while the [11] paper surveys the methods used at the University of California at Berkeley.

Kuh's [5] special issue on routing deals exclusively with routing. The papers and references included there give a very good sample of what is used in practice today. The books and the general papers certainly cover most of the published references before 1983.

Before we review the current methods in circuit layout, we need to understand the practical aspects. First of all, the goals of a designer are a) to build something that works; b) to meet the deadline; c) to have a good design.

The goals of a) and b) are much more important than c). Thus, all CAD algorithms have to be fast and easily implementable. Due to the real-time constraint, an $O(n^5)$ algorithm is usually of no value to the industry, and the comparison of two algorithms $O(n^2 \log n)$ and $O(n^3)$ is based on $n = 1000$ or $10\,000$ not on $n = \infty$.

We do not use the word "optimum design" in c) because there is no well-defined optimum design. A good design is usually a well-balanced design among several criteria: e.g., reliability, cost, speed, size, power consumption, and so on. The list of criteria is very long and it is impossible to define an objective function that we should minimize or maximize. Even if we should concentrate on one criteria, the problem is usually NP-complete. (In short, a problem is NP-complete means that the problem is very hard and probably no polynomial algorithm exists. See the book by Garey and Johnson [24] as well as the Journals of Algorithms which lists updated information on NP-complete problems [34].)

Thus, all algorithms currently in use are heuristic algorithms, these algorithms reflect the experiences and the judgments of the engineer. There is no error bounds on the algorithm.

To get an error bound, we have to first define an optimum design and build some kind of a mathematical model. We realize that reality is too complicated to be represented by a simple mathematical model. However, we can gain the understanding of reality through the mathematical model.

In this section, we divide our discussions on circuit layout into four subsections, namely: a) wirability analysis, partitioning, and placement; b) routing; c) layout approaches; and d) module generation.

A. Wirability Analysis, Partitioning, and Placement

In the gate-array design, we first estimate if all the modules can be put on a single chip and wired according to the wiring list. This problem is called wirability. The Heller *et al.* paper (see page 62 of this book) discusses this problem. There are certain relationships between the number of gates, average number of terminals per gate, the wiring length, etc. One such relationship is called the Rent's rule, originated in the design of computer logic and used extensively in PCB (printed circuit board) design. Basically we first estimate the interconnection requirements between four parts of the chip; say, upper left, upper right, lower left, and lower right. Then each part is partitioned recursively into four parts again. This is used to establish an upper bound on the expected average placement distance needed (see the paper by Donath on page 57).

More recently, Karp and others [61] obtained the worst case channel width needed to route an $n \times n$ array.

We have assumed that the chip has two layers. In reality, we can have more than two layers. A very surprising result is that three layers are enough for any number of two-pin nets. This result is due to Preparata and Lipski [66]. Very few papers have been published on wirability. Most results obtained are based on the worst-case analysis. It is hard to define a typical randomized wiring list L. An approach based on the stochastic analysis of wire distribution is published by El Gamel [50].

The problems of partitioning and placement are interrelated. In the graph model discussed in Section III, we consider each module as a node. Two nodes are connected by an arc if there is a net connecting the two corresponding modules. For k nets connecting the same pair of modules, we associate a capacity $c_{ij} = k$ for the arc connecting nodes v_i and v_j. We also associate a weight w_i with every node v_i. The weight may indicate the area of the module for instance. Thus, the mathematical problem is to partition the nodes into two sets of equal weights such that the total capacity of arcs connecting the two sets is a minimum. For generality, we want to partition the nodes into several sets, with the total weight in each set being bounded and the arc capacity connecting the sets being a minimum. In the paper by Kernighan and Lin on page 76, they first partition the nodes into two sets of equal weights and then interchange the nodes between the two sets to improve the objective function. We shall call this kind of technique local improvement or local relaxation. The simplest kind of local improvement is to pairwise exchange two nodes. A slightly general procedure is to exchange λ nodes at a time until no improvement can be made. Kernighan and Lin also discussed other approaches such as network flow and clustering.

Along the same line of local improvement, Goto (see page 123 of this book) proposed a three-way interchange of modules instead of two-way interchanges and, more generally, any "musical chair" interchange of modules which improves the objective function. Two papers related to this partitioning problem should be mentioned. Karmarker and Karp [35] have proposed a way of partitioning a set of numbers w_i into two subsets and minimizing the difference between the total weights of the two sets. This is a well-known NP-complete problem and their heuristic method is very efficient and with known error bounds. Although they do not consider capacity c_{ij}, their method would be an important tool in solving the partitioning problem. Hu and Ruskey [29] introduced the concept of circular cut in a network. Roughly speaking, this is a modification of minimum cut which considers only the arc capacity c_{ij}. A circular cut is a minimum cut with the total weight of nodes on one side of the cut to be less than a required amount. A combination and improvements of the above methods may solve the partitioning problem to a certain extent. More research is definitely needed, especially in the area of a probabilistic model of wirability.

Let us reexamine the graph model just mentioned. If we have a net connecting five modules, then we would have a complete graph of five nodes (a clique of size five).

A partitioning of the five nodes into two nodes and three nodes requires a cut of capacity six. However, if we use a net connecting the five modules in a row, we need a cut of capacity one. To take care of this situation, Schweikert and Kernighan (see page 81 of this book) proposed a model called the net-cut model where the number of wires needed to connect two sets of modules A and B is equal to the number of nets divided or split into A and B. Following Schweikert and Kernighan, we shall call the first-model, the edge cut model. Several computations and comparisons are made between these two models.

To transform the net-cut model into a graph model, Hu and Moerder (see page 87) propose the following.

Every module is represented by a node. A two-pin net is represented by the arc connecting the two modules. For a net connecting k modules ($k \geq 3$), another kind of node, called star-node (or simply a "star"), is introduced. A star has k arcs, one arc connecting to each module. For example, there are three nets: Net 1 connects modules 1, 2, 3, and 4; Net 2 connects modules 1 and 5; and Net 3 connects modules 4 and 5. The graph representing the edge-cut model is shown in Fig. 8(a) and the graph representing the net-cut model is shown in Fig. 8(b).

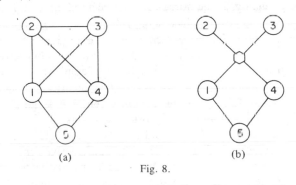

Fig. 8.

The star-node has capacity one. Thus any partition of the four-pin net needs only a cut of capacity one. Now a minimum cut will consist of arcs and star-nodes. The maximum flow algorithm has to be modified (see [23], [27], and [28]).

After partitioning the modules into parts of the chip, we then assign the modules to slots in each part of the chip. This is called the module-placement problem, a gate-assignment in gate-array layout. As we mentioned in Section IV, the objective in the module placement is to have a min-cut placement (M. A. Breuer, see page 105), which is used in partitioning or minimizes a distance function. The minimization of the squares of the total Euclidean distance between the nodes in the slots is called the quadratic assignment problem. Gilmore [52] and Hall (see page 94) both discuss this problem. Another interesting approach using eigen-values of the adjacency matrix of the graph is proposed by Barnes [38]. Since the placement problem can be formulated as an integer programming problem, Widgaja [74] used the branch-and-bound technique to solve the placement problem.

Theoretically, all combinatorial problems can be formulated as integer programming problems, except that the number of rows and columns in the matrix is prohibitive for computations. The column-generating techniques ([21], [22], [26]) have not been popular among the electrical engineers. Hopefully, the progress of solving large zero-one linear programming problems [18] will draw the attention of people in CAD to use the decomposition approach in linear programming.

A recent approach in combinatorial optimization is by using simulated annealing (see Kirkpatrick *et al.* on page 45). The basic idea is derived from a natural analogy with the statistical physics of random systems. The method has been used successfully in solving the gate-array placement problem. Let us assume that we start with a random placement and we choose an objective function, such as the sum of wire lengths. The function has a global minimum with respect to module placement, which unfortunately is impossible to determine in any large problem. The difficulty with the usual iterative improvement techniques based on pairwise exchanges lies in the fact that "accepted" exchanges lead to local minimum which can be far from optimum. The method of simulated annealing, on the other hand, uses a probabilistic measure of "acceptance," thus has the ability of moving uphill in the process of reaching a desirable solution. The price to pay is the large amount of computation usually required.

To conclude our discussion on partitioning and placement, we shall mention briefly other placement methods based on physical analogs. The force-directed method of Quinn and Breuer [67] solves the placement problem using force balancing on a "point" mass which is used to model a module. Hook's law of attraction is used to represent connectivity between modules, and repulsion force is employed between unconnected modules in order to keep them apart. The method usually leads to a good initial placement but requires the solution of large nonlinear algebraic equations.

A recent paper by Cheng and Kuh (see page 115) uses the electric network analogy to solve the constructive placement problem with boundary constraints. A linear resistive network with n nodes is used to model a placement problem with n modules. The connectivity between two modules is represented by the conductance of the resistor between two nodes. There are altogether $n_1 + n_2 = n$ nodes where n_1 represents fixed modules (I–O pads) and n_2 the movable modules whose coordinates are to be determined. The optimization problem amounts to the determination of the n_2 voltages to minimize the total power dissipation. By using only first-order constraints, the problem is reduced to that of finding solutions to a linear resistive network. Scaling, partitioning, and simple relaxation are used to move modules onto slots. The algorithm is fast and has been tested successfully in gate-array placement at the Hughes Aircraft Company. Extensions of the method to modules of irregular shapes and sizes have also been made [43].

B. Routing

We further divide routing into the following:

- Shortest path algorithms and minimum Steiner trees
- Global routing
- Two-layer routing: channel routing and switch-box routing

- Single-layer routing: river routing and single-row routing

1) Shortest Path Algorithms and Minimal Steiner Trees: The shortest path problem can be stated as follows: Given an arbitrary graph G, each arc connecting v_i and v_j has an associated length d_{ij}, find the shortest path between two vertices in the graph. Here, the length of the path is defined as the sum of the lengths of arcs in the path. This problem was solved by Dijkstra [28] in 1959. To find the shortest path between all pairs of vertices, we can use Warshall–Floyd's algorithm [28]. A straightforward implementation of Dijkstra's algorithm needs $O(n^2)$ time, while a straightforward implementation of Warshall–Floyd's algorithm takes $O(n^3)$ time (see, for example, [28]). Since our graph is a grid-graph, $d_{ij} \equiv 1$ for all i, j special algorithms should be used.

Moore [113] suggested that breadth-first-search be used to find the shortest path. Lee [105] also used the breadth-first-search and actually wrote a program. This is probably one of the most publicized CAD tools, commonly referred to as Lee's router or Maze-runner. Several modifications of the algorithm exist (see [83], [88]). This algorithm is guaranteed to find a path with minimum wire-length. Unfortunately, the algorithm is time consuming and picks a shorter path with many vias instead of a longer path with few vias.

In order to minimize the number of vias in the path and to reduce the memory storage and increase speed, another type of algorithm called the line-expansion algorithm was developed (see Hightower [89], [90]; Heyns [141]).

The line-expansion type algorithm is designed to be a fast algorithm to find a path with very few vias. Unfortunately, it does not guarantee to find a path even if such a path exists and its running time is hard to analyze. Recently, Heyns *et al.* [141] suggested an algorithm to generate all possible escape lines from a given line until a solution is found. However, it is not clear that the solution obtained is optimum in terms of some predefined criteria such as shortest wire-length, minimum vias, or a combination of both.

To combine the advantages of both the Maze-runner and the line-expansion algorithm, Hu and Shing (see page 139) proposed an algorithm called the α-β method. It is really a combination of both methods.

The α-β method has the following advantages:

- Like the Lee–Moore method, it is guaranteed to find a path if one exists.
- This method can be used to find the minimum-via path, the shortest wire-length path, or a path which is optimum under a combination of the two criteria.
- The data structure is easy to implement and update.

Here, we define an optimum path as a path of minimum cost, where the cost of the unit wire-length and the cost of a via in the path can be defined arbitrarily by the user.

The basic idea of α-β routing is to associate a cost to every arc of the grid-graph. The cost of an arc is "α" if the arc is of the same direction as the previous arc in a path. The cost of an arc is "β" if the arc is perpendicular to the previous arc in a path. Thus, we can define $\alpha \ll \beta$ if we want a minimum-via path and define $\alpha = \beta$ if we want a shortest wire-length path.

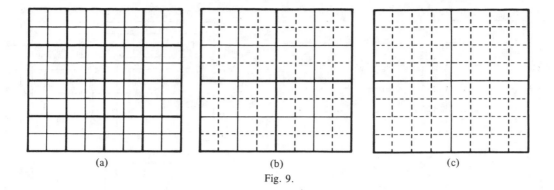

(a)   (b)   (c)

Fig. 9.

If $\beta = 10\alpha$, then a via is equivalent to ten arcs in a straight line.

When a net consists of three or more pins, we need a spanning tree to connect all the pins. When the graph is a grid-graph, we need a rectilinear Steiner tree (see Hanan on page 133). Although the problem of finding a minimum rectilinear Steiner tree is NP-complete, most nets have five pins or fewer. Thus, it is possible to find the minimum rectilinear Steiner tree (see Hwang [94], [95], [96]).

2) Gobal Routing: In global routing the usual approach was to route one net at a time sequentially until all nets are connected. To connect one net, we could use the Maze-runner for example. If some nets cannot be routed at the end, these nets must be routed manually. Obviously, the success of this kind of routing depends on the order we route the nets, and there is no systematic way of rerouting the nets.

Recently, Ting and Tien (see page 155) used the following approach:

- First route every net as if it were the first net to be routed; i.e., pay no attention to conditions at boundary overflows.
- After all nets have been routed, identify the boundaries that are overflowed and the amount of overflow.
- Identify the nets that use these overflow boundaries and form a bipartite graph with one part of the vertices representing the nets and the other part of the vertices representing the overflowed boundaries. An edge connects a net to a boundary if the net uses the boundary. The bipartite graph shows the supply-demand situation among boundaries and nets. And a subset of nets is selected for rerouting. The criteria of selection is "greedy."

Another recent approach (Marek–Sadowska [111]) uses the bottom-up hierarchy and starts the routing with the bottom-level 2×2 cells. As shown in Fig. 9, at each successive level of hierarchy, only 2×2 global cells are routed. The method works extremely fast. Nets not completed are routed with a simple heuristic rerouter at the end. From examples tested, the combination works quite well. Obviously, for difficult problems, a more elaborate rerouter needs to be developed.

An interesting probabilistic approach was used by Vecchi and Kirkpatrick [128]. Basically, they first route the nets randomly, then they change the routing pattern if the new pattern should decrease the objective function

$$F = \sum_{v=1} m_v^{\,2}$$

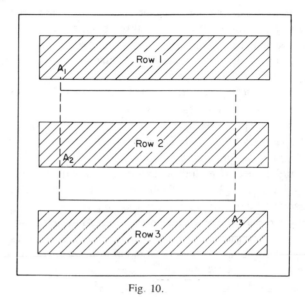

Fig. 10.

where m_v is the number of wires in the vth arc. Since this objective function is convex, a lower value of the objective function indicates that the wires are more uniformly distributed over the whole area. A novel approach used was to change the pattern even if the objective function increases. This happens with a small probability, say 1/100, and changes the pattern if the objective function should decrease with probability, say 99/100.

3) Two-Layer Routing: Consider a chip which has three rows as shown in Fig. 10. Suppose that we want to connect a pin in row 1 to another pin in row 3. We could first use a vertical wire and then use a horizontal wire in channel 2, or first use a horizontal wire in channel 1 and then a vertical wire. If we decide to use a vertical wire first, then the connection of A_2 to A_3 is within channel 2. The problem of connecting pins all within one channel is called channel routing. In routing, we first decide the places in a row the wire is to go through (called vertical assignment), then we route the connections within a channel (channel routing).

Mathematically, the channel routing problem can be stated as follows. Given two parallel horizontal lines at distance $m + 1$ units apart (i.e., there are m tracks between the two horizontal lines), there are numbers $0, 1, 2, \cdots, n$ written on the two horizontal lines. The problem is to connect all nets using two layers and to minimize the required number of tracks m. One such connecting pattern is shown in Fig. 11,

11

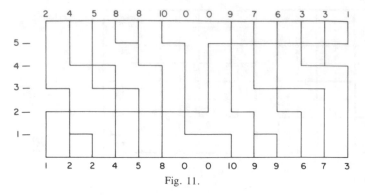

Fig. 11.

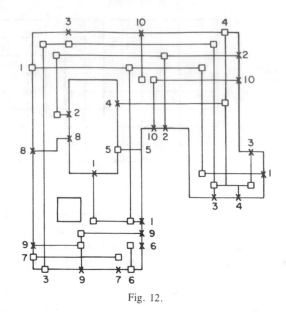

Fig. 12.

where net 1 has two pins—one at the lower left and one at the upper right. Similarly net 2 has three pins, etc.

For channel routing, the reader is referred to [82] and the papers by Yoshimura and Kuh and Burnstein and Pelavin on pages 180 and 191, respectively.

In the paper by Burnstein and Pelavin on page 191, they first partition the channel area into two rows and global cells. Then they use integer programming to route the nets and successfully partition the global cells into smaller cells.

Channel routing has been widely used in automatic layout design because of its high packing density. Given the wiring list of a channel routing problem, we define the maximum density as the maximum intersections of a vertical column with the horizontal net segments over all columns, assuming that each net occupies no more than one horizontal track. As mentioned, several efficient channel routing algorithms are available. By judicious use of doglegs, it is possible to complete the routing of any channel with number of tracks equal to or approaching the maximum density. Thus the maximum density can be used to estimate the required channel width, which is crucial in chip planning.

The more general problem in two-layer detailed routing is to route a *closed*, two-dimensional rectangular region in contrast to an open-ended channel. The pins are on the boundary and can be either fixed in location or floating. This type of two-layer routing problem is called switch-box routing and, at present, there exist some heuristic switch-box routers. Unlike the channel routing problem, there is no prior knowledge of whether a given wiring list can be routed in a given space. If, in addition, the region is rectilinear, and obstacles and internal boundaries exist as shown in Fig. 12, we then have a general two-dimensional, two-layer problem (see the paper by Hsu on page 203).

4) One-Layer Routing: Often in custom chip layout design one-layer routing suffices. Given an open-ended channel with pins on the two boundaries, if routing is to be completed in one layer, the wiring list must satisfy certain topological conditions. If the wiring list can indeed be routed, then, again, we wish to minimize the number of horizontal tracks required. Such a problem is usually referred to as river routing. Hsu (see page 203) presented a simple stack operation to test the routability and carry out the river routing if it is routable. Furthermore, the algorithm can be generalized to the routing of a closed region on one layer. Leiserson and Pinter [108] considered the optimum placement of modules for river routing.

A different type of one-layer routing problem, called single-row routing, has been studied extensively (see Kuh *et al.* (on page 209) and refs. [121] and [124]). The problem originated in multilayer PCB design. Given the positions of pins, which are all on one line (called the reference line), and the wiring list L, connect all the nets and minimize the number of tracks needed on either side of the reference line.

In Fig. 13(a), we have 8 pins at unit distance apart on the reference line (shown as a dashed line). There are four nets which connect the following pins

net 1	connect	v_1 and v_5	
net 2	connect	v_2 and v_6	
net 3	connect	v_3 and v_7	
net 4	connect	v_4 and v_8	

A possible way of connecting the four nets is shown in Fig. 13(a), which requires three tracks on both sides of the reference line. Another arrangement (Fig. 13(b)) connecting the same set of pins requires only two tracks on both sides. The problem is to find the arrangement that minimizes the maximum number of tracks needed in the upper part or lower part of the reference line.

Note that nets 2 and 3 have to cross the reference line once in Fig. 13(a), while nets 1 and 4 have to cross the reference line once in Fig. 13(b). For any wiring list of nets, including k-pin nets, we can always find a feasible arrangement as follows: Put down all the nets from top to bottom in any order. Connect all the pins by the reference line from left to right as shown in Fig. 14(a). Then we can straighten out the reference line and let the nets bend to connect among themselves, as shown in Fig. 14(b). The places of intersections of the reference line and the nets remain the same.

Necessary and sufficient condition for single-row routing are given in the paper by Kuh *et al.* on page 209. However, more work is needed to explore the problem to its fullest extent.

In the future, as VLSI chips become more tightly packed, multilayer routing will be necessary. The problem of intercon-

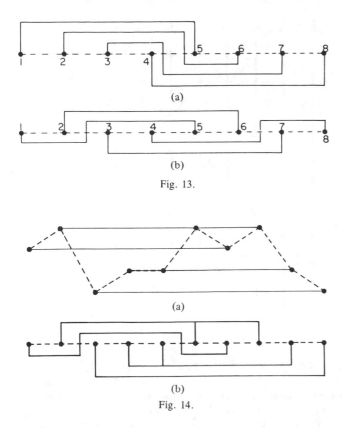

Fig. 13.

Fig. 14.

nection will resemble more that in PCB design. It is conceivable that single row routing will become a principal tool.

C. Layout Approaches

The gate-array chip architecture described in Sections II and III represents a popular semicustom design which is useful for "systems" companies in need of quick production of various kinds of functional chips. A similar layout structure which has rows of standard cells interlaced with routing channels is used extensively for custom chip design, called standard-cell or polycell layout. Standard library cells of equal heights and variable widths are used, typically: 2 or 3-input NOR gates, NAND gates, elementary latches and flip flops, input–output buffers, etc. These are usually manually designed to satisfy design rules for a given technology. The logic and functional specifications prescribe the cells needed as well as the wiring list. The problem is first to place these cells in a row structure to facilitate interconnection and to minimize the total chip area in routing. Similar to the gate-array layout, we also decompose routing into global routing and detailed routing. Different from the gate-array layout, 100 percent routing is guaranteed because channel widths are not fixed. The LTX System of Bell Labs (see the paper by Persky et al. on page 219) is one of the first highly successful systems used in industry based on the standard cell approach.

The standard cell layout approach requires, in addition to well-designed library cells, good algorithms for placement and routing. While there exist excellent two-dimensional placement and channel routing software, more work is needed in two specific areas. One key problem is the linear arrangement of cells in each cell row to minimize the channel density.

Another crucial problem is the feedthrough assignment and the use of equivalent pins in global routing. In this connection, a heuristic approach based on the spanning-tree and Steiner-tree formulation has been proposed [131]. A more general approach to custom chip layout, called the building-block layout, allows modules of irregular shapes and sizes. The modules used include those mentioned in the standard cell library. In addition, functional blocks such as RAM, ROM, PLA, ALU and random logic units are used. The layout problem becomes much more difficult, and, at the present, most custom chips are designed manually. Although there exists a number of automatic layout systems reported in the literature [136], [142], [149], [160], [161], and [165], most are not in production use mainly because the resulting chip area exceeds that of manual design. Thus it is in the building-block layout that major effort is needed.

The problem is again divided into placement and routing. In constructive placement, it is necessary to consider, in addition to module connectivity, how various shapes and sizes of modules fit together. The problem is thus a combination of the familiar bin packing problem and the two-dimensional module placement. Furthermore, a good estimate of routing space is also crucial during the placement phase. Lauther [149] uses the min-cut partitioning algorithm to dissect a chip successively into modules with specified area. The module shape is not considered in the process. In global routing, the aim is to assign nets to channels or other routing regions defined by the placement. The common approach is to use a weighted global routing graph and assign nets based on a Steiner algorithm. The graph is updated each time a net is assigned so that information on channel congestion is taken into account. This is called the incremental assignment method and was first proposed in the NEC "Robin" system [145]. The Berkeley building-block layout (BBL) system (see Chen et al. on page 239) divides the routing regions for detailed routing into two kinds, namely, channels and switchboxes. Space is allocated to regions at the conclusion of globe routing. For channels, it is easy to simply use the maximum density for space allocation. However, there is no way to predict the space required for the switchbox regions. The BBL system does guarantee 100 percent routing because it has the feature to move blocks in order to create additional space. Another special feature is that it allows modules to be rectilinear rather than simply rectangular. This could be important in hierarchical layout design.

In hierarchical layout, some blocks may have fixed area but variable aspect ratio. For example, a particular logic block can be created by using standard cell layout. Thus the number of rows and the length of each row need not be specified at the beginning. This flexibility should offer further challenge in the constructive placement stage.

Finally, we mention briefly the problem of chip planning. A specially interesting approach has been proposed by Heller and others [140]. Given the system architecture in the form of a flow diagram, we first create a planar graph model. By finding the dual of the planar graph, we obtain the chip layout schematic, called floor plan. One main advantage of this approach is that the layout is derived directly from the system architecture. Thus modules which are supposed to be tightly connected are neighbors on the chip. Connections are easily done and often can be made by using butting contacts. This, in essence, amounts to the design philosophy proposed by Mead and Conway [9].

D. Module Generation

As mentioned, standard cells are usually designed manually for a given technology and to satisfy particular design rules. Modules of greater complexity need to be optimally designed to minimize the area usage and to satisfy other specifications, such as the aspect ratio and I–O pin locations. We shall briefly mention two topics here which have played an important role in layout design, namely: *symbolic layout and compaction*, and *gate-matrix layout*.

1) Symbolic Layout and Compaction: The detailed task of IC mask design can be simplified by abstraction. By concentrating on the topology of the layout and without paying attention to the geometric rules at the beginning, layout design engineers can often obtain a symbolic layout directly from the circuit schematic. Symbols are placed on the grid to construct the desired circuit similar to putting tiles on a floor. The term stick layout is used to imply that the symbolic layout design is grid free. Only relative positions of symbols are of consequence. After the layout topology is decided, design rules are introduced according to the given technology; the exact geometrical requirements are then imposed. The term "compaction" is used to signify an optimization process to put devices and wires as close as possible to each other without violating design rules. Thus, cell area is minimized. The program Cabbage described by Hsueh and Pederson (see page 245) is one of the first used extensively in industry for stick layout of modules. It contains a compaction algorithm which is based on minimizing the longest path on a graph. Compaction is done in the horizontal and the vertical directions alternately. "Jogs" or "doglegs" are introduced in the process.

2) Gate-Matrix Layout: In the late 60's Weinberger [164] introduced a novel layout method for MOS complex logic. The method combines the placement and routing into one process. A mathematical formulation of the approach was published in the late 70's (see the paper by Ohtsuki *et al.* on page 249). A generalization to the gate matrix method of layout design was made at the Bell Laboratories and used in the design of the 32-bit CPU Bellmac CMOS chip (see the paper by Lopez and Law on page 259).

The problem can be formulated in terms of linear placement of gates to realize a given logic function. In Fig. 15(a), the vertical lines represent the schematic of NAND gates and the horizontal lines are nets which are connected to the gates as marked. Fig. 15(b) gives another realization of the same logic array with gates t_3 and t_5 interchanged in location. Thus in Fig. 15(a) five tracks are required, while in Fig. 15(b) only three tracks are needed. The second realization happens to be an optimum one in that the required number of tracks is a minimum. Thus the problem is a standard linear placement problem with the objective function equal to the maximum density, i.e., the maximum intersections of a vertical gate with horizontal nets.

The problem can be translated into a graph theoretical

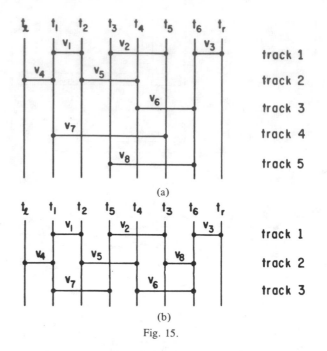

Fig. 15.

problem by associating nets as vertices of a graph and defining an edge (v_i, v_j) if and only if net i and net j intersect by the same vertical gate. The resulting graph is an interval graph, and the clique number is equal to the number of the horizontal tracks. Thus the given gate assignment problem must be first converted to a graph representation according to the gate-net incidence relation. Such a graph is called a connection graph, and it contains only the information on how gates are connected to the nets. By augmenting edges to the connection graph, we obtain an interval graph. Our aim is to add the edges in such a way that the resulting interval graph has the minimum clique number. This problem has been proven to be NP complete [146], and heuristic algorithms are available to obtain good results for gate matrix layout.

V. CONCLUSION AND OPEN PROBLEMS

Since a VLSI system is such a complex system, the designer is forced to use a hierarchical approach or the top-down approach. Thus, in circuit layout we do placement first, then global and detailed routing. Although we do the placement with the routing in mind, we do not have a systematic way of adjusting the placement if the routing cannot be accomplished. The reason is that we do not understand completely even the simplest special case. Also, we do not have enough mathematical tools to solve these problems.

Most algorithms currently in use are heuristic algorithms. The algorithms are efficient in practice but no error bounds are known. The ideas behind these heuristic algorithms can be classified into the following types:

- Greedy algorithms
- Initialization followed by successive local improvements such as pairwise exchange and local relaxation
- Hierarchy approach or decomposition approach
- Models built on physical analogy; e.g., the required connections are rubber bands, the modules are objects in viscous fluids, forces and electric networks, etc.

- Probabilistic approach

From the overall point of view, we have mentioned that more work is needed in chip planning and wirability analysis.

In our discussion we have assumed that the gate-array chip has two layers. If one layer is metal and the other is polysilicon, these two layers should not be treated as the same since the conductivities of metals and polysilicon are different. The current technology has three or more layers: metal, polysilicon, and diffusion. In industry, blue color is used to represent metal, red color to represent polysilicon, and green for diffusion. Very elaborate multicolor graphical software packages are developed to help the designer. Most of these software packages are based on simple mathematical ideas.

The placement problem in building-block layout is a challenging one, especially if we introduce other constraints; for example, some of the modules are preplaced, others may have variable aspect ratio.

In detailed routing, the grid-graph is a mathematical model where the nodes are the positions of pins or the potential positions of vias. For two parallel wires not interfering with each other, they must be placed at λ distance apart. Thus the nodes should be at λ distance apart. However, the clearance between two vias should be more than λ distance apart; in other words, we cannot have two parallel L-shaped wires as shown in Fig. 16.

More generally, we need to consider gridless routing, multilayer routing, interactive and dynamic routing. The latter is of crucial importance since we cannot hope to have totally automatic systems for all possible applications. Recent work on the two-path problem [156] and application of computational geometry seems to be very attractive [158].

Other simple and special cases which have precise mathematical formulation yet need more work in order to obtain the ultimate solutions are

- Linear arrangement to satisfy different objective functions; e.g., total wire length, maximum density, for single-row routing, for multirow routing.
- Two-dimensional placement to satisfy different objective functions.
- Optimum rectilinear Steiner's tree with arcs having different length and Steiner's tree on graph.
- Channel routing and general two-dimensional, two-layer routing which is not limited to HV routing.
- Multilayer routing and optimum decomposition.

Most of the above problems are NP-complete, and we need fast heuristic algorithms with *error bounds*.

Even when all the above problems are solved, the rapid progress in technology will demand refinements or modification of the mathematical models, hence new algorithms.

The VLSI design is a real challenge to scientists, engineers and mathematicians—to create models, to define problems, and to invent algorithms. Eventually, there will be a theory of VLSI design like a branch of geometry. However, the final test of a theory is its capacity to solve the problems which originated it.

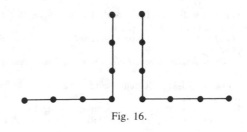
Fig. 16.

BIBLIOGRAPHY

Books and General Surveys on VLSI and Layout

[1] R. F. Ayres, *VLSI, Silicon Compilation and the Art of Automatic Microchip Design*. Engelwood Cliffs, NJ: Prentice-Hall, 1983.

[2] R. K. Brayton, G. D. Hachtel, and A. Sangiovanni-Vincentelli, "A taxonomy of CAD for VLSI," *Proceedings of European Conference on Circuit Theory and Design*, Hague, The Netherlands, 1981.

[3] M. A. Breuer, "Theory and techniques," *Design Automation of Digital Systems*, Vol. 1. Engelwood Cliffs, NJ: Prentice-Hall, 1972, Chs. 4, 5, 6.

[4] M. A. Breuer, A. D. Friedman, and A. Iosupovicz, "A survey of the state of the art of design automation," *Computer*, vol. 14, no. 10, pp. 58–75, Oct. 1981.

[5] E. S. Kuh, Ed., "The special issue on routing and microelectronics," *IEEE Trans. on Computer-Aided Design of Integrated Circuit and Syst.*, vol. CAD-2, no. 4, Oct. 1983.

[6] H. T. Kung, B. Sproull, and G. Steele, *VLSI Systems and Computations*. Rockville, MD: Computer Science Press, 1981.

[7] F. T. Leighton, *Complexity Issues in VLSI*. Cambridge, MA: MIT Press, 1983.

[8] C. E. Leiserson, *Area-Efficient VLSI Computation*. Cambridge, MA: MIT Press, 1983.

[9] C. Mead and L. Conway, *Introduction to VLSI Systems*. Reading, MA: Addison-Wesley, 1980.

[10] S. Muroga, *VLSI System Design*. New York: Wiley-Interscience Publications, 1982.

[11] A. R. Newton, D. O. Pederson, A. L. Sangiovanni-Vincentelli, and C. H. Séquin, "Design aids for VLSI: The Berkeley perspective," *IEEE Trans. Circuits Syst.*, vol. CAS-28, no. 7, July 1981.

[12] C. Rabbat, *Hardware and Software Concepts in VLSI Design*. Princeton: Van Nostrand, 1983.

[13] B. Randell and P. C. Treleaven, Eds., *VLSI Architecture*. Engelwood Cliffs, NJ: Prentice-Hall, 1982.

[14] J. Soukup, "Circuit layout," *Proc. IEEE*, vol. 69, no. 10, pp. 1281–1304, Oct. 1981.

[15] C. D. Thompson, "A complexity theory for VLSI," Computer Science Rep. CMU-CS-80-140, Carnegie-Mellon University, Aug. 1980.

[16] J. D. Ullman, *Computational Aspects of VLSI*. Rockville, MD: Computer Science Press, 1984.

Mathematical Books and Papers

[17] V. Chvatal, *Linear Programming*. San Francisco, CA: Freeman, 1983.

[18] H. Crowder, E. L. Johnson, and M. W. Pedberg, "Solving large scale zero-one linear programming programs," *J. Opt. Soc. Amer.*, vol. 31, no. 5, pp. 803–834, Oct. 1983.

[19] G. B. Dantzig, *Linear Programming and Extensions*. Princeton, NJ: Princeton University Press, 1963.

[20] G. B. Dantzig and R. M. Van Slyke, "Generalized upper bounded techniques for linear programming," *J. Comput. and Syst. Sci.*, vol. 1, pp. 213–226, 1967.

[21] G. B. Dantzig and P. Wolf, "Decomposition principle for linear programs," *J. Opt. Soc. Amer.*, vol. 8, no. 1, pp. 101–111, Jan. 1960.

[22] L. R. Ford and D. R. Fulkerson, "Suggested computations for maximal multi-commodity network flows," *Management Sci.*, vol. 5, no. 1, pp. 97–101, Oct. 1958.

[23] ——, *Flows in Networks*. Princeton, NJ: Princeton University Press, 1962.

[24] M. R. Garey and D. S. Johnson, *Computers and Intractability*. San Francisco, CA: Freeman, 1979.

[25] R. E. Gomory, "Mathematical programming," *Amer. Math. Monthly*, vol. 72, no. 2, pp. 99–110, Feb. 1965.

[26] R. E. Gomory and T. C. Hu, "Multi-terminal network flows," *J. SIAM*, vol. 9, no. 4, pp. 551–570, 1961.

[27] T. C. Hu, *Integer Programming and Network Flows*. Reading, MA: Addison-Wesley, 1969.

[28] ——, *Combinatorial Algorithms*. Reading, MA: Addison-Wesley, 1982.

[29] T. C. Hu and F. Ruskey, "Circular cut in a network," *Math. Oper. Res.*, vol. 5, no. 3, pp. 422–434, Aug. 1980.

[30] T. C. Hu and M. T. Shing, "A decomposition algorithm for circuit routing," Computer Science Report TRCS 84-01, U.C.S.B., Santa Barbara, CA.

[31] ——, "A decomposition algorithm for multi-terminal network flows," Computer Science Report TRCS 83-01, U.C.S.B., Santa Barbara, CA.

[32] ——, "Multi-terminal flows in outplanar networks," *J. Algorithms*, vol. 4, pp. 241–261, 1983.

[33] E. L. Johnson, "Integer programming," presented at the CBMS-NSF Regional Conference on Applied Math, no. 32 SIAM, 1980.

[34] D. S. Johnson, "The NP-completeness column," in *An On-going Guide in Journal of Algorithms*, 1981 to present.

[35] N. Karmarker and R. M. Karp, "The differencing method of set partitioning," UCB/CSD 82/113, Univ. of California, Berkeley, Dec. 1982.

[36] K. Murty, *Linear Programming*. New York: Wiley, 1983.

References on Wirability, Partitioning, and Placement

[37] K. J. Antreich, F. M. Johnnes, and F. H. Kirsch, "A new approach for solving the placement problem using force models," in *Proc. IEEE Int. Symp. Circuits Syst.*, pp. 481–486, 1982.

[38] E. R. Barnes, "An algorithm for partitioning the nodes of a graph," *SIAM J. Discrete Method*, vol. 3, no. 4, Dec. 1982.

[39] M. A. Breuer, "A class of min-cut placement algorithms," in *Proc. 14th Design Automat. Conf.*, pp. 284–290, 1977.

[40] ——, "Min-cut placement," *J. Design Automat. and Fault-Tolerant Comput.*, vol. 1, no. 4, pp. 343–362, Oct. 1977.

[41] R. L. Brooks, A. B. Smith, A. H. Stone, and W. T. Tutte, "The dissection of rectangles into squares," *Duke Math. J.*, vol. 7, pp. 312–340, 1940.

[42] H. R. Charney and D. L. Plato, "Efficient partitioning of components," in *Proc. 5th Design Automat. Workshop*, pp. 16–1 to 16–21, 1968.

[43] C. K. Cheng, "Placement algorithms and applications to VLSI design," Ph.D. thesis, Dept. of EE, Univ. California, Berkeley, CA, 1984.

[44] C. K. Cheng and E. S. Kuh, "Module placement based on network optimization," *IEEE Trans. Computer-Aided Design*, vol. CAD-3, no. 3, pp. 218–225, Oct. 1984.

[45] W. A. Dees and R. J. Smith, "Performance of interconnection rip-up and reroute strategies," in *Proc. 18th Design Automat. Conf.*, pp. 382–390, 1981.

[46] W. E. Donath, "Placement and average interconnection lengths of computer logic," *IEEE Trans. Circuits Syst.*, vol. CAS-26, no. 4, pp. 272–277, Apr. 1979.

[47] M. Feuer, "Connectivity of random logic," *IEEE Trans. Comput.*, vol. C-31, no. 1, pp. 29–33, Jan. 1982.

[48] C. M. Fiduccia and R. M. Mattheyses, "A linear-time heuristic for improving network partitions," in *Proc. 19 Design Automat. Conf.*, pp. 175–181, 1982.

[49] K. Fukunaga, S. Yamada, H. S. Stone, and T. Kasai, "Placement of circuit modules using a graph space approach," in *Proc. ACM IEEE 20th Design Automation Conf.*, pp. 465–471, Miami Beach, FL, Jan. 1983.

[50] A. A. El Gamal, "Two-dimensional stochastic model for interconnections in master-slice integrated circuits," *IEEE Trans. Circuits Syst.*, vol. CAS-28, no. 2, pp. 127–138, Feb. 1981.

[51] A. El Gamal and Z. A. Syed, "A new statistical model for gate array routing," in *Proc. 20th Design Automat. Conf.*, pp. 671–674, 1983.

[52] P. C. Gilmore, "Optimum and suboptimum algorithms for the quadratic assignment problem," *J. SIAM*, vol. 10, no. 2, pp. 305–313, June 1962.

[53] S. Goto, "An efficient algorithm for the two-dimensional placement problem in electrical circuit layout," *IEEE Trans. Circuits Syst.*, vol. CAS-28, no. 1, pp. 12–19, Jan. 1981.

[54] S. Goto, I. Cederbaum, and B. S. Ting, "Suboptimum solution of the back-board ordering with channel capacity constraint," *IEEE Trans. Circuits Syst.*, vol. CAS-24, no. 11, pp. 645–652, 1977.

[55] S. Goto and E. S. Kuh, "An approach to the two-dimensional placement problem in circuit layout," *IEEE Trans. Circuits Syst.*, vol. CAS-25, no. 4, pp. 208–214, 1978.

[56] K. M. Hall, "An *r*-dimension-quadratic placement algorithm," *Management Sci.*, vol. 17, no. 3, pp. 219–229, Nov. 1970.

[57] M. Hanan and J. Kurtzberg, "A review of the placement and quadratic assignment problems," *SIAM Rev.*, vol. 14, no. 2, pp. 324–342, Apr. 1972.

[58] M. Hanan, P. K. Wolff, and B. J. Anguli, "Some experimental results on placement techniques," in *Proc. 13th Design Automation Conf.*, pp. 214–224, 1976.

[59] ——, "Some experimental results on placement techniques," *J. Design Automation and Fault Tolerant Computing*, vol. 2, no. 2, pp. 145–164, May 1978.

[60] W. R. Heller, W. F. Mikhail, and W. E. Donath, "Prediction of wiring space requirements for LSI," *J. Design Automation and Fault Tolerant Computing*, vol. 2, no. 2, pp. 117–144, May 1978.

[61] R. M. Karp, F. T. Leighton, R. L. Rivest, C. D. Thompson, U. Vazivani, and V. Vazivani, "Global wire routing in two-dimensional arrays," in *Proc. 24th Symp. on Foundations of Computer Sci.*, vol. IEEE 83CH1938-0, pp. 453–459, Nov. 1983.

[62] B. W. Kernighan and S. Lin, "An efficient heuristic procedure for partitioning graphs," *Bell Syst. Tech. J.*, vol. 49, no. 2, pp. 291–307, 1970.

[63] F. T. Leighton and C. E. Leiserson, "Wafer-scale integration of systolic arrays," MIT/LCS/TM-236, Feb. 1983.

[64] M. Marek-Sadowska and J. T. Li, "Global router for gate array," *IEEE ICCAD*, pp. 131–132, 1983.

[65] B. T. Preas and W. M. van Cleemput, "Placement algorithms for arbitrarily shaped blocks," in *Proc. 16th Design Automation Conf.*, pp. 474–480, 1979.

[66] F. P. Preparata and W. Lipski, Jr., "Three layers are enough," in *Proc. 23rd IEEE Symp. on Foundations of Computer Sci.*, pp. 350–357, 1982.

[67] N. R. Quinn and M. A. Breuer, "A force directed component placement procedure for printed circuit boards," *IEEE Trans. Circuits Syst.*, vol. CAS-26, no. 6, June 1979.

[68] D. M. Schuler and E. G. Ulrich, "Clustering and linear placement," in *Proc. 9th Design Automation Workshop*, pp. 50–56, 1972.

[69] D. G. Schweikert, "A 2-dimensional placement algorithm for the layout of electrical circuits," in *Proc. Design Automation Conf.*, pp. 408–416, 1976.

[70] D. G. Schweikert and B. W. Kernighan, "A proper model for the partitioning of electrical circuits," in *Proc. 9th Design Automation Workshop*, pp. 57–62, 1972.

[71] L. Steinberg, "The backboard wiring problem: A placement algorithm," *SIAM Rev.*, vol. 3, no. 1, pp. 37–50, Jan. 1961.

[72] J. E. Stevens, "Fast heuristic techniques for placing and wiring printed circuit boards," Ph.D. thesis, CS Dept., Univ. of Illinois, 1972.

[73] K. J. Supowit and E. A. Slutz, "Placement algorithms for custom VLSI," in *Proc. 10th Design Automation Conf.*, pp. 164–170, 1983.

[74] H. Widgaja, "An effective structured approach to finding optimum partitions of networks," *Computing*, vol. 29, pp. 241–262, 1982.

References on Routing

[75] B. S. Baker, S. N. Bhatt, and F. T. Leighton, "An approximate algorithm for Manhattan routing," MIT/LCS/TM-238, Feb. 1983.

[76] D. J. Brown and R. L. Rivest, "New lower bounds on channel width," in *Proc. of CMU Conf. VLSI Syst. and Computations*, 1981.

[77] M. Burstein and R. Pelavin, "Hierarchical channel routes," *Integration, The VLSI J.*, vol. 1, no. 1, pp. 21–38, Mar. 1983.

[78] ——, "Hierarchical wire routing," *IEEE Trans. Computer-Aided Design*, vol. CAD-2, no. 4, pp. 223–234, Oct. 1983.

[79] M. C. Chen, "Space-time algorithms: Semantics and methodology," Tech. Report #5090, Calif. Inst. of Tech., Pasadena, CA, 1983.

[80] N. P. Chen, "New algorithms for Steiner tree on graphs," in *Proc. ISCAS*, pp. 1217–1219, 1983.

[81] M. Cutler and Y. Shiloach, "Permutation layout," *Networks*, vol. 8, pp. 253–278, 1978.

[82] D. N. Deutsch, "A dogleg channel router," presented at *Proc. 19th IEEE Design Automation Conf.*, 1976.

[83] J. M. Geyer, "Connection routing algorithm for printed circuit board," *IEEE Trans. Circuit Theory*, vol. CT-18, pp. 95–100, 1971.

[84] E. N. Gilbert and H. O. Pollak, "Steiner minimal trees," *SIAM J. Appl. Math.*, vol. 16, no. 1, pp. 1–29, 1968.

[85] S. L. Hakimi, "Steiner's problem in graphs and its implications," *Networks*, vol. 1, pp. 113–133, 1971.

[86] M. Hanan, "On Steiner's problem with rectilinear distance," *SIAM J. Appl. Math.*, vol. 14, no. 2, pp. 255–265, Mar. 1966.

[87] A. Hashimoto and J. Stevens, "Wire routing by channel assignment within large apertures," in *Proc. 8th Design Automation Workshop*, pp. 155–169, 1971.

[88] S. Heiss, "A path connection algorithm for multi-layer boards," in *Proc. 5th Design Automation Conf.*, pp. 6-1–6-14, 1968.

[89] D. W. Hightower, "A solution to the line-routing problem on the continuous plane," in *Proc. 6th Design Automation Workshop*, pp. 1–24, 1969.

[90] ——, "The interconnection problem: A tutorial," *IEEE Comput.*, vol. 7, no. 4, pp. 18–32, Apr. 1974.

[91] C. P. Hsu, "Theory and algorithms for signal routing in integrated circuit layout," Ph.D. thesis, EECS Dept., UC Berkeley, 1983.

[92] ——, "A new two-dimensional routing algorithm," in *Proc. 19th Design Automation Conf.*, pp. 46–50, June 1982.

[93] ——, "General river routing algorithm," in *Proc. 20th Design Automation Conf.*, pp. 578–583, 1983.

[94] F. K. Hwang, "On Steiner minimal trees with rectilinear distance," *SIAM J. Appl. Math.*, vol. 30, pp. 104–114, 1976.

[95] ——, "The rectilinear Steiner problem," *Design Automation and Fault-Tolerant Computing*, vol. 2, no. 4, pp. 303–310, Oct. 1978.

[96] ——, "An $O(n \log n)$ algorithm for suboptimal rectilinear Steiner trees," *IEEE Trans. Circuits Syst.*, vol. CAS-26, no. 1, pp. 75–77, 1979.

[97] D. L. Johannsen, "Silicon compilation," Tech. Rep. 4530, Calif. Inst. of Tech., Pasadena, CA.

[98] Y. Kajitani, "On via hole minimization on routing on a two-layer board," in *Proc. ICCC*, pp. 295–298, 1980.

[99] R. M. Karp, F. T. Leighton, R. L. Rivest, C. D. Thompson, U. Vazirani, and V. Vazirani, "Global wire routing in two dimensional arrays," in *Proc. 24th Ann. Symp. on Foundations of Computer Science*, pp. 453–459, Nov. 1983.

[100] B. W. Kernighan and S. Lin, "An efficient procedure for partitioning graphs," *Bell Syst. Tech. J.*, pp. 291–307, Feb. 1970.

[101] B. Kernighan, D. Schwiekert, and G. Persky, "An optimal channel-routing algorithm for polycell layouts of integrated circuits," in *Proc. 10th Design Automation Workshop*, pp. 50–59, 1973.

[102] S. Kirkpatrick, C. D. Gelatt, and M. P. Vecchi, "Optimization by simulated annealing," *Science*, vol. 220, no. 4598, pp. 671–680, May 13, 1983.

[103] J. B. Kruskal, "On the shortest spanning subtree of a graph," in *Proc. Amer. Math. Soc.*, vol. 7, pp. 48–50, 1956.

[104] E. S. Kuh, T. Kashiwabara, and T. Fujisawa, "On optimum single-row routing," *IEEE Trans. Circuits Syst.*, vol. CAS-26, no. 6, pp. 361–368, July 1979.

[105] C. Y. Lee, "An algorithm for path connection and its applications," *IRE Trans. Electron. Comput.*, vol. EC-10, pp. 346–365, 1961.

[106] J. H. Lee, N. K. Bose, and F. K. Hwang, "Use of Steiner's problem in sub-optimal routing in rectilinear metric," *IEEE Trans. Circuits Syst.*, vol. CAS-23, pp. 470–1401, 1976.

[107] F. T. Leighton, "Layouts for the shuffle-exchange graph and lower bound techniques for VLSI," M.I.T. Lab for Computer Science Report TR-274, June 1982.

[108] C. E. Leiserson and R. Y. Pinter, "Optimum placement for river routing," in *VLSI Systems and Computations,* H. T. Kung, Ed. Computer Science Press, 1981.

[109] M. Marek-Sadowska, "Global router for gate array," in *Proc. ICCD*, pp. 332–337, Oct. 1984.

[110] M. Marek-Sadowska and E. S. Kuh, "A new approach to routing two-layer printed circuit boards," *Int. J. Circuit Theory and Appl.*, vol. 9, no. 3, pp. 331–341, July 1981.

[111] M. Marek-Sadowska and J. T. Li, "Global router for gate array," in *IEEE ICCAD*, pp. 131–132, 1983.

[112] K. Mikami and Tabushi, "A computer program for optimal routing of printed circuit connections," in *IFIPS Proc.*, pp. 1475–1478, 1968.

[113] E. F. Moore, "The shortest path through a maze," in *Proc. Int. Symp. Theory of Switching, Part II*, pp. 285–290, 1959.

[114] N. Nan and M. Feuer, "A method for the automatic wiring of LSI chips," in *Proc. ISCAS*, pp. 11–16, 1978.

[115] B. Preas and W. M. Van Cleemput, "Routing algorithms for hierarchical IC layout," in *Proc. ISCAS*, pp. 482–485, 1979.

[116] F. Preparata and W. Lipski, "Three layers are enough," in *Proc. of 23rd Annual IEEE Symp. on Foundations of Comput. Sci.*, 1982.

[117] R. C. Prim, "Shortest connecting networks and some generalizations," *Bell Syst. Tech. J.*, vol. 36, pp. 1389–1401, 1957.

[118] R. L. Rivest and C. M. Fiduccia, "A greedy channel router," in *Proc. of 19th IEEE Design Automation Conf.*, pp. 418–424, 1982.

[119] A. L. Rosenberg, "Three-dimensional VLSI: A case study," *J. Ass. Comput. Mach.*, vol. 30, no. 3, pp. 397–416, July 1983.

[120] A. Siegel, "Fast placement for river routing," Ph.D. thesis, Dept. of Computer Science, Stanford University, 1982.

[121] H. C. So, "Some theoretical results on the routing of multilayer printed wiring boards," in *Proc. IEEE Int. Symp. Circuits Syst.*, pp. 296–303, 1974.

[122] J. Soukup, "Fast mazer router," in *Proc. Design Automation Conf.*, pp. 100–101, 1978.

[123] ——, "Global router," in *Proc. 16th Design Automation Conf.*, pp. 481–484, 1979.

[124] B. S. Ting, E. S. Kuh, and A. Sangiovanni-Vincentelli, "Via assignment problem in multilayer printed circuit board," *IEEE Trans. Circuits Syst.*, vol. CAS-26, no. 4, pp. 261–272, Apr. 1979.

[125] B. S. Ting, E. S. Kuh, and I. Shirakawa, "The multilayer routing problem: Algorithms and necessary and sufficient conditions for single-row single-layer case," *IEEE Trans. Circuit Syst.*, vol. CAS-23, no. 12, pp. 768–778, 1976.

[126] M. Tompa, "An optimal solution to a wire-routing problem," in *Proc. 12th ACM Symp. on Theory of Comput.*, pp. 161–176, 1980.

[127] S. Tsukiyama, E. S. Kuh, and I. Shirakawa, "An algorithm for single-row routing with prescribed street congestions," *IEEE Trans. Circuits Syst.*, vol. CAS-27, no. 9, pp. 765–772, Sept. 1980.

[128] M. P. Vecchi and S. Kirkpatrick, "Global wiring by simulated annealing," *IEEE Trans. Computer-Aided Design*, vol. CAD-2, no. 4, pp. 215–222, Oct. 1983.

[129] Y. Y. Yang and O. Wing, "Suboptimal algorithm for a wire routing problem," *IEEE Trans. Circuit Theory*, vol. CT-19, pp. 508–511, 1972.

[130] T. Yoshimura and E. S. Kuh, "Efficient algorithms for channel routing," *IEEE Trans. Computer-Aided Design on Integrated Circuits and Systems*, vol. CAD-1, no. 1, pp. 25–35, Jan. 1982.

References on Layout Procedures and Methods

[131] K. Aoshima and E. S. Kuh, "Multi-channel optimization in gate array layout," in *Proc. 1983 IEEE Symp. Circuits and Syst.*, pp. 1005–1008, 1983.

[132] T. Asano, "An optimal gate placement algorithm for MOS one-dimensional arrays," *J. Digital Syst.*, vol. VI, no. 1, pp. 1–27, 1982.

[133] K. A. Chen, M. Feuer, K. H. Khokhani, N. Nan, and S. Schmidt, "The chip layout problem: An automatic layout procedure," in *Proc. 14th Design Automation Conf.*, pp. 298–302, 1977.

[134] N. P. Chen, C. P. Hsu, and E. S. Kuh, "The Berkeley building-block layout system," F. Anceau and E. J. Aas, Eds., in *Proc. VLSI '83*, pp. 37–44, 1983.

[135] N. P. Chen, C. P. Hsu, E. S. Kuh, C. C. Chen, and M. Takahashi, "BBL: A building-block layout system for custom chip IC design," in *Proc. IEEE Int. Conf. CAD*, pp. 40–41, 1983.

[136] B. W. Colbry and J. Soukup, "Layout aspects of the VLSI microprocessor design," in *Proc. Int. Symp. Circuits and Syst.*, pp. 1214–1228, 1982.

[137] W. A. Dees and R. J. Smith, "Performance of interconnection rip-up and reroute strategies," in *Proc. 18th Design Automation Conf.*, pp. 382–390, 1981.

[138] R. L. Donze and G. Sporzynski, "Masterimage approach to VLSI design," *IEEE Comput.*, pp. 18–25, 1983.

[139] A. E. Dunlop, "SLIP: Symbolic layout of integrated circuits with compaction," *IEEE Trans. Computer-Aided Design*, vol. 10, no. 6, pp. 387–391, Nov. 1978.

[140] W. R. Heller, G. Sorkin, and K. Maling, "The planar package planner for system designers," in *Proc. 19th Design Automation Conf.*, pp. 253–260, 1982.

[141] W. Heyns, W. Sansen, and H. Beke, "A line-expansion algorithm for the general routing problem with a guaranteed solution," in *Proc. 17th Design Automation Conf.*, pp. 243–249, 1980.

[142] C. S. Horng and M. Lie, "An automatic/interactive layout planning system for arbitrarily-sized rectangular building blocks," in *Proc. 18th Design Automation Conf.*, pp. 293–300, 1981.

[143] M. Y. Hsueh and D. O. Pederson, "Computer-aided layout of LSI circuit building-blocks," in *Proc. IEEE Int. Symp. Circuits and*

Syst., pp. 474–477, 1979.

[144] S. M. Kang, R. H. Krambeck, H-F. S. Law, and A. D. Lopez, "Gate matrix layout of random control logic in a 32-bit CMOS CPU chip adaptable to evolving logic design," *IEEE Trans. Computer-Aided Design*, vol. CAD-2, no. 1, pp. 18–29, Jan. 1983.

[145] K. Kani, H. Kawanishi, and A. Kishimoto, "ROBIN: A building block LSI routing program," in *Proc. IEEE ISCAS*, pp. 658–661, 1976.

[146] T. Kashiwabara and T. Fujisawa, "NP-completeness of the problem of finding a minimum-clique-number interval graph containing a given graph as a subgraph," in *Proc. IEEE ISCAS*, pp. 657–659, 1979.

[147] K. H. Khokhani and A. M. Patel, "The chip layout problem: A placement procedure for LSI," in *Proc. 14th Design Automation Conf.*, pp. 291–297, 1977.

[148] B. S. Landman and R. L. Russo, "On a pin versus block relationship for partitions of logic graphs," *IEEE Trans. Comput.*, vol. C-20, pp. 1469–1479, Dec. 1971.

[149] U. Lauther, "A min-cut placement algorithm for general cell assemblies based on a graph representation," in *Proc. 16th Design Automation Conf.*, pp. 1–10, June 1979.

[150] J. T. Li, "Algorithms for gate matrix layout," *IEEE ISCAS*, pp. 1013–1016, May 1983.

[151] J. T. Li, C. K. Cheng, M. Turner, E. S. Kuh, and M. Marek-Sadowska, "Automatic layout of gate arrays," in *Proc. Custom Integrated Circuit Conf.*, pp. 518–521, May 1984.

[152] A. D. Lopez and H-F. S. Law, "A dense gate matrix layout method for MOS VLSI," *IEEE Trans. Elec. Devices*, vol. ED-27, pp. 1671–1675, Aug. 1980.

[153] K. Maling, S. H. Meuller, and W. R. Heller, "On finding most optimal rectangular package plans," in *Proc. 19th Design Automation Conf.*, pp. 663–670, 1982.

[154] T. Matsuda, T. Fujita, K. Takamizawa, H. Mizumura, H. Nakamura, F. Kitajima, and S. Goto, "LAMDA, an integrated master-slice LSI CAD system," *Integration, The VLSI J.*, vol. 1, no. 1, pp. 53–69, Apr. 1983.

[155] T. Ohtsuki, "The two disjoint path problem and wire routing design," in *Proc. Symp. on Graph Theory and Applications at Tohoku Univ.*, pp. 257–267, 1980 (Lecture Notes in Computer Science 108, Springer, Berlin).

[156] ——, "Minimum dissection of rectilinear regions," in *IEEE Proc. ISCAS*, pp. 1210–1213, 1982.

[157] T. Ohtsuki, H. Mori, T. Kashiwabara, and T. Fujisawa, "On minimal augmentation of a graph to obtain an interval graph," *J. Comput. Syst. Sci.*, vol. 22, no. 1, pp. 60–97, Feb. 1981.

[158] T. Ohtsuki, H. Mori, E. S. Kuh, T. Kashiwabara, and T. Fujisawa, "One-dimensional logic gate assignment and interval graphs," *IEEE Trans. Circuits and Syst.*, vol. CAS-26, no. 9, pp. 675–684, Sept. 1979.

[159] G. Persky, D. N. Deutsch, and D. G. Schweikert, "LTX—A system for the directed automatic design of LSI circuits," in *Proc. 13th Design Automation Conf.*, pp. 399–407, 1976.

[160] B. T. Preas and C. W. Gwyn, "Methods for hierarchical automatic layout of custom LSI circuit masks," in *Proc. 15th Design Automation Conf.*, pp. 206–212, 1978.

[161] R. I. Rivest, "The PI (placement and interconnect) system," in *Proc. 19th Design Automation Conf.*, pp. 475–481, June 1982.

[162] I. Shirakawa, N. Okuda, T. Harada, S. Tani, and H. Ozaki, "A layout system for the random logic portion of an MOS LSI chip," *IEEE Trans. Comput.*, vol. C-30, no. 8, pp. 572–581, Aug. 1981.

[163] B. N. Tien, B. S. Ting, J. Cheam, K. Chow, and S. C. Evans, "GALA—An automatic layout system for high density CMOS gate arrays," in *Proc. 21st Design Automation Conf.*, pp. 657–662, 1984.

[164] A. Weinberger, "Large-scale integration of MOS complex logic: A layout method," *IEEE Solid-State Circuits*, vol. SC-2, pp. 182–190, Dec. 1967.

[165] M. Wiesel and D. A. Mlynski, "An efficient channel model for building block LSI," *IEEE Proc. ISCAS*, pp. 118–121, 1981.

[166] O. Wing, "Automated gate matrix layout," in *Proc. IEEE ISCAS*, pp. 681–685, May 1982.

[167] ——, "Interval-graph-based circuit layout," *IEEE Int. Conf. Computer-Aided Design*, pp. 84–85, 1983.

Part II
General

Circuit Layout

JIRI SOUKUP, SENIOR MEMBER, IEEE

Abstract—This paper gives a general overview of circuit layout, taking a unified approach to various styles of integrated circuits, printed circuit boards, and hybrid circuits. A lot of attention is given to the layout of large and complicated circuits, in particular, to the layout of very-large-scale-integration (VLSI) chips. Though the paper is an overview, and one could almost say a tutorial, it is intended for readers with some basic knowledge of what a circuit layout is and what some of the basic problems are. The main subjects discussed are: assignment of gates, placement methods, loose routing, final routing, and problems associated with the implementation of a hierarchical system. The emphasis is on new, not widely published methods, and on methods that seem to have potential for solving some of the current problems. Practical examples illustrate this rather personal account of circuit layout and suggest where we may go from here.

I. INTRODUCTION

THE PROBLEM of layout—as presented in this paper—is a problem of placing box-like modules in a plane and connecting them by wires according to a given set of rules. Of course, when logical design is iterated with physical design, the shapes of the modules may change. If interpreted in this general way, the layout of metal–oxide–silicon integrated circuits (MOS/IC's) is very much like the layout of printed circuit boards (PCB's) or hybrid (or thick film) circuits, see Fig. 1. It is important to remember this basic concept; though most examples used within the paper are from the IC layout, the methods are equally applicable to other circuits.

This paper begins by giving examples that demonstrate the similarity among the layouts of different types of electrical circuits. The paper shows how today's complex circuits require a more structured approach to both circuit design and layout and how better organization can be achieved through a structured hierarchical design, such as developed at California Institute of Technology.

The notion of hierarchy has become quite popular, but not everybody understands what it means. At the end of the paper, examples of technical difficulties associated with the implementation of a truly hierarchical system are given.

The paper discusses the traditional division of layout into placement and routing and describes typical algorithms. For both placement and routing, numerous algorithms are available. The paper concentrates on those considered to be the most practical. Some algorithms are explained in more detail. These algorithms, which are not widely published or recognized, have the potential for solving some of the current pressing problems. For other algorithms, refer to the list of selected papers.

The proper placement and selection of shapes seems to be key in hierarchical design. The paper describes details of loose

Manuscript received June 10, 1981.
The author is with Bell Laboratories, Murray Hill, NJ 07974.

routing, which has not been discussed in other publications as much as placement and final routing, and describes the procedures used in manual layout. For an example of the manual style of routing, see Fig. 2.

Results obtained from the most fully automated layout systems are still quite discouraging when compared to manually packed designs. And still we know that automation is needed for its speed, and for getting useable chips without redesign cycles.

In general, the paper is rather subjective: It describes one person's view and experience, and it should be read that way. The references, again, reflect the author's personal opinion, and they are a selection, rather than a compilation, of all published papers.

II. MAPPING LOGICAL ONTO PHYSICAL

The design of any electrical digital circuit consists of two major phases. In the first phase, the designer builds the circuit from abstract blocks, such as NAND or NOR gates, registers, and latches. The result of the first phase is usually a schematic diagram or a computer file that describes the type of blocks used and the way they are to be connected. The circuit description is abstract (logical) in nature, and although some blocks may be given performance parameters, such as output power or delay, very little is known about the physical device that will actually be used.

In the second phase, the designer builds the circuit from chosen modules, which are fully determined in size, shape, and internal structure.

The first major step of the layout, sometimes called *gate assignment*, is the transition from the first phase to the second phase. Let's look at examples of gate assignment in different technologies.

1) In the first example, the design calls for a transistor-to-transistor logic (TTL) circuit to be implemented as a PCB. Some of the building blocks correspond to whole IC packages, such as "up/down counter" (74192) or "4-bit arithmetic unit" (74181). Other blocks are only a part of an IC package. For example, circuit 7400 contains four two-input NAND gates, and circuit 7428 contains four two-input NOR buffers. If the design includes 22 NAND gates and 17 NOR buffers, there is a multiple choice of assigning the gates and buffers to six 7400 circuits and five 7428 circuits. Also, some gates and buffers will not be used—we call them spare: two NAND gates and three NOR buffers. Such gates are called spare, because they can be used for later repairs or modifications of the board.

2) The second example is the design of a hybrid circuit, with 4 fully specified TTL packages, and 10 resistors, ranging from 500 Ω to 500 kΩ. The resistors can be printed with three different inks, 1, 10, 100 kΩ/□. (After printing, the cir-

Reprinted from *Proc. IEEE*, vol. 69, pp. 1281–1304, Oct. 1981.

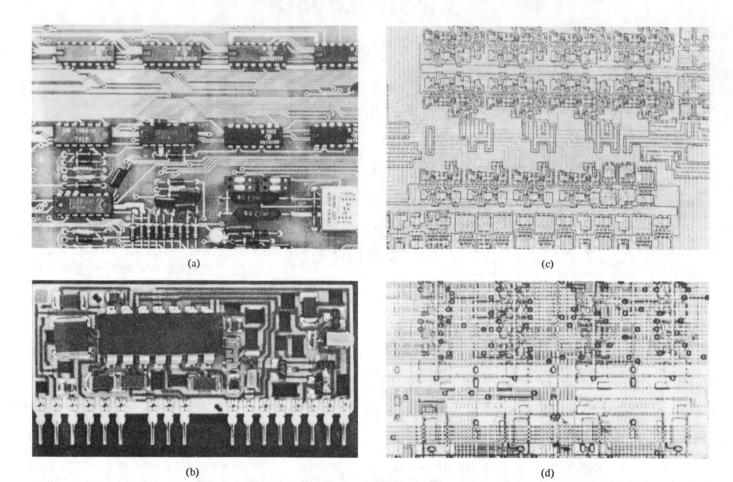

Fig. 1. Layouts for different technologies. (a) Printed circuit board. (b) Hybrid printed circuit. (c) Enlarged detail of a MOS silicon chip. (d) Silicon chip—gate array layout style (this is a detail from the IBM chip shown in Fig. 33). Note that in all four cases, the designer has to place modules and connect them by a conductive path. ((a), (b), and (c) were reprinted from *Telesis*, the technical magazine of Bell-Northern Research Ltd., from the article "Electronic circuit layout algorithm," by J. Soukup. (d) was obtained from IBM.)

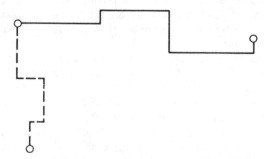

Fig. 2. A typical style of human wiring on two layers.

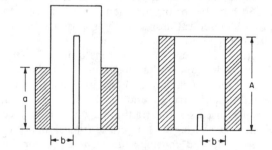

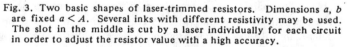

Fig. 3. Two basic shapes of laser-trimmed resistors. Dimensions a, b are fixed $a < A$. Several inks with different resistivity may be used. The slot in the middle is cut by a laser individually for each circuit in order to adjust the resistor value with a high accuracy.

cuit is baked in a furnace to obtain the resistors.) Each printing operation adds to the cost of the circuit. Note that all resistors can be printed with two inks, causing some resistors to be bigger in size. There is a tradeoff between the cost of printing and the size of the circuit. The assignment step consists of two parts. In the first, the economy of various alternatives, such as minimizing the resistor area or number of print operations, must be determined. The result of this part is a selection of the inks (one, two, or three) to be used for the resistors and, particularly which ink should be used for

each resistor. Then in the second part of the assignment step, the shape of all resistors can be determined based on standards for laser-trimmed resistors (see Fig. 3).

3) The third example is a simple case of a complementary MOS (CMOS) circuit that was designed using standard library cells. Most building blocks correspond to library cells on a one-to-one basis, except that several library cells are available for each logic block (or gate). Such cells have identical logic,

but their output power and sizes are different. Also, certain cells may be available in several shapes or terminal combinations designed to link cells into registers for counters. The assignment of gates to cells is straightforward. The size is determined by the required power, and the shape is mainly determined by the adjacent logic.

Of the three technologies mentioned, only the gate assignment for the printed boards requires more automatic processing. The usual assignment algorithm is based on the following rules:

1) Assign all IC's that contain only one logical block or gate.
2) If any IC package has an available gate, use it for the gate that has the most connections in common with gates already assigned to this IC package.
3) If there is no partially used IC package, take the next unused package and assign the first gate randomly.

If there are any spare gates, this greedy algorithm tends to leave them together in one IC package. However, from the viewpoint of future repairs, it is better to distribute spare gates across the board. For that reason, some assignment algorithms start with distributing spare gates, and only then proceed with the three steps described earlier.

Also, the random assignment of the first gate for each new IC package can cause an unfavorable partitioning and long leads between packages. However, the author is not aware of any PCB layout system based on a more intelligent gate assignment. Instead, most systems start with a crude gate assignment, and then refine it after placement when the position of the terminals can be taken into consideration as well. This gate optimization is typically based on a pair-wise swapping of gates and terminals.

Note that integer programming methods for the assignment problem (AP) as defined in Operations Research, may be applicable to the gate assignment as well. For more on the AP-methods, see Section V.

III. LAYOUT: PLACING AND CONNECTING MODULES

A typical layout problem can be described as a puzzle: many modules are to be placed in a given area so that they do not overlap. At the same time, certain points, called terminals, must be connected by mutually noninterfering wires (according to a given wiring list). The modules and the area are usually (but not always) rectangular in shape, and the wires must run with certain clearance between the modules.

There can be several different types of terminals (see Fig. 4). Electrically equivalent terminals are terminals connected inside the module. They represent different points of connecting to the same electrical terminal. Feedthroughs (sometimes also called jumpers) are a special case of electrically equivalent terminals—they are pieces of metal or polysilicon (depending on the technology) with no logic attached. Feedthroughs can also be viewed as prearranged partial or full connections for certain nets.

Fixed terminals, the most common terminals, are used for fully designed modules. Edge-assigned and floating terminals (see Fig. 4) are important in the top-down design phase, where only an approximation of some modules is used. Terminals of such modules are free to move (within some limits).

The connections can usually run on two or more layers without interference, although long superposed wires on different layers should be avoided, unless purposely used as capacitance. When working on two layers, horizontal connections are used on one layer, and the vertical connections

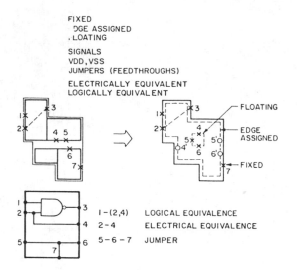

Fig. 4. Different types of terminals, and terminals for a composite module.

are used on the other. However, I have observed that the human designer does not follow this rule rigidly on densely packed interactively designed boards or chips. Most connections generally follow one direction on one layer, then use a contact window or a via hole, and continue in the perpendicular direction on the other layer. The resulting path has sideways jogs and bends, as shown in Fig. 2. A similar characteristic in layout can be observed for both PCB and MOS layouts in Fig. 1.

In the MOS layout, the resistivity and capacitance per unit path length are not the same for different layers. The metal layer is used whenever possible. The prime purpose of a polysilicon layer in routing is to cross under the metal. It should not extend far beyond this because polysilicon connections, and even long metal connections, have a detrimental effect on the performance of the circuit. Similar problems occur (though not that severe) even in two-metal MOS technology, where the electrical properties of connections on the two metal layers are not identical. On the other hand, main routing layers on PCB's usually have the same properties; the connections between the layers decrease reliability and increase the production cost. In fact, the speed of many chips today is limited more by the delay along the connections than by the physical speed of the basic devices, such as the transistor. This is especially so for large-scale-integration (LSI) layouts, where, for example, in polycell layouts up to 80 percent of the area is used for communication between the modules. For example, note the relative amount of the routing area in the layout shown in Fig. 5(a).

Current layouts involve so many modules and connections that it is a difficult task to place and connect them. For example, a typical printed circuit board may have 200 modules and 1200 connections, while the number of modules and connections on an LSI circuit may be in the thousands. Photographs at the end of the paper show the layout style of some advanced chip designs: Bell Laboratories (Fig. 32), IBM (Fig. 33), Intel (Fig. 34), Hewlett Packard (Fig. 35), and National Semiconductor (Fig. 36).

Because of the close coupling between circuit performance and layout, it is essential that the designer can simulate any part of the circuit at any time, using wiring capacitances and other electrical parameters of the layout. On the other hand,

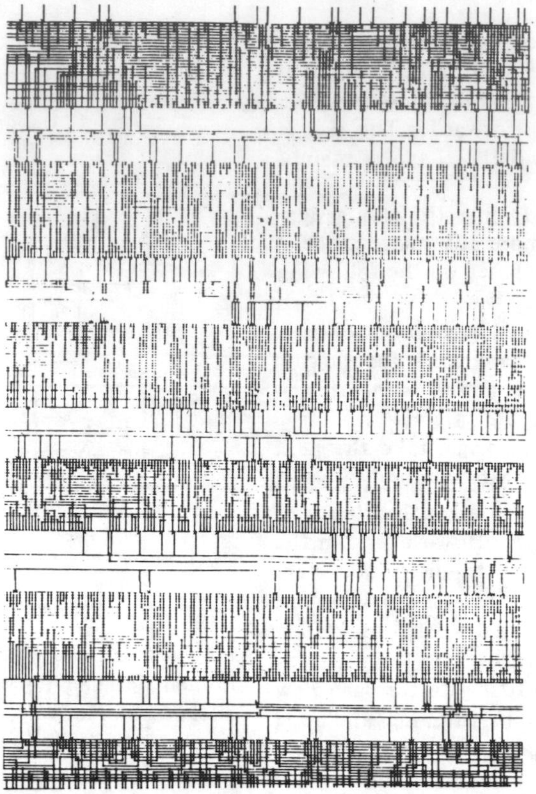

(a)

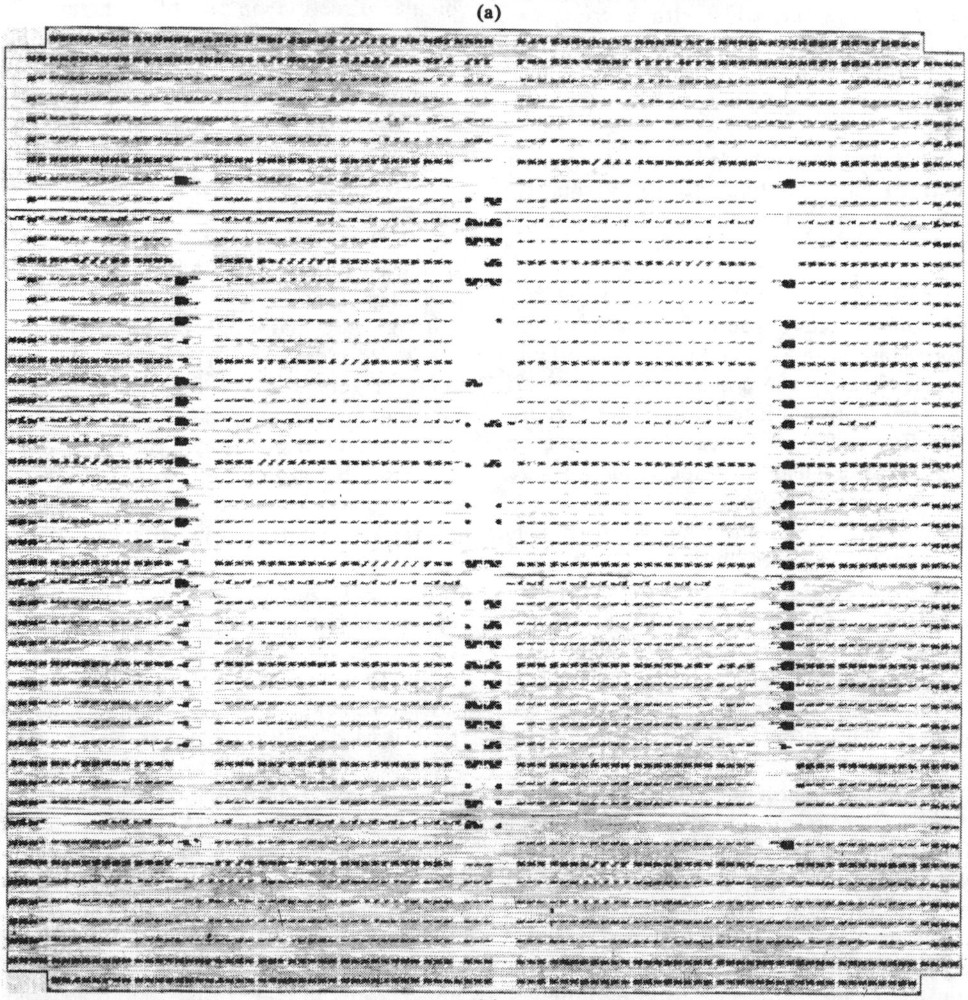

(b)

Fig. 5. On complex silicon layouts, the routing consumes a larger area than the modules. (a) Standard cell layout in rows (Bell Laboratories). (b) Gate array layout (metal layer of the IBM chip shown in Fig. 33, reprinted from *Electronics,* Oct. 9, 1980. Copyright © 1985, McGraw-Hill, Inc. All right reserved.

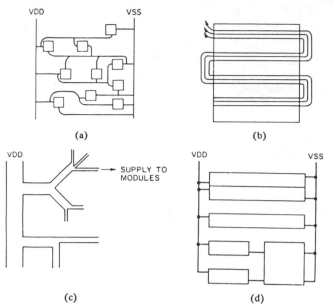

(a)

(b)

(c)

VDD   → SUPPLY TO MODULES

(d)

VDD   VDD   VSS

Fig. 6. (a) For one VDD- and one VSS terminal per module, VDD and VSS can always be connected on one layer. (b) Snake-like connections are used for multiple voltage supplies or when modules have several terminals for power and ground. (c) The track width may gradually decrease on power and ground nets. (d) Method of feeding VDD and VSS to rows of standard cells.

the layout algorithms (for both placement and routing) should consider a limit on the length of important nets, such as clock lines.

Power and ground (VDD and VSS) on IC's are usually designed all in metal. As shown in Fig. 6(a), this can always be done, providing each module has only one VDD and one VSS terminal. If two supply voltages are used, such as +5 V and +10 V, and if only one metal layer is available for routing, either all of the modules must be connected in a snake-like manner (see Fig. 6(b)), or else wide diffusion bridges must be used to cross under the metal.

In practice, the connections for power and ground are conveniently prearranged so that an automated system cannot miss anything in this critical task. In designs based on standard cell rows, such as in Fig. 5(a), each cell has two rails, one for power and one for the ground. In this manner, power and ground connect automatically within the row just by the butting of cells. Only the rows have to be supplied power and ground, and this can be done either by feeding ground to all the rows from the left and the power from the right (see Fig. 6(d)), or by connecting rows into a long snake (see Fig. 6(b)).

On PCB's and hybrid circuits, the power distribution schemes include wide parallel power or ground frames around the circuit. In the case of the PCB's one frame is on each side of the board. When the technology permits four or more routing layers, the usual method of increasing the circuit density is to use two layers for power and ground only. Large ground-connected layers also have the function of protection against noise and crosstalk between nets. The board shown in Fig. 1(a) uses this method of distributing power and ground.

The power and ground connections may decrease in width along the path from the source to the end points of the distribution net, see Fig. 6(c). The current density is the main factor to be considered. High current density may cause overheating, and it may burn the connections. On IC's, high current may also cause a directional diffusion of metal ions, resulting in an open circuit.

The general tendency in layout is a move toward more intelligent systems. For example, a system that performs an on-line rule check can detect rule violations right within the interactive layout. (The automatic layout should not cause any violations.) A rule check avoids tedious loops in the design process, which are necessary if only a postprocessor-type rule checker is available. What we want is an intelligent tool. Systems that draw rectangles and lines are not enough any more.

In order to provide on-line rule checking in real life layouts, sufficiently general modules are needed. Fig. 7 shows such modules used in Bell Laboratories new layout system. The module boundary is a polygon composed of horizontal and vertical lines. The terminals reflect the true size of physical features, and they may extend from, be flush with, or be inside the boundary.

This paper avoids the question of how to generate basic modules. It simply assumes they are given. Somewhere before the layout, there is assumed to be a set of specialized programs that generates gate matrix blocks or PLA's according to a given specification. Modules used as standard cells or memories are assumed to come from libraries or to be custom designed interactively on systems like APPLICON or CALMA. Stick diagrams and programs for compacting the mask layout also fall into this category. All we need are modules with their boundaries, the terminals, and the wiring list.

IV. CAN IT BE DONE?

The problem of layout is extremely complex. If, for a given set of connections, the modules are reduced to points, and we require the placement with the minimum wire length, we get the following assignment problem:

Let a_{ij} be a variable describing the assignment of module i to location j

$$a_{ij} = 0 \text{ or } 1, \quad \begin{cases} i = 1, \cdots, n \\ j = 1, \cdots, m. \end{cases}$$

Each module must be placed, and at any location there cannot be more than one module:

$$\sum_{j=1}^{m} a_{ij} = 1, \qquad \text{for all } i = 1, \cdots, n$$

$$\sum_{i=1}^{m} a_{ij} = 0 \text{ or } 1, \qquad \text{for all } j = 1, \cdots, m.$$

The objective is to minimize the wire length

$$\sum_{(r,s) \in G} \left(\left| \sum_{j=1}^{m} (x_j a_{rj} - x_j a_{sj}) \right| + \left| \sum_{j=1}^{m} (y_j a_{rj} - y_j a_{sj}) \right| \right) = \min$$

where G is the set of all connections (only two point nets are assumed), and x_j, y_j are coordinates of location j.

This problem is difficult enough for n in the order of hundreds, and still this representation is too simple for what we really need. (I am assuming that the problem is nondeterministic polynomial-time (NP) complete,[1] but I am not certain that this problem has been proved to be NP complete or not.)

[1] The time for solving such problem grows exponentially with the number of variables. In practice, for larger problems of this type, only an approximate solution can be found.

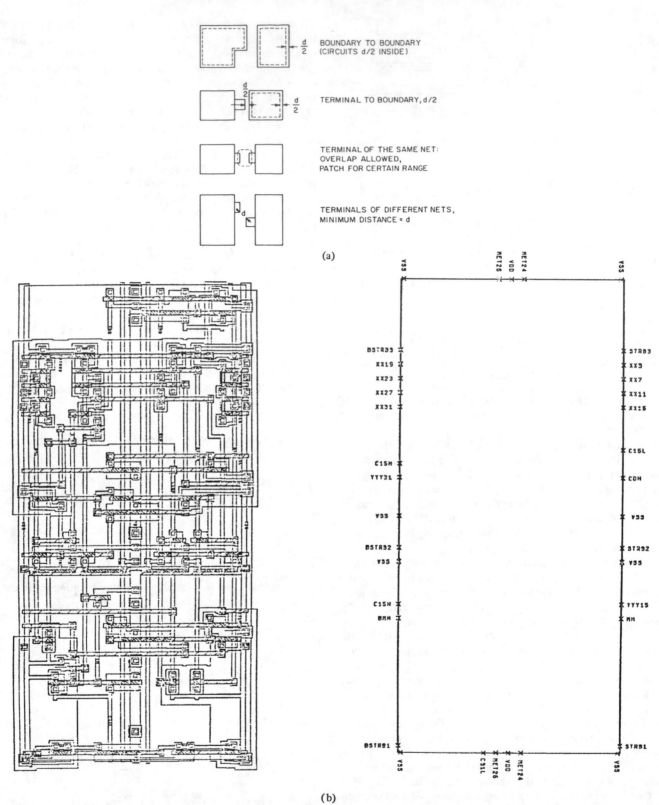

Fig. 7. (a) A module can have a more general shape than just a rectangle; also, the on-line rule check is performed on a realistic terminal representation. (b) Gate matrix block and its representation in the layout (such a representation is generated automatically).

The problem of finding the optimal interconnection of multipoint nets is NP complete. The optimum net is a minimum Steiner tree (see Fig. 10(d)). The optimum choice of channels in a complicated layout is yet a more complicated assignment problem than that of the placement just shown. Conditions for continuity of the connections and the limit on the number of connections in individual channels make it more complex. The channel assignment (sometimes called loose routing or

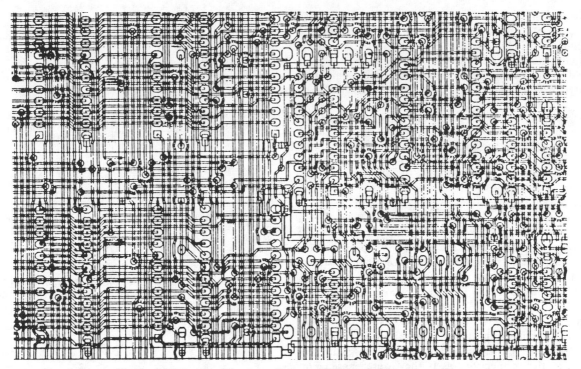

Fig. 8. An example of dense routing on a PCB (placement and routing produced automatically on the CPS system, Bell Northern Research).

also global routing) depends partially on the shapes of individual connections and on their interaction. Placement, loose routing, and final routing are mutually dependent, and still each one is too difficult to solve alone.

This explains why the task of layout is traditionally performed as the placement, the loose routing, and the final routing (or the track assignment) steps. At each step, mainly heuristic methods are used, and the divide-and-conquer approach or the procedures used in interactive designs are frequently applied.

When working on the placement, a simplified model is used to estimate the routing. When assigning nets to channels in loose routing, the connections are also simplified. For example, connections may be represented as straight trunks with links to terminals on both sides of the channel. Only during the final routing does everything become real. Frequently not all of the routing can fit into the available area. A new placement is required: new loose routing, new final routing, sometimes a tedious interactive cleanup, where a designer tries to squeeze the connections into an area that is too small to accommodate them. (Imagine trying to insert an additional connection into a dense layout, such as the one shown in Fig. 8.)

I am aware of only two unique attempts to perform both placement and routing simultaneously. One is the commercially available system GAELIC-COMPEDA [27]; the other is the experimental global router developed at Bell-Northern Research (BNR).

The COMPEDA system simulates the procedure of the manual hand-packed design.[2] The program starts from the bottom of a rectangular area, adding modules one by one.

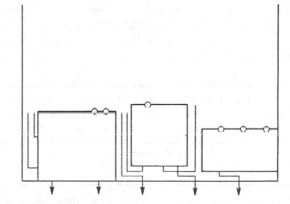

Fig. 9. COMPEDA method of layout. After adding each block, the bottom boundary line moves up, and new temporary terminals are made for unfinished nets.

With each module, all the unfinished connections are extended to the new bottom line, which rises gradually up, like porridge in a pot, until all modules are placed, and all the nets are connected (see Fig. 9). Though this approach can be quite efficient in some cases, it has the same disadvantage as the manually packed layout. Its placement is only as good as the initial placement (see section five). The iterative improvement of the placement is very difficult and sometimes impossible.

The global router paper [1] describes a global method for final routing. The idea is to expand areas around the terminals, with each area assigned a different priority. It is similar to pumping air into balloons placed at terminal locations. When balloons associated with terminals of the same signal touch, they spark a connection, stick together, and their priority drops. The computer algorithm operates on a grid, changing dynamically both the extension of the areas and their priority.

This idea of global routing can be expanded even more. If

[2] This description is based more on an analysis of the test runs with the COMPEDA system than on any published material; see also [27].

the balloons are fixed to free floating modules, the balloons push away not only other balloons, but also whole modules—causing simultaneous changes in placement and loose routing. Unfortunately, the improvements in layout were not as great as we would need to compensate for the enormous computing time required for running this router. The Fortran code of the router is available free of charge from the BNR.

In a way, Mead's approach to layout [18] also combines placement and routing. In his design style, most modules are designed considering the overall context; small modules fit into large modules. Most of the placement is predetermined by the overall system design, and most routing is done through the butting of modules.

It is interesting to watch how quickly circuit designers adjust to the improvements in the quality of available design tools. The demand for the amount of circuitry to fit one card or a given area of a chip is always just a bit too high. This continuous challenge is what keeps many of us in this business. But this kind of work is also continuously frustrating in the sense that we are always trying to perform an impossible task.

From this point of view, methods for estimating 'wireability' are of extreme importance. The number of modules and connections on boards and chips is getting so great, that we can look at the connections as if they were chosen randomly. Assuming that the logic is random in character (it has no regular pattern such as those that might appear in memories or registers), useful parameters may be obtained through the statistical analysis of existing layouts. Heller *et al.* [14] have developed generally valid formulas to estimate the wireability of new designs even before the layout actually starts. Their method can determine the probability of being able to finish with few connections left. The results can be used as an acceptance criterion for new circuit designs. Their method not only protects the layout people and the system, but it also saves the circuit designers money and time that they may otherwise spend trying to accomplish something impossible.

V. Placement

Good placement is the key ingredient to a successful layout, but it is also where we have to face defeat most frequently. This computer-aided design (CAD) area is still under development.

When placing the modules, we need a model for individual nets. The model should resemble the final routing. The model should also be easy to compute because it will be used over and over. Fig. 10 shows some typical net representations. My favorite is the straight trunk representation, which has the advantage of being close to the usual net appearance and is easy to calculate (see Fig. 10(c)).

The true objective of the placement phase is to achieve completely automatic routing in a small compact circuit. Such an objective is not mathematically well defined. Unless we try the final routing, we never know whether we have succeeded or not. We replace the true goal by a simplified objective that is easy to enumerate on the computer. When we choose this objective, we hope that by improving it, we also improve routability.

For many years, the minimum wire length of all combined connections had been the placement objective. We know today that such an objective leads to crowded unroutable channels, especially in the middle of the layout. A dramatic improvement can be achieved when the peak of the crossing

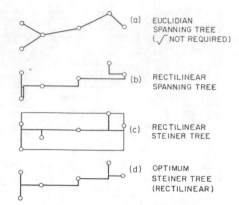

Fig. 10. Net representation in placement. The rectilinear Steiner tree based on a single trunk is a realistic model, and it is also easy to calculate.

count is minimized. Fig. 11 shows an example of the definition of the crossing count and a fairly distributed density after using this placement improvement method. I have had a good experience with the following objective:

$$\max_{v,h}(c_i) + g \sum_{v,h} c_i^2 = \min$$

where g is a sufficiently small, positive number, and c_i is the number of connections crossed by the grid line i; note that there are horizontal (h) and vertical (v) grid lines.

Note the similarity between the objective function for minimizing the peak of the crossing count and the function to minimize the wire length. The wire length function can be written also in this form:

$$\sum_{v,h} c_i = \min.$$

The routing density in Fig. 12 is a realistic estimate of the channel width when using a channel router as the main routing algorithm. Current channel routers generate routing with density equal to or very close to the channel density. The placement objective for channel-routed regions typically has three objectives:

1) to avoid cycle constraints between nets; otherwise the simple router will not work;
2) to minimize the channel density;
3) to minimize the span over which the maximum density occurs.

The algorithms for placement can be divided into two groups: constructive (initial) placement and interactive improvement of placement.

A. Constructive Placement

The epitaxial growth algorithm is the most usual form of constructive placement. The algorithm is a reflection of the human interpretation of the problem, and it is quite simple to code [15], [19].

The algorithm assumes some seeded[3] placement. The seeds are either manually preplaced modules or fixed I/O points,

---

[3] Seeded placement is similar to growing crystals. Adjacent modules are placed around the seeds, causing modules to cluster like crystals.

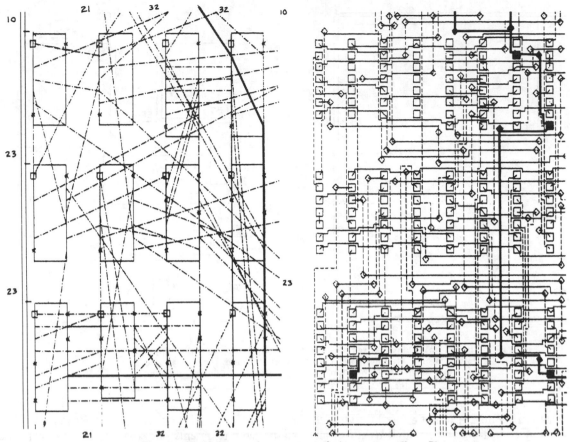

Fig. 11. Crossing count is a measure of placement quality. Placement and its crossing count is at the left, the final routing is at the right. (Reprinted from *Telesis*, the technical magazine of Bell-Northern Research Ltd., from the article "Electronic circuit layout algorithm," by J. Soukup.)

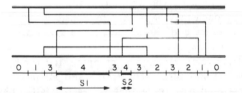

Fig. 12. The definition of the channel density is based on counting the number of nets crossing any interval. The secondary measure is the span of the max. density, in this example span equals $s_1 + s_2$. (Can you find routing using 4 tracks only?)

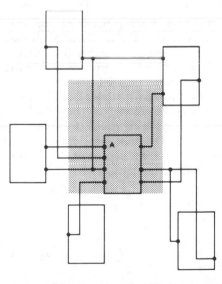

Fig. 13. For every module there is an equilibrium rectangle where the total pull from all other modules is zero. A zero pull is equivalent to the minimum length of the wire for all connected signals. (Reprinted from *Telesis*, the technical magazine of Bell-Northern Research Ltd., from the article "Electronic circuit layout algorithm," by J. Soukup.)

such as the connector on a PCB or I/O pads on a silicon chip. The manually preplaced seed is either an important module, such as a microprocessor on a PCB, or a special module having a central position in a functional block of logic. By planting such a seed, the designer determines indirectly the placement of the whole block.

The algorithm finds the next unplaced module with the maximum number of connections to the placed modules. Then it moves the module into the best available position, finds the next unplaced module with the maximum number of connections, and so on, until all the modules are placed. The crossing count does not make much sense when many modules are still unplaced. Therefore, the wire length is used to determine the best position, usually by trying all available positions. The size of the module, the number of its terminals, and other factors may be used to weigh the number of connections when choosing the next module to be placed.

29

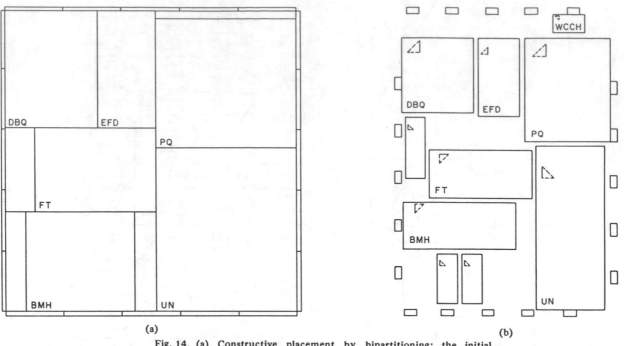

(a)   (b)

Fig. 14. (a) Constructive placement by bipartitioning; the initial stage, where modules have correct areas, but not the shapes yet. (b) Placement after switching to true shapes of modules, and then iterating to improve the packing.

The best position for a module is found either by trying all available positions, and comparing the required length of all connections, or by placing the module directly into the zero-force rectangle shown in Fig. 13.

The epitaxial growth placement algorithm is easy to implement on a grid (typically on printed boards), or in a row (for a row-organized LSI layout). But for a continuous placement of modules of variable sizes, recently designed systems favor bipartitioning algorithms [2], [3], [20], [26]. Such algorithms are based on finding the minimum cut through the circuit in such way that the total size of all modules in each set is about equal, and the number of connections between the two sets is minimized. Then each set is partitioned again into two halves, and so on, until the remaining modules are small in size or single basic modules (see Fig. 14).

The partitioning was based on heuristic rules, and only recently have attempts been made to apply integer programming techniques. The new development started with the invention of new efficient algorithms for solving the assignment problem (AP) in operations research literature:

$$\text{Minimize} \quad \sum_{i,j} c_{ij} \cdot a_{ij}$$

$$\text{subject to} \quad \sum_{i=1}^{n} a_{ij} = 1, \quad \sum_{j=1}^{m} a_{ij} = 1, \quad a_{ij} = 0 \text{ or } 1.$$

For example, the algorithm reported by Hung and Rom [22] is of the order of $O(n^3)$, and it takes about 1.5 s of central processing unit (CPU) time on the IBM 370/158 to find the optimum assignment of 100 objects.

Note that the general advantage of bipartitioning is that it minimizes the peak of the crossing count. Therefore, it is better than the seeded initial placement described above.

A reasonably good initial placement can also be obtained from the following physical model. Let us assume that all

modules are small objects connected with elastic bands. Each elastic band represents one connection to be made or a part of a multipoint net (multipoint nets cause some problems in this model). The bands pull the modules together with a force that is proportional to the distance. After picking up four modules and placing them in the corners of the available area, all other modules have an equilibrium position, which can be found by solving a system of $(n - 4)$ variables for $(n - 4)$ linear equations, where n is the number of modules.

This approach may be refined by considering the true shapes of the modules and the shape of the given area. Also, the elastic bands may be attached at true terminal locations, and their attractive force may be compensated for by a repulsive force of the modules. The frame and all modules generate negative (repulsive) force that will affect all other modules. This force increases as the distance gets shorter; and the force approaches infinity when two modules touch. In this way, the overlap of modules is automatically avoided. The modules spread in the given area, and no modules have to be fixed in the corners. Assuming that the modules have negligible mass and that they float in a heavy oil,[4] we get a system of differential equations of the first order. The steady state of the system is easy to find by simulating the movement of all modules in small time intervals.

Many people who believe in a random initial placement argue that the iterative placement improvement (see below) will change the placement so much that any placement is really good enough to start with. Another argument is that if we start with a good initial placement, we may be enforcing a local optimum from which the improvement algorithm can find no exit. The third argument is that to avoid local minima in the improvement phase, it is useful to try at least several

[4] Modules floating in heavy oil move with speed proportional to the force. When there is no mass, the system does not oscillate or overshoot.

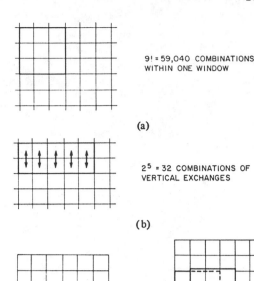

9! = 59,040 COMBINATIONS
WITHIN ONE WINDOW

(a)

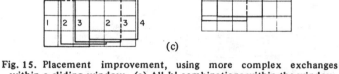

2^5 = 32 COMBINATIONS OF
VERTICAL EXCHANGES

(b)

(c)

Fig. 15. Placement improvement, using more complex exchanges within a sliding window. (a) All $k!$ combinations within the window. (b) Two pair exchanges within a double row of length k. (c) Moving the window with an overlap.

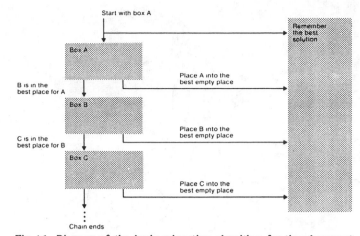

Fig. 16. Diagram of the basic relaxation algorithm for the placement improvement. This method is able to get over some local minima. (Reprinted from *Telesis*, the technical magazine of Bell-Northern Research Ltd., from the article "Electronic circuit layout algorithm," by J. Soukup.)

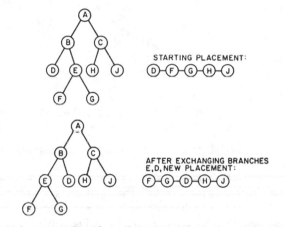

STARTING PLACEMENT:

AFTER EXCHANGING BRANCHES E,D, NEW PLACEMENT:

Fig. 17. Improvement of a linear placement by exchanging two branches of the clustering tree.

initial placements. The constructive algorithms provide only one placement each, but we can generate an almost infinite number of random placements to start [15].

B. Placement Improvement

The purpose of this phase is to improve the placement by applying small local changes, such as the pair-wise exchange of modules. Typically, there is a large number of trials and it is essential that the objective (crossing count or the wire length) is calculated on an incremental basis. For example, when two modules are moved, the original connections attached to the two modules are subtracted from the count. Then the move is made, and the count for those several connections is added to the total.

There are many ways to shuffle the modules around efficiently. For example:

1) $k(k-1)/2$ pairwise exchanges may be done only in a certain neighborhood of each module (Fig. 15(a)),

2) all $k!$ of placement combinations may be tried for groups of k modules (for example by using a window of k modules, and sliding over the given area) (Fig. 15(a) and (c)).

3) for a double row of $2k$ modules, all 2^k vertical exchanges may be tried (Fig. 15(b)); the sliding window may be used again.

The placement of modules in a single row is a special case. This placement can be improved by employing a tree representation of a clustering tree (Fig. 17). (As the reader may have recognized, the problem of placement has a lot in common with partitioning and clustering.) Any two branches at the same level of hierarchy may be exchanged, and this exchange is much more powerful than exchanging simple modules.

All of these algorithms are essentially greedy, and that is their major weakness. They easily get stuck in a local minimum. For this reason, the relaxation algorithm, which does not stop when placement ceases to improve, usually achieves better placements. Fig. 16 shows the idea of the algorithm: starting from an arbitrary module the program looks first for the best[5] available (empty) location. The value of the objective function is recorded, and the program places the module into the best position, regardless of whether the position is occupied. The module that occupied that position is the next to be moved. It is first moved to the best available position, and the value of the objective function is recorded. Then it is moved into the absolutely best position, regardless of whether it is occupied. This is repeated until either the best position is available (there is no next module to move) or until a module, which was already moved, would move again. (There would be a danger of getting into a loop.) When the chain stops, the subsolution with the best value of the objective function is chosen as the next improvement. (This, of course, can also be the original placement, which means there will be no change). The next chain will start with a different module, and so on.

A more general description of the relaxation algorithm was published by Goto [4]. The simple chain algorithm described

---

[5] The word best is being used here for any position inside the rectangle of the equilibrium pull (Fig. 13).

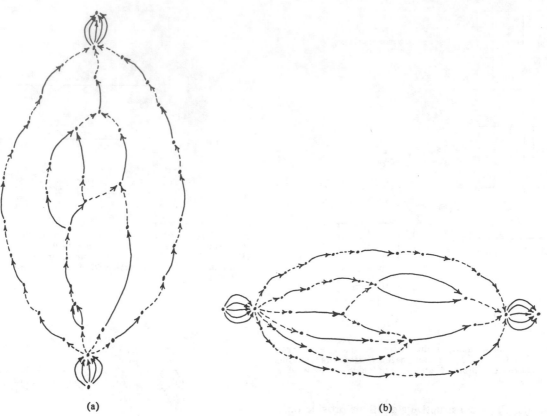

(a)   (b)

Fig. 18. Position graph for the placement from Fig. 14(b). (a) Vertical position graph. (b) Horizontal position graph.

above is a special case of Goto's algorithm for parameters $\varepsilon = 0$, $\lambda = \infty$.

All of the algorithms based on the exchange of modules become difficult to control if the modules are of different sizes. In fact, the exchange algorithms are practically useless in problems, such as the one shown in Fig. 14(b). For such cases, a relatively new method of improvement was developed by Preas and Lauther [2], [3]. The method describes the relative positions of modules in the form of two position graphs, one for the horizontal constraints, one for the vertical constraints. The graph has one node for each module or, in some versions, one node for each channel between the modules (Fig. 18). The length of the critical path in the two graphs determines the overall x and y dimensions required for the placement. A set of simple rules transforms changes in the placement into changes in the position graph.

The position graph has an advantage that it can take into account not only the size of the modules, but also the width of the channels between the modules. Thanks to this property, the feedback between routing (or loose routing) and placement becomes simple. Values assigned to the edges of the position graph include the required channel width, and the placement automatically adjusts to the routing requirements.

C. Considering Shapes

The placement algorithms, as described earlier, have one common deficiency: they do not consider shapes. The result can be seen in Figs. 19(a) and (b). The two figures represent two versions of a layout for the same circuit. The layout in Fig. 19(a) was produced using a two-level optimization. First,

natural clusters (logical units) were manually detected in the circuit. Then each cluster was hand designed and packed as much as possible. In the second optimization, automatic placement and routing was used, including bipartitioning, improvements based on the position graph, loose routing, and final routing using a Lee-type algorithm. Fig. 19(b) shows an independently done manual design that requires only 45 percent of the area (see [23]). How is that possible?

My explanation is that when the automatic layout was generated (Fig. 19(a)), no attention was given to the shape of the modules. The hand-packed design is in neat rows, taking full advantage of matching heights and of power and ground distributing rails. The automatic layout was shortsighted in optimizing each cluster, without considering the whole layout. This result is even more interesting because the manual layout (Fig. 19(b)) followed the strategy of rows typical for some automatic systems [19].

The influence of shape is a new factor we will have to consider in layout. This applies even more for general (non-rectangular) modules, such as L shapes or E shapes. For such modules, matching shapes may be more important than a large number of common connections.

VI. Loose Routing

When PCB's and IC's had a small numbers of modules, there was no need for loose routing. With the size of the circuits used today, loose routing has become an important part of any layout system. One of the key reports on loose routing is the paper from IBM [5] on routing gate-array chips.

Loose routing is a preliminary planning stage for the final

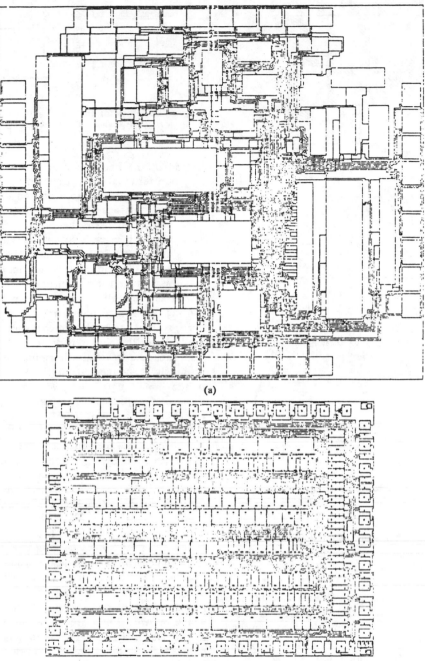

(a)

(b)

Fig. 19. Comparing the automatic and hand-packed layout. (a) Automatic layout (an earlier version). (b) Carefully handpacked layout, requiring only 45 percent of the area.

routing. The loose router decides through which channels individual connections will run. The main consideration is the flow through narrow or important channels. Some connections may be routed around to avoid critical bottlenecks.

The preliminary stage of loose routing involves the definition of the routing area (Fig. 20). The definition must fit the algorithm for loose routing, but must also provide a fair representation of the space required for the final routing. The definition of the routing area has a direct impact on all parts of the layout: on the adjustment of placement when routing does not have enough space, on data organization, and on algorithms of both loose routing and final routing.

I am aware of five ways to represent the routing area.

1) The BNR system cuts the area into numerous small rectangles. It then applies a generalized Lee algorithm to expand from rectangle to rectangle (Fig. 21 and paper [23]).

2) The PI System as being developed at M.I.T. creates larger rectangles using only the shorter of the two possible edges for each corner (Fig. 20(b)).

3) A new layout system at Bell Laboratories divides the routing area into similar rectangles that are then combined (automatically or interactively) into channels. The loose router operates on these channels, considering the available channel density.

4) The CYCLOP system developed at Stanford University and Sandia Laboratories uses a system of horizontal and verti-

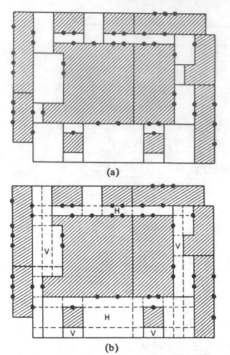

(a)

(b)

Fig. 20. (a) Dividing the routing area into rectangular blocks, then defining the routing channels (Bell Laboratories). (b) Dividing the same routing area into rectangles (M.I.T.).

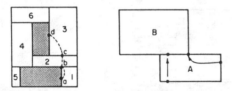

Fig. 21. Rectangle to rectangle expansion is the basic idea used in loose routing.

cal lines through channels. The line segments do not carry all of the information available in the two other systems. However, the channel graph immediately tells the sequence in which the channels must be routed.

5) The method used by Wiesel and Mlynski [24] divides each routing layer into a different set of rectangles. Rectangles on one layer are long in the horizontal direction, and rectangles on the other layer are long in the vertical direction. In both loose and final routing, a connection can change layer only in the area where horizontal and vertical rectangles overlap.

The algorithm for breaking the routing area into rectangles does not take too much computer time [23], [24]; after sorting the boundary segments of all modules by their lower x and y coordinates, the rectangles can be obtained by a single scan over the segment data. However, if the channel router is to be used as the main algorithm for final routing, it is most important to avoid loops between horizontal and vertical channels. If possible, no routing area should have fixed terminals on two sides that complete the rectangle. We want typical channels and no switchbox areas (see the next paragraph for more details). Switchbox areas cannot be routed by the channel router and, for them, we cannot guarantee that all of the connections will be finished in final routing. See Fig. 20(a) for an example of how the routing channels are created.

A loose router [23] operates on a principle that is a general-

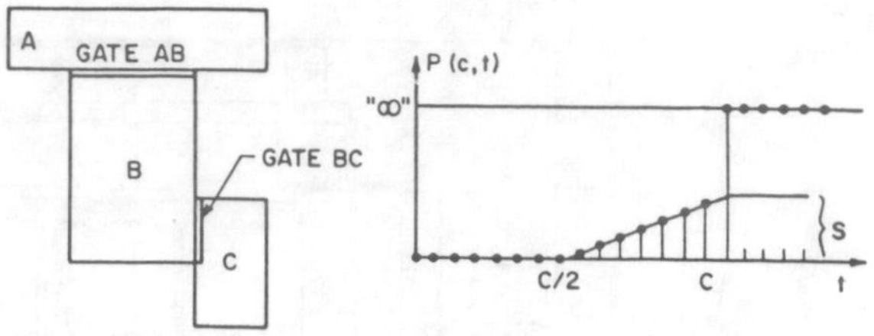

Fig. 22. Definition of the 'gate' between two channels, and the progressive penalty for crossing crowded gates.

ization of the Lee-expansion algorithm. When connecting a net, one terminal expands from one rectangle to all adjacent rectangles, generating points called source points. Each source point expands again into all adjacent rectangles, and so on, until at least one of the target terminals is reached. Each source point is given a cost value relating to the distance from the starting terminal. In fact, the expansion process only stops when the minimum cost from all target terminals (there may be several for a multiterminal net) is smaller than the lowest cost of all active source points.[6] In the process of routing, the source points are kept in a partially sorted binary heap, so that the minimum-cost source point is always directly available for the next expansion.

In the case of a multiterminal net, after the first pair of terminals is connected, both the terminals and the source points along the path are given a zero starting cost for the next expansion sequence. This helps to generate nets with additional branching points (Steiner nets).

The main control mechanism of the loose router is the system of penalties for routing through critical areas. The penalties depend on the algorithm used for the final routing. For example, a penalty for overflowing channel density is required for the channel router, and a penalty for the gate[7] capacity overflow (see Fig. 22) is required for a Lee-type maze router.

The loose router may run in two passes. In the first, all connections are routed without penalties, based on the shortest distance (or the smallest resistance) only. The first pass discovers the critical channels or gates, and it also monitors overflow. The second pass may have several iterations with progressive penalties in critical channels. The sequence in which individual nets are routed may also dynamically change in the second pass, giving some nets priority in choosing the path. Fig. 22 shows an example of the gate penalty that may be used.

The result of loose routing is a decomposition of final routing into a number of smaller problems, one for each rectangle or channel. These subproblems are small enough to be handled in a short time and without memory overflow even on a relatively small computer system.

Note that several attempts have been made to develop a global router that would combine loose routing and final routing (track assignment) into a single step:

1) The global router, based on expansion of the area around the terminals, has been mentioned previously [1].

---

[6] Active source points are source points in the front wave. These points are kept as sources of future expansion.

[7] In order to be consistent with paper [23], we use the word gate for the interval common to two adjacent routing rectangles. This gate has nothing to do with logical gates used in electrical circuits; it is simply a routing bottleneck between two routing rectangles.

2) Vintr developed another form of global routing (see Section VII and [13]).
3) Any form of the ripout router, such as the one developed in American Microsystems (AMI) [6], is an attempt to combine loose routing and final routing.

VII. FINAL ROUTING

The final routing of connections is a most exciting game. It is in final routing that all of the layout effort suddenly assumes its physical appearance. The task is often difficult or impossible (when the placement is bad), and because its result is so vividly physical and convincing, usually more recognition is obtained for any finished routing then for a superior, really smart placement.

Historically, there have been three types of routers:

1) grid expansion (by C. Y. Lee [17]);
2) channel router (by A. Hashimoto and J. Stevens [28]);
3) linear expansion (by D. Hightower [16]).

Fig. 23, from one of the first reports published by D. Hightower, explains how the router finds a connection through a maze of obstructions. The router runs vertical and horizontal expansion lines from the two terminals to be connected. Then, for each line, it finds the longest perpendicular escape line. If there is a multiple choice, the escape line nearest to the starting terminal is taken. This process is repeated until expansion lines from both sets (one set for each terminal) intersect, generating a connection. In most cases, the algorithm generates the path with the minimum number of bends, that is, the path with the minimum number of contact windows or vias.

The Hightower algorithm is gridless in principle, and it is quite fast for simple mazes. However, for complicated mazes, it becomes slow, needs a large stack of data, and does not guarantee the connection, whenever it exists.

The Lee algorithm is based on expanding a wave from one point to another. At each step, grids on a diamond-shaped front wave are expanded one step further. Each grid point is marked with a traceback code. The traceback code stores the direction to the source of the expansion. The points of the wave front are kept in a stack. The traceback code requires 3 bits of memory per grid for multilayer routing (see Fig. 24(a)). The algorithm guarantees minimum path length using rectilinear distance (a path composed from horizontal and vertical line segments).

The Lee algorithm requires a large memory for large and dense layouts. Its original version was slow for long connections. But the router guarantees a connection if it exists, regardless of how complicated the maze may be. The speed improves as the area gets more congested.

The channel router is a special router designed for routing in an area with no inside obstructions and with terminals on two opposite sides. The connections may exit from the channel through the two remaining sides, but the router determines the exact place. The first paper published on channel routing was by Hashimoto and Stevens [28]. The dogleg router [7] that is still used in Bell Laboratories has been one of the most successful implementations of this router. Its idea is simple: Straight trunk connections are used whenever possible. If an interference can be detected at the next terminal, a dogleg is used to switch to the track on the opposite side of the channel (Fig. 25). The algorithm assigns one track at a time; and it is greedy since it fills one track as densely as possible, even if the connections come from the opposite side of the channel. The algorithm avoids the directional bias by assign-

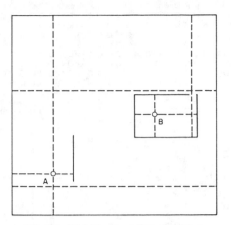

Fig. 23. Hightower routing algorithm.

ing the top track from the left, then the bottom track from the right, then the top track from the right, and the bottom track from the left. Several parameters can be chosen: one specifies the track selection sequence, and the other the minimum length of the trunk without doglegs. The algorithm generates routing with the density almost equal to the mathematical minimum.

The channel router is gridless, and it always makes all of the connections—at least in the topological sense. If the channel is not wide enough, the routing may overflow the channel, but the router implements all of the connections. It is very fast, and it has been the main routing algorithm for routing IC's for many years. The router has two limitations:

1) Terminals in the channel may create a constraint loop (Fig. 25(c)); the placement program checks for such loops, and it avoids them at any cost.
2) The given terminals must be on opposite sides of the channel; if the terminals are given on adjacent sides, the channel router cannot be used (with some exceptions for the three-sided channel routing problem).

Though the Hightower router has been used extensively, especially for the routing of printed boards, no major improvement in its algorithm had been made until last year, when Heyns et al. [21] combined the line expansion router and the grid expansion router into a new algorithm (Fig. 26). Their new router expands from a line like the Hightower router, but then it fills an area like the Lee router. However, the area is not kept in memory; only its boundary segments are remembered, as in the Hightower router. The character of the resulting path is similar to the Hightower router, and in most cases, the path has the minimum number of bends. In my opinion, this router has a great potential for today's applications. It finds a path, whenever it exists, and it does not need a grid. It is fast, and it can even use some penalty functions similar to those described below for the Lee-type routers.

Two approaches were suggested to improve the speed of the Lee algorithm [8], [9], both using the same idea. If the direction of the target point is known, expand more in that direction, unless that direction is blocked. The expansion does not proceed all around the wave front. Grids toward the target have priority in being expanded. The Rubin router achieves this by sorting the wave front using the key that is a total of the true distance from the starting point plus the estimated distance to the target point (equal to air rectilinear distance). The arrow router achieves a similar effect by a simple control over the loading and unloading of the stack. Points expanded toward the target are next to be expanded. The speed improve-

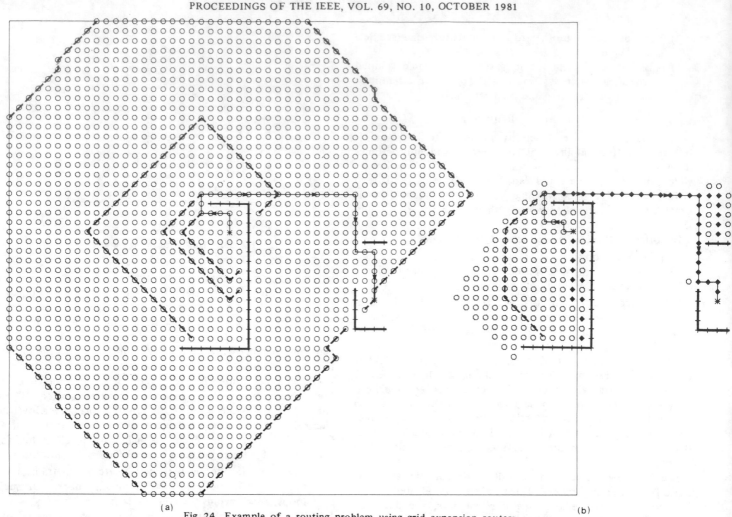

(a)   (b)

Fig. 24. Example of a routing problem using grid expansion routers. (a) Basic Lee algorithm. (b) 'Arrow' router enters fewer grids, and it is faster. (Reprinted from *Telesis*, the technical magazine of Bell-Northern Research Ltd., from the article "Electronic circuit layout algorithm," by J. Soukup.)

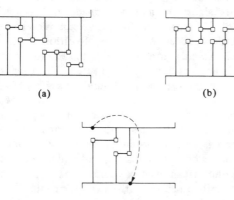

(a)   (b)

(c)

Fig. 25. (a) Channel routing with straight trunks. (b) When doglegs are used, fewer tracks are needed. (c) Most channel routers fail on a terminal constraint loop.

Fig. 26. Basic idea of the new router by Heyns, Sansen, and Beke. The router expands from a line over an area, but it records only boundary line segments for future expansion.

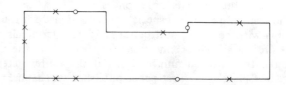

Fig. 27. An example of the three-sided channel, for which we need a good router today. (x and ○ denote terminals on different layers.)

ment is dramatic, in the order of 20–50 times for typical printed boards. Other properties of the Lee-type router remain the same, in particular the guaranteed connection and a large memory required.

The Lee router has proved to be an enormously flexible and efficient tool when more than a few bits of memory are used per grid. A set of penalty functions may control the router to generate the minimum resistance path with a minimum number of bends, allowing small side jogs, running tracks along existing structures, and running wide tracks for power and ground. The router can intelligently, not rigidly, use

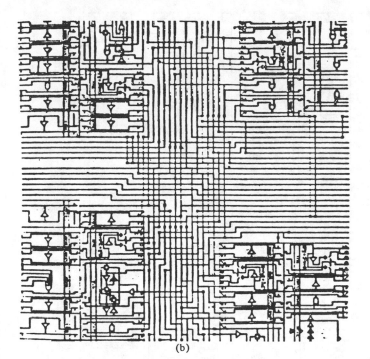

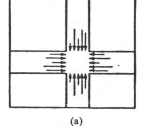

(a)

(b)

Fig. 28. Switchbox routing problem. (a) A general representation.
(b) An example of switchbox routing on a manually routed chip.

preferred layers. For this reason, a large effort was invested in the refinements of the Lee router. It may also be because it is so graphically illustrative. The author still rates it as the best router when smart routing is required.

In the area of the channel routing, cleaner algorithms of a more mathematical nature have been developed. Yoshimura and Kuh [10] developed an algorithm based on constraint trees, and they have results slightly better than those of the dogleg router [7]. The improvement, however, in the number of tracks is only minor. Both channel routers operate very close to what is the mathematical lower bound for the channel density.

Roberts, currently with G.T.E., showed that the dogleg router algorithm can be modified to reduce drastically the number of contact windows in the channel.

I think that a new channel router will soon emerge; a router that will permit fixed terminals on three sides of the channel (Fig. 27), with only one side for the algorithm-defined exits. Such a router will be extremely useful for channel routing of general layouts, such as in Fig. 28, where it is impossible to apply two-sided channels as we use them today.

Another router, which we desperately need, is a router for the switchbox problem (see Fig. 28(a)). Switchbox is a rectangular routing area with no obstruction inside and with signals entering from all four directions, possibly on several layers. The switchbox problem is more difficult than the channel routing problem, but it is easier than the general routing problem because there are no obstructions inside, and the terminals are on the perimeter only. Lee-type algorithms are not suitable for solving this problem; they do not check ahead to avoid unnecessary blocking of other terminals, as the channel router does.

One possible candidate for the switchbox problem is Akers's router, though it is believed that it has never been published [11]. The idea is to expand from the outside boundary in the inward direction (see Fig. 29). The shortest possible

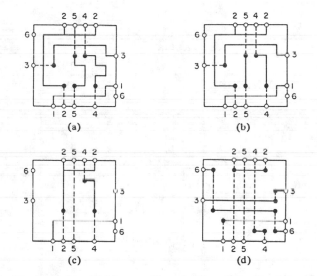

Fig. 29. A simple switchbox problem and its solution by Akers's and Lee's routers. (a) With Akers's router, connections are implemented in this order: 2, 1, 4, 2, 3, 5. There is a problem finishing net 6. (b) Akers's cleanup phase. The U shapes are removed on nets 5 and 4. (c) Lee router with a preference for one layer. A small number of changes between layers is used as the secondary objective. Problems to implement nets 6 and 3. (d) For a regular Lee router based on horizontal and vertical layers, the problem is easy to solve. All nets are connected. Note that terminals may not be always available on both layers.

connections along the outside wall are done on the main routing layer first. When such a connection blocks some terminals, partial connections are extended directly inward on a secondary routing layer, but only as much as is needed for the access to the inside routing area. The wall of routed connections closes more and more, until (hopefully) all the connections are completed, and some unused space is left, usually in the center area. At that time, most connections appear crooked and strange. They are long and run around

37

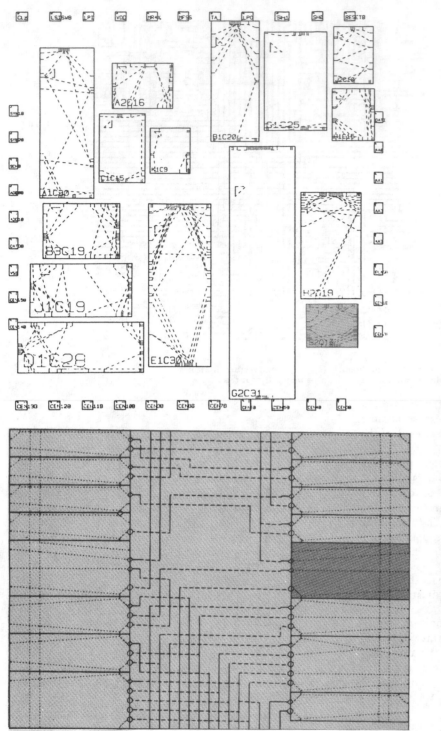

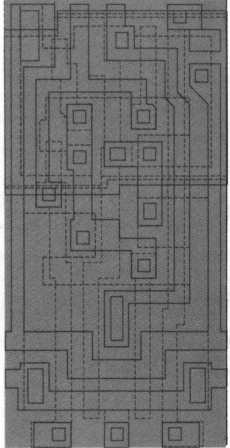

Fig. 30. Several levels of hierarchy in a chip design. (Reprinted from *Telesis*, the technical magazine of Bell-Northern Research Ltd., from the article "Electronic circuit layout algorithm," by J. Soukup.)

the wall, instead of being the straight, short lines we would like them to be. This can be corrected by the postprocessing algorithm, which tries all connections repeatedly, searching for a U-shape on the main routing layer. If the U-shape does not have any other connections inside, it is removed and replaced by a straight line. The result is a dense, intelligent routing, with the main routing layer used whenever possible. (In MOS layouts, metal is the main routing layer, polysilicon is the secondary layer.)

Two new methods for the switchbox problem are being tested at M.I.T., as a part of the PI project. One method is fast and simple, the other more elaborate and more time expensive. The two routers are the prime routers in the M.I.T. layout system. The system requires a switchbox router because all routing rectangles are two-dimensional regions and not channels (see Fig. 20(b)). (Details on the M.I.T. layout system have not been published and are expected to appear by the Summer of 1981.)

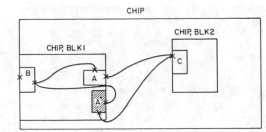

Fig. 31. A change in the placement of module A may change the net list at some levels of the hierarchy. This example shows that physical and logical data are sometimes difficult to separate.

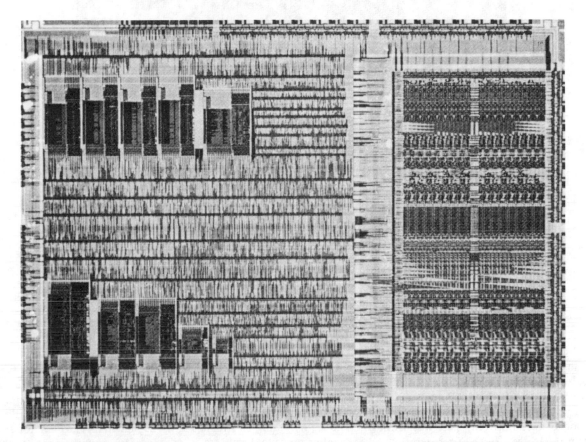

Fig. 32. Bell Laboratories full 32-bit CPU implemented in CMOS. Chip size 10.5 × 13.9 mm (3.5-m rules, 100 000 transistors). Laid out by 5 designers in 3 months, using various automated design tools.

Recently Sadowska and Kuh published a new method for routing [12]. The algorithm proceeds from left to right in a kind of a scanning motion, and it expands all connections from left to right, keeping the planar arrangement as long as possible. (The connections at this phase are more or less horizontal, but not straight, horizontal lines.) When the target terminals are to be reached, the algorithm changes the layer to a set of planar, more or less vertical connections toward the target terminals. The claim of a 100-percent connection rate may be true in the topological sense, but may not be valid under real conditions. For example, there may not always be enough space between all terminals and via holes for a planar connection. On the other hand, this router is unique in that it generates connections very similar to those in manually routed difficult boards. People are usually efficient in solving complex tasks. Is this the beginning of a new era in our understanding of the routing problem?

At various companies, systems simulating the human way of routing were implemented. Handling the final cleanup, and making the last few last connections on which the automatic router failed proved to be a major job. Programs were written to provide some help. The paper cited in reference [6] describes such a ripout router with an excellent performance record. The principle of all similar algorithms is to detect connections that cause obstructions for other nets. Such obstructions are temporarily removed, and other nets are routed. The original net is rerouted only after that. Even if the rerouting fails (being blocked now), the number of unfinished nets may have decreased.

Though obviously quite useful, such programs do not solve the problem; they only alleviate it. Some experienced designers insist that a few last connections with a really tangled routing are worse than more unfinished connections with more space to handle them. The important property of these routers is

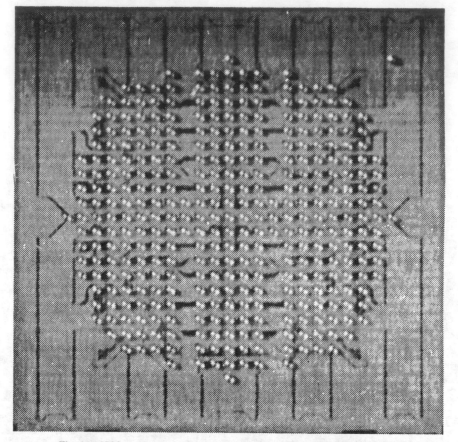

Fig. 33. IBM gate array experimental chip with 7640 logical cells, bipolar technology (7 mm square, contains 11 000 connections). Only 65 percent of available cells are occupied. Using an automated layout system, it took two months to generate the layout (less then 5 h of running time on an IBM 370/168). There was only a modest increase in required computing time, compared to much simpler chips designed before. (Photo courtesy of IBM.)

that they fall halfway between loose routing and final routing; they can run without the loose router.

Vintr's router [13] falls into the same category, except that instead of ripping out unwanted connections, it implements even the good connections only part way through. The algorithm runs in iterations. In the first iteration, all connections are found using a standard router, such as the Lee or Rubin router. However, for each connection, only 10 percent of the path from each end is really implemented. The first iteration ends with little tails hanging from every terminal. In the second iteration, prior to finding a connection for a signal, all of the tails of that signal are canceled. A new connection is found that may (and it frequently does) take a different course. Now 20 percent of the tails is recorded from each side. After 5 iterations, all of the connections should be made. Note that though this router is very efficient, it cannot guarantee all the connections.

In my opinion, Vintr's router is the best global router today. It does not resolve all conflicts, but it is practical and fast. This router is a cure for the most frequent cause of unroutable nets on PCB's: the blocking of pins. Once a net has a tail on every pin, it is very unlikely that access to any of its terminals would be completely blocked by other connections. Vintr's router is also commercially available in a stand-alone version on a small intelligent terminal.

Most routers, with the exception of the Lee-type routers, are designed for one or two layers. For more than two layers, various schemes, such as routing layers in pairs, are used. Multilayer routing has become important with a silicon process using two metals and one silicon layer. I am not aware of any layout system that generates connections on three layers efficiently. An interesting idea was published by Chaudhuri [25]. His routing is based on a hexagonal grid with lines at 0°, 120°, and 240°. The result looks like a honeycomb. Because three systems of lines are used, the router can connect on three layers. Another advantage is that the minimum Steiner trees can be implemented for multipoint nets (using Euclidean distance).

Several studies have been made on the design of a chip performing a fast parallel Lee-type algorithm. My unpublished study, done at Bell-Northern Research, indicates that with today's technology, it is impossible to place the whole router on one chip. When splitting the router into an array of chips, the number of pins becomes a limiting factor; only a matrix of 4 X 4 grid points can be placed on one chip. The estimated cost of the router, composed from such building blocks, is in the range of one-half million dollars for typical printed circuit boards or chips. Also, although hardware routers may provide speed and economy, they may have problems competing with the flexibility and portability of routers designed in soft-

Fig. 34. INTEL APX 43201 Instruction Decode Unit. The chip contains 110 000 devices in HMOS technology. (Photo courtesy of INTEL.)

ware. Note that the routing chip can work asynchronously (at least inside) simplifying the logic and number of required gates.

VIII. HIERARCHICAL LAYOUT

Note that our definition of the hierarchical layout system includes the California Institute of Technology layout style as a subset. A new layout system at Bell Laboratories allows two modes of combining modules:

1) The connectivity information is maintained, both inside and outside of the new composed module.
2) Modules are combined just as building blocks (blindly), without any connectivity information and terminal checking.

The user can move from one level to any other level of the hierarchy. However, from the currently chosen level, all upper levels appear invisible and all lower levels are represented by module boundaries and terminals. The problem is reduced to a practical, manageable size. The interactive design can be done on a screen, and automatic algorithms run quickly [30].

As explained in the beginning of this paper, the introduction of a hierarchy into chip layout became necessary because of the amount of data to be handled, the requirements for a better organization of the design team, and the potential for reusing older layouts.

The first problem is: How can we generate the hierarchy? The first idea may be to use an automatic partitioning algorithm. The best results so far have been obtained with the natural (inherited) hierarchy specified by the designer (for example, by marking an area in the schematic diagram). However, it happens frequently that circuits, which are close together in the schematic diagram, should not be close in the layout, and vice versa. For such designs we may use a different way of specifying the hierarchy (for example, just by a list of names). Because the requirement for recording the hierarchy is relatively new, most programs for the entry and manipulation of schematic diagrams are not designed to record the natural hierarchy of the circuit, and pass it to the layout system. In Bell Laboratories the language used for the circuit description is based on hierarchy, and it provides an entry medium for the layout system. Also, computer programs for the interactive changes of the hierarchy are available both as a stand-alone version and as a part of the layout system.

Fig. 30 is an example of a chip placement with several levels of hierarchy. As shown here (the figure does not represent a real chip), parts of the hierarchy may use some of the older design styles, such as rows of standard cells.

A truly hierarchical system needs a complex data base (or we may call it data structure). Various maintenance programs are required to keep the data base consistent. Because no problems have been reported in literature on the difficulties of handling hierarchical data, the question arises: Do all those people who use this popular word, hierarchy, really understand what they are getting into? Let us point out a few of the problems.

1) If the circuit calls for several superblocks of the same type (for example, several registers of the same size), what

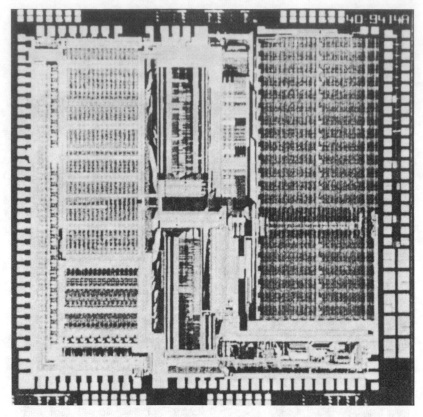

Fig. 35. 450 000 transistor, 32-bit CPU chip developed by the Hewlett-Packard Company. (I feel that the layout was done mostly by hand; photo courtesy of the Hewlett-Packard Corp.)

happens to the boundaries of other registers, if one register and its boundary are changed?

2) Terminals from a lower level of hierarchy have to some-time propagate several levels up, including electrical equivalences, jumpers, or capacitances. How can we do all that consistently, and what can we do if an error is detected? What are the rules for propagating terminals that fall inside the parent module?

3) When a channel is to be rerouted, what happens to other channels? Can the loose router run with some channels already completed?

4) Fig. 31 shows that a change in the physical layout may change the logical circuit description. (When turning module A, as shown in Fig. 31, the signal that was internal to hierarchy level CHIP.BLK1 must now be considered also in the parent level CHIP.) Such a case may occur frequently for rows of standard cells. How will we handle that?

5) We need algorithms for the automatic generation of a composite module boundary; but we need them even for such irregular cases as overlapping modules, or modules in clusters that do not touch each other.

6) Are the design rules (and the rules for checking the correctness of the layout) the same for all levels of a hierarchy?

7) If a jumper is connected to a net, do we list it in the data base as one of the terminals? If so, what happens if we move the module and disconnect the jumper? And if we don't, how do we trace the capacitance associated with the net?

8) What is a module without a boundary? Or, what happens to a module, which partially protrudes from its parent boundary?

9) It is a generally accepted practice to organize the data base so that we store only one module-type description (such as "NAND-gate"), and so that the connectivity list deals with modules, their unique identification names, and a pointer to the module-type. If we find in the layout that two physical terminals touch, what is a fast way of finding out which nets they belong to? (A physical terminal cannot have a pointer to the signal because the module-type is stored only once. Searching through all of the physical terminals may take a long time for modules with many terminals.)

Some of these problems become even more difficult when dealing with blocks of layout that are only partially designed. One of the basic features of a new layout system in Bell Laboratories is the availability of both bottom-up and top-down designs. The top-down design includes operations with modules not yet fully designed; such modules may only be estimated in size, and commands are available to estimate or adjust both area and shape. Top-down design is actually a planning phase that has been done traditionally on a piece of paper or in the designer's head. Enormous potentials of layout improvement seem to be possible, but these improvements cannot be used until we incorporate the top-down design process into our CAD systems.

ACKNOWLEDGMENT

This paper reflects the author's experience in the Ingegrated Circuit Laboratory, Bell Laboratories, Murray Hill, NJ, and also the experience in several CAD-related departments of Bell-Northern Research, Ottawa, Ont., Canada. Some figures were reprinted from *Telesis*, the technical magazine of Bell-Northern

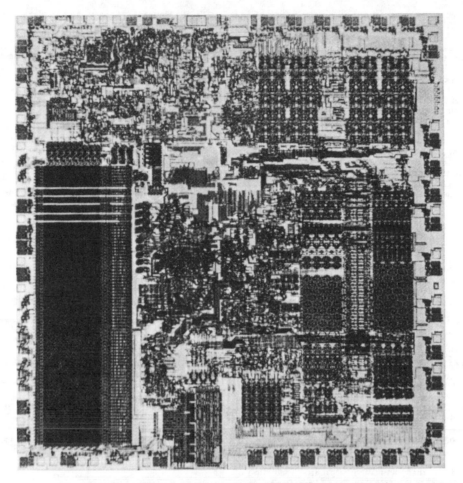

Fig. 36. National Semiconductor CPU chip NS16032 with 60K transistors. Most of the layout was done by hand interactively on the CALMA GDS system. Note the hand-packed area of random logic in the middle of the chip, where regular structures were avoided in order to save the chip area. ROM sections were generated automatically using a macroexpansion capability. This HMS01 chip has an area of 290 square mils, with typical gate size 3.5 µm, and a metal-to-metal pitch of 11 µm. (Photo courtesy of National Semiconductor.)

Research. The paper would have been impossible to complete without numerous personal contacts. Thanks to those who read the paper and suggested improvements: S. Annunziata, D. N. Deutsch, A. E. Dunlop, H. K. Gummel, W. R. Heller, D. C. Schmidt, and D. G. Schweikert.

BIBLIOGRAPHY AND REFERENCES

References are divided into three groups: In the first group, major sources of important publications are given. The second group provides references from the main body of the paper. In the third group, some additional sources of information are given. For someone new to this field, the best way to start would be to look through the *Proceedings of the Design Automation Conference* from the last two or three years.

Major Sources of Publications on Layout

[A] Design Automation Conference (IEEE and ACM); the conference is held each year in June. No. 10-Portland 1973, No. 11-Denver 1974, No. 12-Boston 1975, No. 13-San Francisco 1976, No. 14-New Orleans 1977, No. 15-Las Vegas 1978, No. 16-San Diego 1979, No. 17-Minneapolis 1980, No. 18-Nashville 1981.

[B] IEEE International Symposium on Circuits and Systems; the symposium is held each year in July, and it has a session on Design Automation. New York 1978, Tokyo 1979, Houston 1980, Chicago 1981.

[C] *Journal of Digital Systems*, published quarterly. Note that until 1979 the journal was called the Journal of Design Automation and Fault-Tolerant Computing.

References from the Main Body of the Paper

[1] J. Soukup, "Global router," in *Proc. 16th Design Automation Conf.* (San Diego, CA), pp. 481–484, June 1979. (The paper describes a routing that combines loose routing and final routing into one algorithm.)

[2] B. T. Preas and C. W. Gwyn, "Methods for hierarchical automatic layout of custom LSI circuit masks," in *Proc. 15th Design Automation Conf.* (Las Vegas, NV), pp. 206–212, June 1978. (The first paper combining hierarchical layout and placement based on position graphs.) See also, *J. Des. Automat. Fault-Tolerant Comput.*, vol. 3, no. 1, pp. 41–58, 1978.

[3] U. Lauther, "A min-cut placement algorithm for general cell assemblies based on graph representation," in *Proc. 16th Design Automation Conf.* (San Diego, CA), pp. 1–10, June 1979. (Another version of hierarchical layout and placement based on position graphs. The paper describes the first results obtained at Siemens, Germany.)

[4] S. Goto, "A two-dimensional placement algorithm for the master-slice LSI layout problem, in *Proc. 16th Design Automation Conf.* (San Diego, CA), pp. 11–17, June 1979. (The most general version of the relaxation algorithm for the placement improvement, and some statistics on experimental results.)

[5] N. Nan and M. Feuer, "A method for automatic wiring of LSI chips," in *Proc. IEEE Int. Symp. Circuits and Systems*, (New York), pp. 11-15, 1978. (Method of loose routing applied to a grid of mutually crossing channels.)

[6] H. Bollinger, "A mature DA system for PC layout," in *Proc. 1st Int. Printed Circuit Conf.* (New York), pp. 85-99, 1979. (An interesting description of the rip-out router as used in A.M.I.; a layout system with an excellent performance record.)

[7] D. N. Deutsch, "A "DOGLEG" channel router," in *Proc. 13th Design Automation Conf.* (San Francisco, CA), pp. 425-433, June 1976. (The original dogleg router, as used at Bell Laboratories.)

[8] F. Rubin, "The Lee connection algorithm," *IEEE Trans. Comput.*, vol. C-23, pp. 907-914, 1974. (The paper describes a method of speed improvement for the Lee router, using the information about the location of the target. The algorithm is based on sorting the points of the expansion wave.)

[9] J. Soukup, "Fast maze router," in *Proc. 15th Design Automation Conf.* (Las Vegas, NV), pp. 100-102, June 1978. (Speed improvement of the Lee algorithm based on simple control of the expansion wave.)

[10] T. Yoshimura and E. S. Kuh, "Efficient algorithms for channel routing," Memo. UCB/ERL M80/43, Aug. 11, 1980, Electronics Research Laboratory, College of Engineering, Univ. California, Berkeley. (A better approach to channel routing than the simple heuristics of the original dogleg router.)

[11] Private communication from S. Akers, who works with the General Electric Co., Syracuse, NY. (The router mentioned in this paper is only my interpretation of Akers's verbal presentation in the IEEE Symposium on Network Routing, Columbia Univ., New York, 1976.)

[12] M. M. Sadowska and E. S. Kuh, "A new approach to two layer routing," in *Proc. IEEE Int. Symp. Circuits and Systems* (Chicago, IL), p. 122, 1981. (Only an abstract appears in the conference proceedings, but the full text of the paper can be obtained from the Dep. Electrical Engineering, Univ. California, Berkeley. This paper is on the router that runs in a similar topology as used by human designers. Preferred horizontal and vertical layers are used, but not strictly enforced.)

[13] J. Vintr is the inventor of the algorithm that was reported by R. Dutta at the 1980 CANDE workshop on hardware for CAD, Univ. Michigan, Ann Arbor. There were no proceedings published.

[14] W. R. Heller, W. F. Mikhail, and W. E. Donath, "Prediction of wiring space requirements for LSI," in *Proc. 14th Design Automation Conf.* (New Orleans, LA), pp. 32-42, 1977. (This is the original paper on wireability, developed in IBM.) Heller later presented several papers on the subject of wireability, but the full text does not appear in the proceedings, for example, 1979 IEEE Inter. Symposium on Circuits and Systems, Tokyo. Full text and mathematical analysis was published in the *J. Des. Automat. Fault-Tolerant Comput.*, vol. 2, no. 2, pp. 117-144, May 1978.

[15] M. Hanan, P. K. Wolff, and B. J. Anguli, "Some experimental results on placement techniques," in *Proc. 13th Design Automation Conf.* (San Francisco, CA), pp. 214-224, 1973. (This is one of the best papers published on placement. The same paper was published later in an extended version in the *J. Des. Automat. Fault-Tolerant Comput.*, vol. 2, no. 2, pp. 145-164, May 1978.

[16] D. Hightower, "A solution to the line routing problem on the continuous plane," in *Proc. Design Automation Workshop*, pp. 1-24, 1969. (The original publication of the line router; other papers and reports of experiences were published later.)

[17] C. Y. Lee, "An algorithm for path connections and its application," *IRE Trans. Electron. Comput.*, pp. 346-365, Sept. 1961. (The classical paper on the grid based routing using the expansion wave.)

[18] C. Mead and L. Conway, *Introduction to VLSI*. Reading, MA: Addison Wesley, 1980. (This is the main textbook on silicon design used by many universities today. It describes the layout approach used at California Institute of Technology. Anyone seriously interested in VLSI design should read it.)

[19] D. G. Schweikert, "A 2-dimensional placement algorithm for the layout of electrical circuits," in *Proc. Design Automatic Conf.* (San Francisco, CA), pp. 408-416, 1976. (The paper describes basic methods of the initial placement and the iterative improvement.) Look also at G. Persky, D. N. Deutsch, and D. G. Schweikert, "LTX—A minicomputer-based system for automatic LSI layout," *J. Des. Automat. Fault-Tolerant Comput.*, vol. 1, no. 3, pp. 217-256, May 1977. (This is the best published description of the LTX system. The paper describes details of placement algorithms, net assignment, and the dogleg channel router.)

[20] M. Breuer, "Min-cut placement," *J. Des. Automat. Fault Tolerant Comput.*, vol. 1, no. 4, pp. 343-362, Oct. 1977. (A basic paper on placement based on bipartitioning.)

[21] W. Heyns, W. Sansen, and H. Beke, "A line expansion algorithm for the general routing problem with a guaranteed solution," in *Proc. 17th Design Automation Conf.* (Minneapolis, MN), pp. 243-249, June 1980. (The router that combines grid and line expansion. It is gridless, and it does not require as much memory as Lee-type routers.)

[22] M. Hung and W. O. Rom, "Solving the assignment problem by relaxation," *Operations Res.*, vol. 28, no. 4, pp. 969-982, Aug. 1980. (One of the recently published methods on the Assignment Problem.)

[23] J. Soukup and J. C. Royle, "On hierarchical routing," *J. Digital Systems*, vol. 5, no. 3, Sept. 1981. (Details of an implementation of a hierarchical layout system with an emphasis on loose routing. A Lee-type router is assumed for final routing.)

[24] M. Wiesel and D. A. Mlynski, "An efficient channel model for building block LSI," in *Proc. 1981 IEEE Int. Symp. Circuits and Systems* (Chicago, IL), pp. 118-121, Apr. 1981. (A method for dividing the routing area into perpendicular channels that are different on different layers; these channels are used for both loose and channel routing.)

[25] P. P. Chaudhuri, "An ecological approach to wire routing," in *Proc. 1979 IEEE Int. Symp. Circuits and Systems* (Tokyo, Japan), pp. 854-857, July 1979. (The routing method based on the hexagonal grid.)

[26] D. G. Schweikert and B. W. Kernighan, "Partitioning circuit layout," in *Proc. 9th Annu. Design Automation Workshop*. (This is the same as DA Conf.) (Dallas, TX), June 1972. (A classical paper on partitioning; important and frequently referenced in papers on placement.)

[27] K. J. Loosemore, "Automated layout of integrated circuits," in *Proc. 1979 IEEE Int. Symp. Circuits and Systems*. (Tokyo, Japan), pp. 665-668, 1979. (A good description of the GEALIC-COMPEDA system.)

[28] A. Hashimoto, and J. Stevens, "Wire routing by optimizing channel assignment within large appertures," in *Proc. 8th Design Automation Workshop*, pp. 155-169, 1971. (The first paper on channel routing; the same as DA Conf.)

[29] J. Soukup, "Electronic circuit layout algorithm," *Telesis* (the technical magazine of Bell-Northern Research Ltd.), vol. 7, no. 2, 1980.

[30] R. H. Krambeck, D. E. Blahut, H. F.-S. Law, B. W. Colbry, H. C. So, M. Harrison, and J. Soukup, "Top down design of a one chip 32 bit CPU," VLSI '81 Conf., Edinburgh, Scotland, Aug. 1981.

Additional Bibliography

[a] B. W. Kernighan and S. Lin, "An efficient procedure for partitioning graphs," *Bell Syst. Tech. J*, pp. 291-307, Feb. 1970.

[b] M. Breuer and K. Shamsa, "A hardware router," *J. Digital Syst.*, vol. 4, no. 4, pp. 393-408, Winter 1980.

[c] F. Rubin "A comparison of wire spreading metrics for printed circuit routing," *J. Des. Automat. Fault-Tolerant Comput.*, vol. 1, no. 3, pp. 231-240, July 1978.

[d] K. Sato and T. Nagai, "A method of specifying the relative locations between blocks in a routing program for building block LSI," in *Proc. 1979 IEEE Int. Symp. Circuits and Systems* (Tokyo, Japan), pp. 673-676, 1979.

[e] J. Soukup and U. W. Stockburger, "Routing in theory and practice," in *Proc. 1st Annu. Conf. Computer Graphics in CAD/CAM Systems* (M.I.T.), pp. 126-146, Apr. 1979.

[f] H. Shiraishi and F. Hirose, "Efficient placement and routing techniques for master slice LSI," in *Proc. 17th Design Automation Conf.* (Minneapolis, MN), pp. 458-464, June 1980.

[g] B. T. Preas and W. M. van Cleemput, "Placement algorithms for arbitrarily shaped blocks," in *Proc. 16th Design Automation Conf.* (San Diego, CA), pp. 474-480, 1979.

[h] K. A. Chen, M. Feuer, K. H. Khokhani, N. Nan, and S. Schmidt, "The chip layout problem: An automatic wiring procedure," in *Proc. Design Automation Conf.* (New Orleans, LA), pp. 298-302, 1977.

[i] E. B. Eichelberger and T. W. Williams, "A logic design structure for LSI testability," in *Proc. 14th Design Automation Conf.* (New Orleans, LA), pp. 462-468, 1977.

[j] J. R. Allen, "A topologically adaptable cellular router," in *Proc. 13th Design Automation Conf.* (San Francisco, CA), pp. 161-167, June 1976.

[k] B. R. Rau, "A new philosophy for interconnection on multilayer boards," in *Proc. 13th Design Automation Conf.* (San Francisco, CA), pp. 225-231, 1976.

[l] M. Hanan, "Net wiring for large scale integrated circuits," IBM Research Rep. RC-1375, Feb. 1965. (The report describes a fast algorithm for a Steiner representation of nets. Compare with Fig. 10.)

SCIENCE

Optimization by Simulated Annealing

S. Kirkpatrick, C. D. Gelatt, Jr., M. P. Vecchi

In this article we briefly review the central constructs in combinatorial optimization and in statistical mechanics and then develop the similarities between the two fields. We show how the Metropolis algorithm for approximate numerical simulation of the behavior of a many-body system at a finite temperature provides a natural tool for bringing the techniques of statistical mechanics to bear on optimization.

We have applied this point of view to a number of problems arising in optimal design of computers. Applications to partitioning, component placement, and wiring of electronic systems are described in this article. In each context, we introduce the problem and discuss the improvements available from optimization.

Of classic optimization problems, the traveling salesman problem has received the most intensive study. To test the power of simulated annealing, we used the algorithm on traveling salesman problems with as many as several thousand cities. This work is described in a final section, followed by our conclusions.

Combinatorial Optimization

The subject of combinatorial optimization (1) consists of a set of problems that are central to the disciplines of computer science and engineering. Research in this area aims at developing efficient techniques for finding minimum or maximum values of a function of very many independent variables (2). This function, usually called the cost function or objective function, represents a quantitative mea-

sure of the "goodness" of some complex system. The cost function depends on the detailed configuration of the many parts of that system. We are most familiar with optimization problems occurring in the physical design of computers, so examples used below are drawn from that context. The number of variables involved may range up into the tens of thousands.

The classic example, because it is so simply stated, of a combinatorial optimization problem is the traveling salesman problem. Given a list of N cities and a means of calculating the cost of traveling between any two cities, one must plan the salesman's route, which will pass through each city once and return finally to the starting point, minimizing the total cost. Problems with this flavor arise in all areas of scheduling and design. Two subsidiary problems are of general interest: predicting the expected cost of the salesman's optimal route, averaged over some class of typical arrangements of cities, and estimating or obtaining bounds for the computing effort necessary to determine that route.

All exact methods known for determining an optimal route require a computing effort that increases exponentially

Summary. There is a deep and useful connection between statistical mechanics (the behavior of systems with many degrees of freedom in thermal equilibrium at a finite temperature) and multivariate or combinatorial optimization (finding the minimum of a given function depending on many parameters). A detailed analogy with annealing in solids provides a framework for optimization of the properties of very large and complex systems. This connection to statistical mechanics exposes new information and provides an unfamiliar perspective on traditional optimization problems and methods.

with N, so that in practice exact solutions can be attempted only on problems involving a few hundred cities or less. The traveling salesman belongs to the large class of NP-complete (nondeterministic polynomial time complete) problems, which has received extensive study in the past 10 years (3). No method for exact solution with a computing effort bounded by a power of N has been found for any of these problems, but if such a solution were found, it could be mapped into a procedure for solving all members of the class. It is not known what features of the individual problems in the NP-complete class are the cause of their difficulty.

Since the NP-complete class of problems contains many situations of practical interest, heuristic methods have been developed with computational require-ments proportional to small powers of N. Heuristics are rather problem-specific: there is no guarantee that a heuristic procedure for finding near-optimal solutions for one NP-complete problem will be effective for another.

There are two basic strategies for heuristics: "divide-and-conquer" and iterative improvement. In the first, one divides the problem into subproblems of manageable size, then solves the subproblems. The solutions to the subproblems must then be patched back together. For this method to produce very good solutions, the subproblems must be naturally disjoint, and the division made must be an appropriate one, so that errors made in patching do not offset the gains

S. Kirkpatrick and C. D. Gelatt, Jr., are research staff members and M. P. Vecchi was a visiting scientist at IBM Thomas J. Watson Research Center, Yorktown Heights, New York 10598. M. P. Vecchi's present address is Instituto Venezolano de Investigaciones Científicas, Caracas 1010A, Venezuela.

obtained in applying more powerful methods to the subproblems (4).

In iterative improvement (5, 6), one starts with the system in a known configuration. A standard rearrangement operation is applied to all parts of the system in turn, until a rearranged configuration that improves the cost function is discovered. The rearranged configuration then becomes the new configuration of the system, and the process is continued until no further improvements can be found. Iterative improvement consists of a search in this coordinate space for rearrangement steps which lead downhill. Since this search usually gets stuck in a local but not a global optimum, it is customary to carry out the process several times, starting from different randomly generated configurations, and save the best result.

There is a body of literature analyzing the results to be expected and the computing requirements of common heuristic methods when applied to the most popular problems (1–3). This analysis usually focuses on the worst-case situation—for instance, attempts to bound from above the ratio between the cost obtained by a heuristic method and the exact minimum cost for any member of a family of similarly structured problems. There are relatively few discussions of the average performance of heuristic algorithms, because the analysis is usually more difficult and the nature of the appropriate average to study is not always clear. We will argue that as the size of optimization problems increases, the worst-case analysis of a problem will become increasingly irrelevant, and the average performance of algorithms will dominate the analysis of practical applications. This large number limit is the domain of statistical mechanics.

Statistical Mechanics

Statistical mechanics is the central discipline of condensed matter physics, a body of methods for analyzing aggregate properties of the large numbers of atoms to be found in samples of liquid or solid matter (7). Because the number of atoms is of order 10^{23} per cubic centimeter, only the most probable behavior of the system in thermal equilibrium at a given temperature is observed in experiments. This can be characterized by the average and small fluctuations about the average behavior of the system, when the average is taken over the ensemble of identical systems introduced by Gibbs. In this ensemble, each configuration, defined by the set of atomic positions, $\{r_i\}$, of the system is weighted by its Boltzmann probability factor, $\exp(-E(\{r_i\})/k_B T)$, where $E(\{r_i\})$ is the energy of the configuration, k_B is Boltzmann's constant, and T is temperature.

A fundamental question in statistical mechanics concerns what happens to the system in the limit of low temperature—for example, whether the atoms remain fluid or solidify, and if they solidify, whether they form a crystalline solid or a glass. Ground states and configurations close to them in energy are extremely rare among all the configurations of a macroscopic body, yet they dominate its properties at low temperatures because as T is lowered the Boltzmann distribution collapses into the lowest energy state or states.

As a simplified example, consider the magnetic properties of a chain of atoms whose magnetic moments, μ_i, are allowed to point only "up" or "down," states denoted by $\mu_i = \pm 1$. The interaction energy between two such adjacent spins can be written $J\mu_i\mu_{i+1}$. Interaction between each adjacent pair of spins contributes $\pm J$ to the total energy of the chain. For an N-spin chain, if all configurations are equally likely the interaction energy has a binomial distribution, with the maximum and minimum energies given by $\pm NJ$ and the most probable state having zero energy. In this view, the ground state configurations have statistical weight $\exp(-N/2)$ smaller than zero-energy configurations. A Boltzmann factor, $\exp(-E/k_B T)$, can offset this if $k_B T$ is smaller than J. If we focus on the problem of finding empirically the system's ground state, this factor is seen to drastically increase the efficiency of such a search.

In practical contexts, low temperature is not a sufficient condition for finding ground states of matter. Experiments that determine the low-temperature state of a material—for example, by growing a single crystal from a melt—are done by careful annealing, first melting the substance, then lowering the temperature slowly, and spending a long time at temperatures in the vicinity of the freezing point. If this is not done, and the substance is allowed to get out of equilibrium, the resulting crystal will have many defects, or the substance may form a glass, with no crystalline order and only metastable, locally optimal structures.

Finding the low-temperature state of a system when a prescription for calculating its energy is given is an optimization problem not unlike those encountered in combinatorial optimization. However, the concept of the temperature of a physical system has no obvious equivalent in the systems being optimized. We will introduce an effective temperature for optimization, and show how one can carry out a simulated annealing process in order to obtain better heuristic solutions to combinatorial optimization problems.

Iterative improvement, commonly applied to such problems, is much like the microscopic rearrangement processes modeled by statistical mechanics, with the cost function playing the role of energy. However, accepting only rearrangements that lower the cost function of the system is like extremely rapid quenching from high temperatures to $T = 0$, so it should not be surprising that resulting solutions are usually metastable. The Metropolis procedure from statistical mechanics provides a generalization of iterative improvement in which controlled uphill steps can also be incorporated in the search for a better solution.

Metropolis et al. (8), in the earliest days of scientific computing, introduced a simple algorithm that can be used to provide an efficient simulation of a collection of atoms in equilibrium at a given temperature. In each step of this algorithm, an atom is given a small random displacement and the resulting change, ΔE, in the energy of the system is computed. If $\Delta E \leq 0$, the displacement is accepted, and the configuration with the displaced atom is used as the starting point of the next step. The case $\Delta E > 0$ is treated probabilistically: the probability that the configuration is accepted is $P(\Delta E) = \exp(-\Delta E/k_B T)$. Random numbers uniformly distributed in the interval $(0,1)$ are a convenient means of implementing the random part of the algorithm. One such number is selected and compared with $P(\Delta E)$. If it is less than $P(\Delta E)$, the new configuration is retained; if not, the original configuration is used to start the next step. By repeating the basic step many times, one simulates the thermal motion of atoms in thermal contact with a heat bath at temperature T. This choice of $P(\Delta E)$ has the consequence that the system evolves into a Boltzmann distribution.

Using the cost function in place of the energy and defining configurations by a set of parameters $\{x_i\}$, it is straightforward with the Metropolis procedure to generate a population of configurations of a given optimization problem at some effective temperature. This temperature is simply a control parameter in the same units as the cost function. The simulated annealing process consists of first "melting" the system being optimized at a high effective temperature, then lowering the temperature by slow stages until the system "freezes" and no further changes occur. At each temperature, the simulation must proceed long enough for the system to reach a steady state. The sequence of temperatures and the number of rearrangements of the $\{x_i\}$ attempted to reach equilibrium at each temperature can be considered an annealing schedule.

Annealing, as implemented by the Metropolis procedure, differs from iterative improvement in that the procedure need not get stuck since transitions out of a local optimum are always possible at nonzero temperature. A second and more important feature is that a sort of adaptive divide-and-conquer occurs. Gross features of the eventual state of the system appear at higher temperatures; fine details develop at lower temperatures. This will be discussed with specific examples.

Statistical mechanics contains many useful tricks for extracting properties of a macroscopic system from microscopic averages. Ensemble averages can be obtained from a single generating function, the partition function, Z,

$$Z = \text{Tr} \exp\left(\frac{-E}{k_B T}\right) \quad (1)$$

in which the trace symbol, Tr, denotes a sum over all possible configurations of the atoms in the sample system. The logarithm of Z, called the free energy, $F(T)$, contains information about the average energy, $<E(T)>$, and also the entropy, $S(T)$, which is the logarithm of the number of configurations contributing to the ensemble at T:

$$-k_B T \ln Z = F(T) = <E(T)> - TS \quad (2)$$

Boltzmann-weighted ensemble averages are easily expressed in terms of derivatives of F. Thus the average energy is given by

$$<E(T)> = \frac{-d\ln Z}{d(1/k_B T)} \quad (3)$$

and the rate of change of the energy with respect to the control parameter, T, is related to the size of typical variations in the energy by

$$C(T) = \frac{d <E(T)>}{dT}$$
$$= \frac{[<E(T)^2> - <E(T)>^2]}{k_B T^2} \quad (4)$$

In statistical mechanics $C(T)$ is called the specific heat. A large value of C signals a change in the state of order of a system, and can be used in the optimization context to indicate that freezing has be-

gun and hence that very slow cooling is required. It can also be used to determine the entropy by the thermodynamic relation

$$\frac{dS(T)}{dT} = \frac{C(T)}{T} \quad (5)$$

Integrating Eq. 5 gives

$$S(T) = S(T_1) - \int_T^{T_1} \frac{C(T')\,dT}{T} \quad (6)$$

where T_1 is a temperature at which S is known, usually by an approximation valid at high temperatures.

The analogy between cooling a fluid and optimization may fail in one important respect. In ideal fluids all the atoms are alike and the ground state is a regular crystal. A typical optimization problem will contain many distinct, noninterchangeable elements, so a regular solution is unlikely. However, much research in condensed matter physics is directed at systems with quenched-in randomness, in which the atoms are not all alike. An important feature of such systems, termed "frustration," is that interactions favoring different and incompatible kinds of ordering may be simultaneously present (9). The magnetic alloys known as "spin glasses," which exhibit competition between ferromagnetic and antiferromagnetic spin ordering, are the best understood example of frustration (10). It is now believed that highly frustrated systems like spin glasses have many nearly degenerate random ground states rather than a single ground state with a high degree of symmetry. These systems stand in the same relation to conventional magnets as glasses do to crystals, hence the name.

The physical properties of spin glasses at low temperatures provide a possible guide for understanding the possibilities of optimizing complex systems subject to conflicting (frustrating) constraints.

Physical Design of Computers

The physical design of electronic systems and the methods and simplifications employed to automate this process have been reviewed (11, 12). We first provide some background and definitions related to applications of the simulated annealing framework to specific problems that arise in optimal design of computer systems and subsystems. Physical design follows logical design. After the detailed specification of the logic of a system is complete, it is necessary to specify the precise physical realization of the system in a particular technology.

This process is usually divided into several stages. First, the design must be partitioned into groups small enough to fit the available packages, for example, into groups of circuits small enough to fit into a single chip, or into groups of chips and associated discrete components that can fit onto a card or other higher level package. Second, the circuits are assigned specific locations on the chip. This stage is usually called placement. Finally, the circuits are connected by wires formed photolithographically out of a thin metal film, often in several layers. Assigning paths, or routes, to the wires is usually done in two stages. In rough or global wiring, the wires are assigned to regions that represent schematically the capacity of the intended package. In detailed wiring (also called exact embedding), each wire is given a unique complete path. From the detailed wiring results, masks can be generated and chips made.

At each stage of design one wants to optimize the eventual performance of the system without compromising the feasibility of the subsequent design stages. Thus partitioning must be done in such a way that the number of circuits in each partition is small enough to fit easily into the available package, yet the number of signals that must cross partition boundaries (each requiring slow, power-consuming driver circuitry) is minimized. The major focus in placement is on minimizing the length of connections, since this translates into the time required for propagation of signals, and thus into the speed of the finished system. However, the placements with the shortest implied wire lengths may not be wirable, because of the presence of regions in which the wiring is too congested for the packaging technology. Congestion, therefore, should also be anticipated and minimized during the placement process. In wiring, it is desirable to maintain the minimum possible wire lengths while minimizing sources of noise, such as cross talk between adjacent wires. We show in this and the next two sections how these conflicting goals can be combined and made the basis of an automatic optimization procedure.

The tight schedules involved present major obstacles to automation and optimization of large system design, even when computers are employed to speed up the mechanical tasks and reduce the chance of error. Possibilities of feedback, in which early stages of a design are redone to solve problems that became apparent only at later stages, are greatly reduced as the scale of the overall system being designed increases. Op-

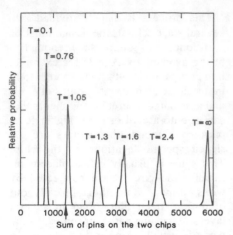

Fig. 1. Distribution of total number of pins required in two-way partition of a microprocessor at various temperatures. Arrow indicates best solution obtained by rapid quenching as opposed to annealing.

timization procedures that can incorporate, even approximately, information about the chance of success of later stages of such complex designs will be increasingly valuable in the limit of very large scale.

System performance is almost always achieved at the expense of design convenience. The partitioning problem provides a clean example of this. Consider N circuits that are to be partitioned between two chips. Propagating a signal across a chip boundary is always slow, so the number of signals required to cross between the two must be minimized. Putting all the circuits on one chip eliminates signal crossings, but usually there is no room. Instead, for later convenience, it is desirable to divide the circuits about equally.

If we have connectivity information in a matrix whose elements $\{a_{ij}\}$ are the number of signals passing between circuits i and j, and we indicate which chip circuit i is placed on by a two-valued variable $\mu_i = \pm1$, then N_c, the number of signals that must cross a chip boundary is given by $\Sigma_{i>j}(a_{ij}/4)(\mu_i - \mu_j)^2$. Calculating $\Sigma_i\mu_i$ gives the difference between the numbers of circuits on the two chips. Squaring this imbalance and introducing a coefficient, λ, to express the relative costs of imbalance and boundary crossings, we obtain an objective function, f, for the partition problem:

$$f = \sum_{i>j}\left(\lambda - \frac{a_{ij}}{2}\right)\mu_i\mu_j \qquad (7)$$

Reasonable values of λ should satisfy $\lambda \lesssim z/2$, where z is the average number of circuits connected to a typical circuit (fan-in plus fan-out). Choosing $\lambda \simeq z/2$ implies giving equal weight to changes in the balance and crossing scores.

The objective function f has precisely the form of a Hamiltonian, or energy function, studied in the theory of random magnets, when the common simplifying assumption is made that the spins, μ_i, have only two allowed orientations (up or down), as in the linear chain example of the previous section. It combines local, random, attractive ("ferromagnetic") interactions, resulting from the a_{ij}'s, with a long-range repulsive ("antiferromagnetic") interaction due to λ. No configuration of the $\{\mu_i\}$ can simultaneously satisfy all the interactions, so the system is "frustrated," in the sense formalized by Toulouse (9).

If the a_{ij} are completely uncorrelated, it can be shown (13) that this Hamiltonian has a spin glass phase at low temperatures. This implies for the associated magnetic problem that there are many degenerate "ground states" of nearly equal energy and no obvious symmetry. The magnetic state of a spin glass is very stable at low temperatures (14), so the ground states have energies well below the energies of the random high-temperature states, and transforming one ground state into another will usually require considerable rearrangement. Thus this analogy has several implications for optimization of partition:

1) Even in the presence of frustration, significant improvements over a random starting partition are possible.

2) There will be many good near-optimal solutions, so a stochastic search procedure such as simulated annealing should find some.

3) No one of the ground states is significantly better than the others, so it is not very fruitful to search for the absolute optimum.

In developing Eq. 7 we made several severe simplifications, considering only two-way partitioning and ignoring the fact that most signals connect more than two circuits. Objective functions analogous to f that include both complications are easily constructed. They no longer have the simple quadratic form of Eq. 7, but the qualitative feature, frustration, remains dominant. The form of the Hamiltonian makes no difference in the Metropolis Monte Carlo algorithm. Evaluation of the change in function when a circuit is shifted to a new chip remains rapid as the definition of f becomes more complicated.

It is likely that the a_{ij} are somewhat correlated, since any design has considerable logical structure. Efforts to understand the nature of this structure by analyzing the surface-to-volume ratio of components of electronic systems [as in "Rent's rule" (15)] conclude that the

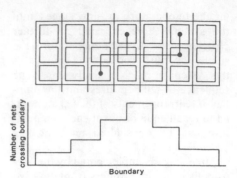

Fig. 2. Construction of a horizontal net-crossing histogram.

circuits in a typical system could be connected with short-range interactions if they were embedded in a space with dimension between two and three. Uncorrelated connections, by contrast, can be thought of as infinite-dimensional, since they are never short-range.

The identification of Eq. 7 as a spin glass Hamiltonian is not affected by the reduction to a two- or three-dimensional problem, as long as $\lambda N \simeq z/2$. The degree of ground state degeneracy increases with decreasing dimensionality. For the uncorrelated model, there are typically of order $N^{1/2}$ nearly degenerate ground states (14), while in two and three dimensions, $2^{\alpha N}$, for some small value, α, are expected (16). This implies that finding a near-optimum solution should become easier, the lower the effective dimensionality of the problem. The entropy, measurable as shown in Eq. 6, provides a measure of the degeneracy of solutions. $S(T)$ is the logarithm of the number of solutions equal to or better than the average result encountered at temperature T.

As an example of the partitioning problem, we have taken the logic design for a single-chip IBM "370 microprocessor" (17) and considered partitioning it into two chips. The original design has approximately 5000 primitive logic gates and 200 external signals (the chip has 200 logic pins). The results of this study are plotted in Fig. 1. If one randomly assigns gates to the two chips, one finds the distribution marked $T = \infty$ for the number of pins required. Each of the two chips (with about 2500 circuits) would need 3000 pins. The other distributions in Fig. 1 show the results of simulated annealing.

Monte Carlo annealing is simple to implement in this case. Each proposed configuration change simply flips a randomly chosen circuit from one chip to the other. The new number of external connections, C, to the two chips is calculated (an external connection is a net with circuits on both chips, or a circuit connected to one of the pins of the original single-chip design), as is the new balance score, B, calculated as in deriving Eq. 7. The objective function analogous to Eq. 7 is

$$f = C + \lambda B \qquad (8)$$

where C is the sum of the number of external connections on the two chips and B is the balance score. For this example, $\lambda = 0.01$.

For the annealing schedule we chose to start at a high "temperature," $T_0 = 10$, where essentially all proposed circuit flips are accepted, then cool exponentially, $T_n = (T_1/T_0)^n T_0$, with the ratio $T_1/T_0 = 0.9$. At each temperature enough flips are attempted that either there are ten accepted flips per circuit on the average (for this case, 50,000 accepted flips at each temperature), or the number of attempts exceeds 100 times the number of circuits before ten flips per circuit have been accepted. If the desired number of acceptances is not achieved at three successive tempera-

tures, the system is considered "frozen" and annealing stops.

The finite temperature curves in Fig. 1 show the distribution of pins per chip for the configurations sampled at $T = 2.5$, 1.0, and 0.1. As one would expect from the statistical mechanical analog, the distribution shifts to fewer pins and sharpens as the temperature is decreased. The sharpening is one consequence of the decrease in the number of configurations that contribute to the equilibrium ensemble at the lower temperature. In the language of statistical mechanics, the entropy of the system decreases. For this sample run in the low-temperature limit, the two chips required 353 and 321 pins, respectively. There are 237 nets connecting the two chips (requiring a pin on each chip) in addition to the 200 inputs and outputs of the original chip. The final partition in this example has the circuits exactly evenly distributed between the two partitions. Using a more complicated balance score, which did not penalize imbalance of less than

100 circuits, we found partitions resulting in chips with 271 and 183 pins.

If, instead of slowly cooling, one were to start from a random partition and accept only flips that reduce the objective function (equivalent to setting $T = 0$ in the Metropolis rule), the result is chips with approximately 700 pins (several such runs led to results with 677 to 730 pins). Rapid cooling results in a system frozen into a metastable state far from the optimal configuration. The best result obtained after several rapid quenches is indicated by the arrow in Fig. 1.

Placement

Placement is a further refinement of the logic partitioning process, in which the circuits are given physical positions (*11, 12, 18, 19*). In principle, the two stages could be combined, although this is not often possible in practice. The objectives in placement are to minimize

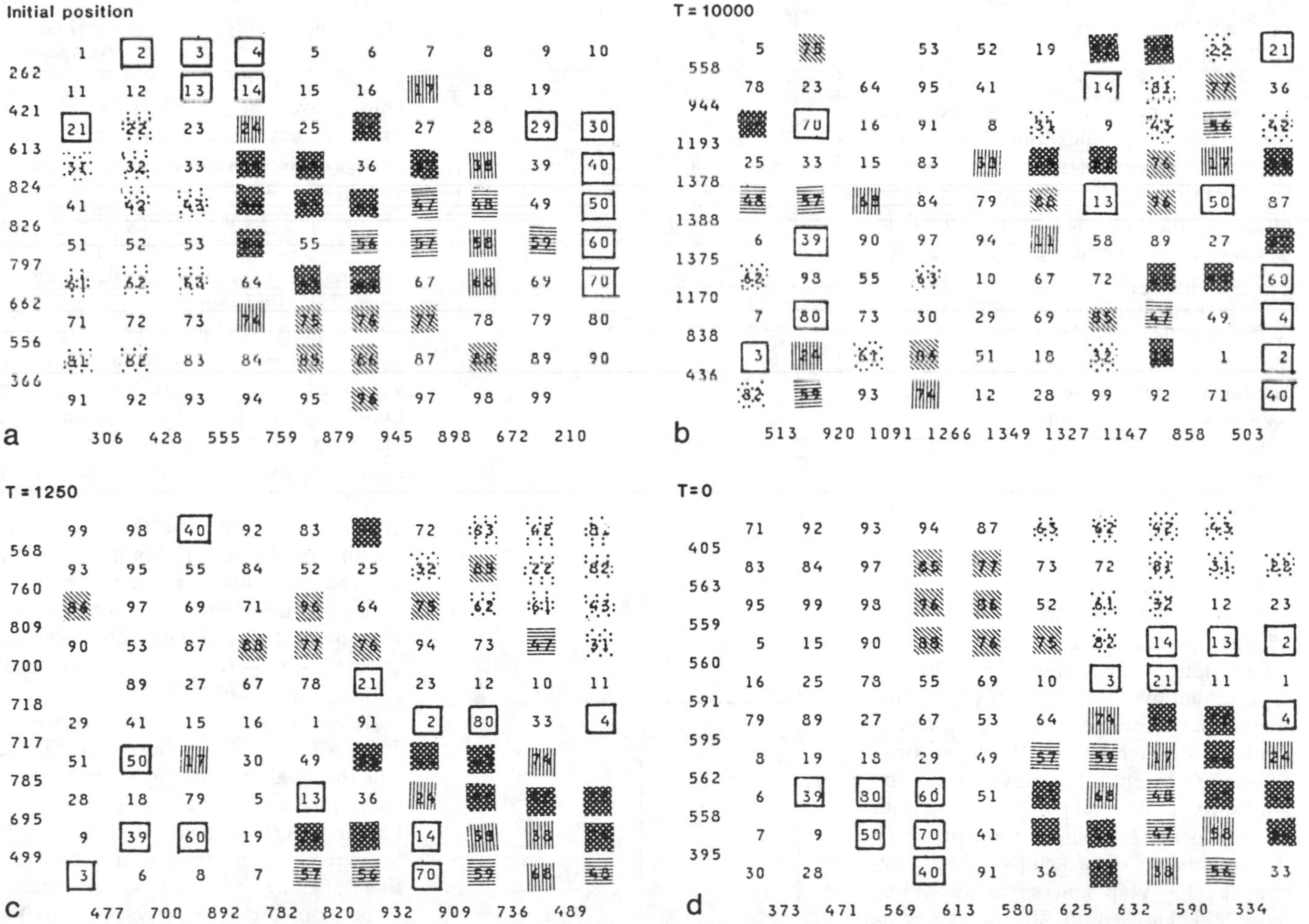

Fig. 3. Ninety-eight chips on a ceramic module from the IBM 3081. Chips are identified by number (1 to 100, with 20 and 100 absent) and function. The dark squares comprise an adder, the three types of squares with ruled lines are chips that control and supply data to the adder, the lightly dotted chips perform logical arithmetic (bitwise AND, OR, and so on), and the open squares denote general-purpose registers, which serve both arithmetic units. The numbers at the left and lower edges of the module image are the vertical and horizontal net-crossing histograms, respectively. (a) Original chip placement; (b) a configuration at $T = 10,000$; (c) $T = 1250$; (d) a zero-temperature result.

signal propagation times or distances while satisfying prescribed electrical constraints, without creating regions so congested that there will not be room later to connect the circuits with actual wire.

Physical design of computers includes several distinct categories of placement problems, depending on the packages involved (20). The larger objects to be placed include chips that must reside in a higher level package, such as a printed circuit card or fired ceramic "module" (21). These chip carriers must in turn be placed on a backplane or "board," which is simply a very large printed circuit card. The chips seen today contain from tens to tens of thousands of logic circuits, and each chip carrier or board will provide from one to ten thousand interconnections. The partition and placement problems decouple poorly in this situation, since the choice of which chip should carry a given piece of logic will be influenced by the position of that chip.

The simplest placement problems arise in designing chips with structured layout rules. These are called "gate array" or "master slice" chips. In these chips, standard logic circuits, such as three- or four-input NOR's, are preplaced in a regular grid arrangement, and the designer specifies only the signal wiring, which occupies the final, highest, layers of the chip. The circuits may all be identical, or they may be described in terms of a few standard groupings of two or more adjacent cells.

As an example of a placement problem with realistic complexity without too many complications arising from package idiosyncrasies, we consider 98 chips packaged on one multilayer ceramic module of the IBM 3081 processor (21). Each chip can be placed on any of 100 sites, in a 10 × 10 grid on the top surface of the module. Information about the connections to be made through the signal-carrying planes of the module is contained in a "netlist," which groups sets of pins that see the same signal.

The state of the system can be briefly represented by a list of the 98 chips with their x and y coordinates, or a list of the contents of each of the 100 legal locations. A sufficient set of moves to use for annealing is interchanges of the contents of two locations. This results in either the interchange of two chips or the interchange of a chip and a vacancy. For more efficient search at low temperatures, it is helpful to allow restrictions on the distance across which an interchange may occur.

To measure congestion at the same time as wire length, we use a convenient intermediate analysis of the layout, a net-crossing histogram. Its construction is summarized in Fig. 2. We divide the package surface by a set of natural boundaries. In this example, we use the boundaries between adjacent rows or columns of chip sites. The histogram then contains the number of nets crossing each boundary. Since at least one wire must be routed across each boundary crossed, the sum of the entries in the histogram of Fig. 2 is the sum of the horizontal extents of the rectangles bounding each net, and is a lower bound to the horizontal wire length required. Constructing a vertical net-crossing histogram and summing its entries gives a similar estimate of the vertical wire length.

The peak of the histogram provides a lower bound to the amount of wire that must be provided in the worst case, since each net requires at least one wiring channel somewhere on the boundary. To combine this information into a single objective function, we introduce a threshold level for each histogram—an amount of wire that will nearly exhaust the available wire capacity—and then sum for all histogram elements that exceed the threshold the square of the excess over threshold. Adding this quantity to the estimated length gives the objective function that was used.

Figure 3 shows the stages of a simulated annealing run on the 98-chip module. Figure 3a shows the chip locations from the original design, with vertical and horizontal net-crossing histograms indicated. The different shading patterns distinguish the groups of chips that carry out different functions. Each such group was designed and placed together, usually by a single designer. The net-crossing histograms show that the center of the layout is much more congested than the edges, most likely because the chips known to have the most critical timing constraints were placed in the center of the module to allow the greatest number of other chips to be close to them.

Heating the original design until the chips diffuse about freely quickly produces a random-looking arrangement, Fig. 3b. Cooling very slowly until the chips move sluggishly and the objective function ceases to decrease rapidly with change of temperature produced the result in Fig. 3c. The net-crossing histograms have peaks comparable to the peak heights in the original placement, but are much flatter. At this "freezing point," we find that the functionally related groups of chips have reorganized from the melt, but now are spatially separated in an overall arrangement quite different from the original placement. In the final result, Fig. 3d, the histogram peaks are about 30 percent less than in the original placement. Integrating them, we find that total wire length, estimated in this way, is decreased by about 10 percent. The computing requirements for this example were modest: 250,000 interchanges were attempted, requiring 12 minutes of computation on an IBM 3033.

Between the temperature at which clusters form and freezing starts (Fig. 3c) and the final result (Fig. 3d) there are many further local rearrangements. The functional groups have remained in the same regions, but their shapes and relative alignments continue to change throughout the low-temperature part of the annealing process. This illustrates that the introduction of temperature to the optimization process permits a controlled, adaptive division of the problem

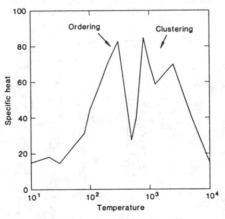

Fig. 4. Specific heat as a function of temperature for the design of Fig. 3, a to d.

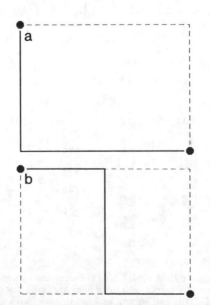

Fig. 5. Examples of (a) L-shaped and (b) Z-shaped wire rearrangements.

through the evolution of natural clusters at the freezing temperature. Early prescription of natural clusters is also a central feature of several sophisticated placement programs used in master slice chip placement (22, 23).

A quantity corresponding to the thermodynamic specific heat is defined for this problem by taking the derivative with respect to temperature of the average value of the objective function observed at a given temperature. This is plotted in Fig. 4. Just as a maximum in the specific heat of a fluid indicates the onset of freezing or the formation of clusters, we find specific heat maxima at two temperatures, each indicating a different type of ordering in the problem. The higher temperature peak corresponds to the aggregation of clusters of functionally related objects, driven apart by the congestion term in the scoring. The lower temperature peak indicates the further decrease in wire length obtained by local rearrangements. This sort of measurement can be useful in practice as a means of determining the temperature ranges in which the important rearrangements in the design are occurring, where slower cooling with be helpful.

Wiring

After placement, specific legal routings must be found for the wires needed to connect the circuits. The techniques typically applied to generate such routings are sequential in nature, treating one wire at a time with incomplete information about the positions and effects of the other wires (11, 24). Annealing is inherently free of this sequence dependence. In this section we describe a simulated annealing approach to wiring, using the ceramic module of the last section as an example.

Nets with many pins must first be broken into connections—pairs of pins joined by a single continuous wire. This "ordering" of each net is highly dependent on the nature of the circuits being connected and the package technology. Orderings permitting more than two pins to be connected are sometimes allowed, but will not be discussed here.

The usual procedure, given an ordering, is first to construct a coarse-scale routing for each connection from which the ultimate detailed wiring can be completed. Package technologies and structured image chips have prearranged areas of fixed capacity for the wires. For the rough routing to be successful, it must not call for wire densities that exceed this capacity.

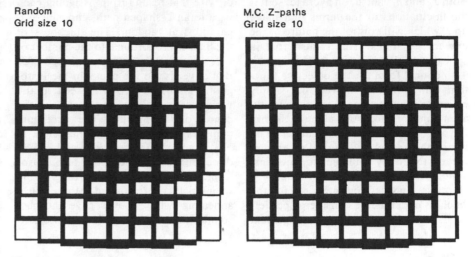

Random
Grid size 10

M.C. Z-paths
Grid size 10

Fig. 6 (left). Wire density in the 98-chip module with the connections randomly assigned to perimeter routes. Chips are in the original placement.   Fig. 7 (right). Wire density after simulated annealing of the wire routing, using Z-shaped moves.

We can model the rough routing problem (and even simple cases of detailed embedding) by lumping all actual pin positions into a regular grid of points, which are treated as the sources and sinks of all connections. The wires are then to be routed along the links that connect adjacent grid points.

The objectives in global routing are to minimize wire length and, often, the number of bends in wires, while spreading the wire as evenly as possible to simplify exact embedding and later revision. Wires are to be routed around regions in which wire demand exceeds capacity if possible, so that they will not "overflow," requiring drastic rearrangements of the other wires during exact embedding. Wire bends are costly in packages that confine the north-south and east-west wires to different layers, since each bend requires a connection between two layers. Two classes of moves that maintain the minimum wire length are shown in Fig. 5. In the L-shaped move of Fig. 5a, only the essential bends are permitted, while the Z-shaped move of Fig. 5b introduces one extra bend. We will explore the optimization possible with these two moves.

For a simple objective function that will reward the most balanced arrangement of wire, we calculate the square of the number of wires on each link of the network, sum the squares for all links, and term the result F. If there are N_L links and N_W wires, a global routing program that deals with a high density of wires will attempt to route precisely the average number of wires, N_W/N_L, along each link. In this limit F is bounded below by N_W^2/N_L. One can use the same objective function for a low-density (or high-resolution) limit appropriate for de-

tailed wiring. In that case, all the links have either one or no wires, and links with two or more wires are illegal. For this limit the best possible value of F will be N_W/N_L.

For the L-shaped moves, F has a relatively simple form. Let $\epsilon_{iv} = +1$ along the links that connection i has for one orientation, -1 for the other orientation, and 0 otherwise. Let a_{iv} be 1 if the ith connection can run through the vth link in either of its two positions, and 0 otherwise. Note that a_{iv} is just ϵ_{iv}^2. Then if $\mu_i = \pm 1$ indicates which route the ith connection has taken, we obtain for the number of wires along the vth link,

$$n_v = \sum_i \frac{a_{iv}(\epsilon_{iv}\mu_i + 1)}{2} + n_v(0) \quad (9)$$

where $n_v(0)$ is the contribution from straight wires, which cannot move without increasing their length, or blockages.

Summing the n_v^2 gives

$$F = \sum_{ij} J_{ij}\mu_i\mu_j + \sum_i h_i\mu_i + \text{constants} \quad (10)$$

which has the form of the Hamiltonian for a random magnetic alloy or spin glass, like that discussed earlier. The "random field," h_i, felt by each movable connection reflects the difference, on the average, between the congestion associated with the two possible paths:

$$h_i = \sum_v \epsilon_{iv} [2n_v(0) + \sum_j a_{jv}] \quad (11)$$

The interaction between two wires is proportional to the number of links on which the two nets can overlap, its sign depending on their orientation conventions:

$$J_{ij} = \sum_v \frac{\epsilon_{iv}\epsilon_{jv}}{4} \quad (12)$$

Both J_{ij} and h_i vanish, on average, so it is the fluctuations in the terms that make up F which will control the nature of the low-energy states. This is also true in spin glasses. We have not tried to exhibit a functional form for the objective function with Z-moves allowed, but simply calculate it by first constructing the actual amounts of wire found along each link.

To assess the value of annealing in wiring this model, we studied an ensemble of randomly situated connections, under various statistical assumptions. Here we consider routing wires for the 98 chips on a module considered earlier. First, we show in Fig. 6 the arrangement

of wire that results from assigning each wire to an L-shaped path, choosing orientations at random. The thickness of the links is proportional to the number of wires on each link. The congested area that gave rise to the peaks in the histograms discussed above is seen in the wiring just below and to the right of the center of the module. The maximum numbers of wires along a single link in Fig. 6 are 173 (x direction) and 143 (y direction), so the design is also anisotropic. Various ways of rearranging the wiring paths were studied. Monte Carlo annealing with Z-moves gave the best solution, shown in Fig. 7. In this exam-

ple, the largest numbers of wires on a single link are 105 (x) and 96 (y).

We compare the various methods of improving the wire arrangement by plotting (Fig. 8) the highest wire density found in each column of x-links for each of the methods. The unevenness of the density profiles was already seen when we considered net-crossing histograms as input information to direct placement. The lines shown represent random assignment of wires with L-moves; aligning wires in the direction of least average congestion—that is, along h_i—followed by cooling for one pass at zero T; simulated annealing with L-moves only; and annealing with Z-moves. Finally, the light dashed line shows the optimum result, in which the wires are distributed with all links carrying as close to the average weight as possible. The optimum cannot be attained in this example without stretching wires beyond their minimum length, because the connections are too unevenly arranged. Any method of optimization gives a significant improvement over the estimate obtained by assigning wire routings at random. All reduce the peak wire density on a link by more than 45 percent. Simulated annealing with Z-moves improved the random routing by 57 percent, averaging results for both x and y links.

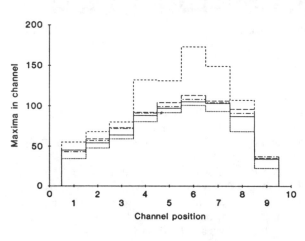

Fig. 8. Histogram of the maximum wire densities within a given column of x-links, for the various methods of routing.

Traveling Salesmen

Quantitative analysis of the simulated annealing algorithm or comparison between it and other heuristics requires problems simpler than physical design of computers. There is an extensive literature on algorithms for the traveling salesman problem (3, 4), so it provides a natural context for this discussion.

If the cost of travel between two cities is proportional to the distance between them, then each instance of a traveling salesman problem is simply a list of the positions of N cities. For example, an arrangement of N points positioned at random in a square generates one instance. The distance can be calculated in either the Euclidean metric or a "Manhattan" metric, in which the distance between two points is the sum of their separations along the two coordinate axes. The latter is appropriate for physical design applications, and easier to compute, so we will adopt it.

We let the side of the square have length $N^{1/2}$, so that the average distance between each city and its nearest neighbor is independent of N. It can be shown that this choice of length units leaves the optimal tour length per step independent of N, when one averages over many

Fig. 9. Results at four temperatures for a clustered 400-city traveling salesman problem. The points are uniformly distributed in nine regions. (a) $T = 1.2$, $\alpha = 2.0567$; (b) $T = 0.8$, $\alpha = 1.515$; (c) $T = 0.4$, $\alpha = 1.055$; (d) $T = 0.0$, $\alpha = 0.7839$.

instances, keeping N fixed (25). Call this average optimal step length α. To bound α from above, a numerical experiment was performed with the following "greedy" heuristic algorithm. From each city, go to the nearest city not already on the tour. From the Nth city, return directly to the first. In the worst case, the ratio of the length of such a greedy tour to the optimal tour is proportional to $\ln(N)$ (26), but on average, we find that its step length is about 1.12. The variance of the greedy step length decreases as $N^{-1/2}$, so the situation envisioned in the worst case analysis is unobservably rare for large N.

To construct a simulated annealing algorithm, we need a means of representing the tour and a means of generating random rearrangements of the tour. Each tour can be described by a permuted list of the numbers 1 to N, which represents the cities. A powerful and general set of moves was introduced by Lin and Kernighan (27, 28). Each move consists of reversing the direction in which a section of the tour is traversed. More complicated moves have been used to enhance the searching effectiveness of iterative improvement. We find with the adaptive divide-and-conquer effect of annealing at intermediate temperatures that the subsequence reversal moves are sufficient (29).

An annealing schedule was determined empirically. The temperature at which segments flow about freely will be of order $N^{1/2}$, since that is the average bond length when the tour is highly random. Temperatures less than 1 should be cold. We were able to anneal into locally optimal solutions with $\alpha \leq 0.95$ for N up to 6000 sites. The largest traveling salesman problem in the plane for which a proved exact solution has been obtained and published (to our knowledge) has 318 points (30).

Real cities are not uniformly distributed, but are clumped, with dense and sparse regions. To introduce this feature into an ensemble of traveling salesman problems, albeit in an exaggerated form, we confine the randomly distributed cities to nine distinct regions with empty gaps between them. The temperature gives the simulated annealing method a means of separating out the problem of the coarse structure of the tour from the local details. At temperatures, such as $T = 1.2$ (Fig. 9a), where the small-scale structure of the paths is completely disordered, the longer steps across the gaps are already becoming infrequent and steps joining regions more than one gap are eliminated. The configurations studied below $T = 0.8$ (for instance, Fig. 9b) had the minimal number of long steps,

but the detailed arrangement of the long steps continued to change down to $T = 0.4$ (Fig. 9c). Below $T = 0.4$, no further changes in the arrangement of the long steps were seen, but the small-scale structure within each region continued to evolve, with the result shown in Fig. 9d.

Summary and Conclusions

Implementing the appropriate Metropolis algorithm to simulate annealing of a combinatorial optimization problem is straightforward, and easily extended to new problems. Four ingredients are needed: a concise description of a configuration of the system; a random generator of "moves" or rearrangements of the elements in a configuration; a quantitative objective function containing the trade-offs that have to be made; and an annealing schedule of the temperatures and length of times for which the system is to be evolved. The annealing schedule may be developed by trial and error for a given problem, or may consist of just warming the system until it is obviously melted, then cooling in slow stages until diffusion of the components ceases. Inventing the most effective sets of moves and deciding which factors to incorporate into the objective function require insight into the problem being solved and may not be obvious. However, existing methods of iterative improvement can provide natural elements on which to base a simulated annealing algorithm.

The connection with statistical mechanics offers some novel perspectives on familiar optimization problems. Mean field theory for the ordered state at low temperatures may be of use in estimating the average results to be obtained by optimization. The comparison with models of disordered interacting systems gives insight into the ease or difficulty of finding heuristic solutions of the associated optimization problems, and provides a classification more discriminating than the blanket "worst-case" assignment of many optimization problems to the NP-complete category. It appears that for the large optimization problems that arise in current engineering practice a "most probable" or average behavior analysis will be more useful in assessing the value of a heuristic than the traditional worst-case arguments. For such analysis to be useful and accurate, better knowledge of the appropriate ensembles is required.

Freezing, at the temperatures where large clusters form, sets a limit on the energies reachable by a rapidly cooled spin glass. Further energy lowering is possible only by slow annealing. We

expect similar freezing effects to limit the effectiveness of the common device of employing iterative improvement repeatedly from different random starting configurations.

Simulated annealing extends two of the most widely used heuristic techniques. The temperature distinguishes classes of rearrangements, so that rearrangements causing large changes in the objective function occur at high temperatures, while the small changes are deferred until low temperatures. This is an adaptive form of the divide-and-conquer approach. Like most iterative improvement schemes, the Metropolis algorithm proceeds in small steps from one configuration to the next, but the temperature keeps the algorithm from getting stuck by permitting uphill moves. Our numerical studies suggest that results of good quality are obtained with annealing schedules in which the amount of computational effort scales as N or as a small power of N. The slow increase of effort with increasing N and the generality of the method give promise that simulated annealing will be a very widely applicable heuristic optimization technique.

Dunham (5) has described iterative improvement as the natural framework for heuristic design, calling it "design by natural selection." [See Lin (6) for a fuller discussion.] In simulated annealing, we appear to have found a richer framework for the construction of heuristic algorithms, since the extra control provided by introducing a temperature allows us to separate out problems on different scales.

Simulation of the process of arriving at an optimal design by annealing under control of a schedule is an example of an evolutionary process modeled accurately by purely stochastic means. In fact, it may be a better model of selection processes in nature than is iterative improvement. Also, it provides an intriguing instance of "artificial intelligence," in which the computer has arrived almost uninstructed at a solution that might have been thought to require the intervention of human intelligence.

References and Notes

1. E. L. Lawlor, *Combinatorial Optimization* (Holt, Rinehart & Winston, New York, 1976).
2. A. V. Aho, J. E. Hopcroft, J. D. Ullman, *The Design and Analysis of Computer Algorithms* (Addison-Wesley, Reading, Mass., 1974).
3. M. R. Garey and D. S. Johnson, *Computers and Intractability: A Guide to the Theory of NP-Completeness* (Freeman, San Francisco, 1979).
4. R. Karp, *Math. Oper. Res.* **2**, 209 (1977).
5. B. Dunham, *Synthese* **15**, 254 (1963).
6. S. Lin, *Networks* **5**, 33 (1975).
7. For a concise and elegant presentation of the basic ideas of statistical mechanics, see E. Shrödinger, *Statistical Thermodynamics* (Cambridge Univ. Press, London, 1946).
8. N. Metropolis, A. Rosenbluth, M. Rosenbluth, A. Teller, E. Teller, *J. Chem. Phys.* **21**, 1087 (1953).
9. G. Toulouse, *Commun. Phys.* **2**, 115 (1977).

10. For review articles, see C. Castellani, C. DiCastro, L. Peliti, Eds., *Disordered Systems and Localization* (Springer, New York, 1981).
11. J. Soukup, *Proc. IEEE* **69**, 1281 (1981).
12. M. A. Breuer, Ed., *Design Automation of Digital Systems* (Prentice-Hall, Engelwood Cliffs, N.J., 1972).
13. D. Sherrington and S. Kirkpatrick, *Phys. Rev. Lett.* **35**, 1792 (1975); S. Kirkpatrick and D. Sherrington, *Phys. Rev. B* **17**, 4384 (1978).
14. A. P. Young and S. Kirkpatrick, *Phys. Rev. B* **25**, 440 (1982).
15. B. Mandelbrot, *Fractals: Form, Chance, and Dimension* (Freeman, San Francisco, 1979), pp. 237–239.
16. S. Kirkpatrick, *Phys. Rev. B* **16**, 4630 (1977).
17. C. Davis, G. Maley, R. Simmons, H. Stoller, R. Warren, T. Wohr, in *Proceedings of the IEEE International Conference on Circuits and Computers*, N. B. Guy Rabbat, Ed. (IEEE, New York, 1980), pp. 669–673.

18. M. A. Hanan, P. K. Wolff, B. J. Agule, *J. Des. Autom. Fault-Tolerant Comput.* **2**, 145 (1978).
19. M. Breuer, *ibid.* **1**, 343 (1977).
20. P. W. Case, M. Correia, W. Gianopulos, W. R. Heller, H. Ofek, T. C. Raymond, R. L. Simek, C. B. Steiglitz, *IBM J. Res. Dev.* **25**, 631 (1981).
21. A. J. Blodgett and D. R. Barbout, *ibid.* **26**, 30 (1982); A. J. Blodgett, in *Proceedings of the Electronics and Computers Conference* (IEEE, New York, 1980), pp. 283–285.
22. K. A. Chen, M. Feuer, K. H. Khokhani, N. Nan, S. Schmidt, in *Proceedings of the 14th IEEE Design Automation Conference* (New Orleans, La., 1977), pp. 298–302.
23. K. W. Lallier, J. B. Hickson, Jr., R. K. Jackson, paper presented at the European Conference on Design Automation, September 1981.
24. D. Hightower, in *Proceedings of the 6th IEEE Design Automation Workshop* (Miami Beach, Fla., June 1969), pp. 1–24.
25. J. Beardwood, J. H. Halton, J. M. Hammersley, *Proc. Cambridge Philos. Soc.* **55**, 299 (1959).
26. D. J. Resenkrantz, R. E. Stearns, P. M. Lewis, *SIAM (Soc. Ind. Appl. Math.) J. Comput.* **6**, 563 (1977).
27. S. Lin, *Bell Syst. Tech. J.* **44**, 2245 (1965).
28. _____ and B. W. Kernighan, *Oper. Res.* **21**, 498 (1973).
29. V. Černy has described an approach to the traveling salesman problem similar to ours in a manuscript received after this article was submitted for publication.
30. H. Crowder and M. W. Padberg, *Manage. Sci.* **26**, 495 (1980).
31. The experience and collaborative efforts of many of our colleagues have been essential to this work. In particular, we thank J. Cooper, W. Donath, B. Dunham, T. Enger, W. Heller, J. Hickson, G. Hsi, D. Jepsen, H. Koch, R. Linsker, C. Mehanian, S. Rothman, and U. Schultz.

Part III
Wirability, Partitioning, and Placement

Placement and Average Interconnection Lengths of Computer Logic

WILM E. DONATH, MEMBER, IEEE

Abstract—The length of the interconnections for a placement of logic gates is an important variable in the estimation of wiring space requirements, delay values, and power dissipation. A formula for an upper bound on expected average interconnection length, based on partitioning results, is given for linear and square arrays of gates. This upper bound gives significantly lower interconnection length than the bound based upon random placement. Actual placements give average interconnection lengths of about half the upper bound given by theory.

I. INTRODUCTION

IN the physical layout of computer logic complexes, a critical quantity is the amount of wire required for the interconnections; quite frequently, this length is used as a measure of the quality of the placement [1]. Knowledge about achievable length is important from several points of view.

1) The amount of space required for the interconnections is largely governed by the total length of wire that must be accommodated. This governs the spacing between gates and the size of the logic carrier (chip, card, etc.).

2) Time delay for signals depends on the wiring length, either because the additional capacitance slows the circuit down or because the interconnection acts as a transmission line.

Manuscript received November 11, 1974; revised August 11, 1977.
The author is with the IBM Thomas J. Watson Research Center, Yorktown Heights, NY 10598.

3) Power requirements depend, in part, on the amount of capacitance at the output of a gate which in turn depends on wire length.

It should be pointed out that other factors affect actual wire length; efficiency of the wiring program, details of the accessing of the connections to the gates, and the possibility of taking wireability into account in the placement program itself. Any result regarding wiring length is, therefore, a first cut at the wiring problem.

Several authors were concerned with the problem of *a priori* length estimation; Gilbert [2] developed relationships for average minimal spanning tree lengths of randomly placed nets of n points; while this is of some interest, it should be pointed out that this does not take into account the effect of good placement algorithms. This author [3] developed a lower bound on average placement length for random graphs; however, graphs generated in the design of logic complexes are significantly different from random graphs [4], and therefore this paper is only of theoretical interest.

Sutherland and Oestreicher [5] developed a method for estimating track requirements for PC boards; their method, however, depends on random placement and would, for large arrays, yield excessively large estimates. We are, therefore, interested in finding results for placement lengths which reflect both the characteristics of logic

Reprinted from *IEEE Trans. Circuits Syst.*, vol. CAS-26, pp. 272–277, Apr. 1979.

complexes as they are designed by engineers and the effect of a placement procedure. Given this point of view, we are justified in using as a starting point the empirically observed terminal to gate (i.e., circuit) count relationships known as Rent's Rule [6]; this relationship can also be derived from a stochastic model [4] using a principle of self-simularity [7]. The theoretical model of Rent's Rule relates it closely to the way engineers design logic complexes now; it also uses a hierarchical partitioning scheme, which is an assumption in our model in addition to the Rent Relationship. Graphs, which obey the Rent Relationship, but do not do so on a hierarchical basis, could probably be designed, but it seems unlikely that they would be common.

The Rent Relationship is

$$T = AC^p \qquad (1)$$

where T is the average number of terminals required by a group containing an average of C gates, A is the average number of terminals per gate, and p $(0 \leqslant p \leqslant 1)$ is a constant for a given logic graph. We use this relationship on a hierarchical basis, i.e., it is expected to hold recursively. We then derive a connection length which is an average for this class of graphs over a subset of all possible placements, where the subset is chosen such that relatively few corrections are long. This differs from random placement, where the average overall placements of a graph are computed. Asymptotic results are also derived; these depend on the value of the exponent p in the partitioning relationship.

We find then that the average length $\bar{R}$ per interconnection behaves for large C and placement on square arrays as

$$\bar{R} \sim C^{p-1/2}, \qquad p > \frac{1}{2}$$

$$\bar{R} \sim \log C, \qquad p = \frac{1}{2}$$

$$\bar{R} \sim f(p), \qquad p < \frac{1}{2} \qquad (2)$$

where $f(p)$ is independent of C. For linear placements, the asymptotic law is

$$\bar{R} \sim C^p. \qquad (3)$$

Since highly parallel circuitry [4], [7], [8] has large exponents p (as much as 0.75), while highly serialized circuitry may be designed with low exponents (as little as 0.47), this indicates that wiring space requirements may vary drastically as the character of the logic changes.

We note that, in any case, these results are significantly better upper bounds than the random placement of computer logic on either a square or linear array:

square array: $\qquad \bar{R} = \sqrt{C}/3$

linear placement: $\qquad \bar{R} = (C+1)/3$

so that the theory described here takes into account in a

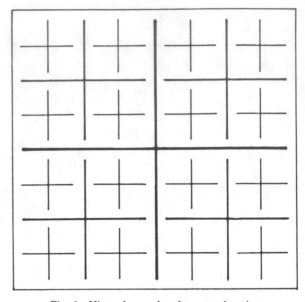

Fig. 1 Hierarchy on the placement locations.

significant way the effect of placement for logic complexes typical of present day design practice.

It is to be noted again that the theoretical results are in the nature of an upper bound; experimental results are needed both to verify the asymptotic trends and the actual improvement possible over the theoretical results given here. The results given in Section III verify the trend with circuit count as well as partitioning exponent p and show the empirical $\bar{R}$ can be about one-half the theoretical upper bound.

II. MATHEMATICAL DERIVATION

The derivation of the upper bound rests basically on the observed average terminal to gate count relationship generalized for a hierarchial structure; we also make an assumption that the number of interconnections is proportional to the terminal count, and that the constant of proportionality does not change with level of hierarchy. Secondly, we impose a hierarchy on the sockets and the hierarchy on the logic gates are parallel, and for our particular purposes, each element of the hierarchy consists of 4 elements of the next lower level of hierarchy (see Fig. 1). This means that there are exactly 4^L (L = number of levels) gates in the logic complex and sockets in the arrays. The placement is then assumed to occur in the following recursive way.

1) The top level element of the hierarchy on the blocks is placed on the top level element of the hierarchy on the placement locations (or sockets). At this stage, the set of elements placed and not expanded consists of one element —the top level element of the hierarchy on the logic blocks. The list of members of the next lower level of the hierarchy of logic gates are placed in a list called the PNX list.

2) Each element E in the PNX list is now treated as follows.

a) Each descendant element of E is placed, at random, on a descendant element of the hierarchical element on which E was placed.

b) It is removed from the PNX list.

c) If the descendant elements of E are also hierarchical elements, they are entered into the PNX list. If they are individual logic gates, nothing more is done.

3) The process terminates when no more elements are in the PNX list.

We note two factors.

1) We can deduce the expected number $\bar{n}_l$ of connections at each level of the hierarchy.

2) We can deduce the average lengths $\bar{r}_l$ of interconnections between sockets belonging to the same $(l+1)$ level hierarchical element, but to different lth level hierarchical elements.

The average length $\bar{R}$ of the interconnections for this process is then given by

$$\bar{R} = \frac{\sum_{l<L} \bar{n}_l \bar{r}_l}{\sum_{l<L} \bar{n}_l}. \tag{4}$$

We note that the partition on the placement location (or sockets) is such that $\bar{r}_l$ increases with increasing l (l=level of hierarchy), while, as we shall see later, the partition on the logic complex is such that $\bar{n}_l$ decreases with increasing level of hierarchy. This leads to an overall lowering of $\bar{R}$; we shall proceed now by deriving expressions for $\bar{n}_k$, and we shall derive expressions for $\bar{R}$ by using expressions for $\bar{r}_k$ for the linear and square case. It is to be noted that this placement procedure can be improved in practice to yield a lower distance, which is one reason why $\bar{R}$ is in the nature of an upper bound. We also ignore connections made to the outside of the array.

In order to derive an expression for n_k, we first consider (1) for the total terminal count of a subcomplex of K gates, which gives us

$$T = AK^p.$$

Suppose we have a total of C gates, divided into groups of K (K may have the value 4^k), we have then a total of

$$T_{\text{total}}(K) = AK^p C/K$$
$$= ACK^{p-1} \tag{5}$$

terminals for complexes of size K. Let $N(K)$ denote the number of interconnections among complexes of size K, and let us assume that this number $N(K)$ is a fraction α of the number of terminals. The value of α is $1/2$ if each net has just two terminals, and is somewhat larger in the normal case, but must be less than 1. Since the range of variation is relatively small for α, it is a good approximation to assume that variations of α are insignificant. We have then

$$N(K) = \alpha ACK^{p-1}. \tag{6}$$

n_k is given by

$$n_k = N(4^k) - N(4^{k+1}) \tag{7}$$

where $N(4^k)$ is the number of connections between groups of size 4^k, which also includes connections between groups of size 4^{k+1}; (7) expresses the fact that these higher level connections must be subtracted out. We have then

$$n_k = \alpha AC(1 - 4^{p-1})4^{k(p-1)}. \tag{8}$$

We shall later derive the expression for $\bar{r}_k$, which are (in terms of circuit spacings) for the

linear case: $\quad \bar{r}_k = \frac{5}{3}4^k \tag{9}$

square case: $\quad \bar{r}_k = \frac{14}{9}2^k - \frac{2}{9}2^{-k}. \tag{10}$

Note that 4^k is the basic unit in the linear case, since 4^k circuits are placed in a row, while 2^k is the basic unit in the square case, where the block is $2^k \times 2^k$. We note, lastly, that

$$C = 4^L.$$

$$L = \log C / \log 4. \tag{11}$$

We find

$$\sum_{k<L} n_k = \alpha AC(1 - 4^{p-1}) \sum_{k=0}^{L-1} 4^{k(p-1)}$$

$$= \alpha AC(1 - 4^{p-1}) \frac{1 - 4^{(p-1)L}}{1 - 4^{p-1}}$$

and, substituting (11)

$$= \alpha AC(1 - C^{p-1}). \tag{12}$$

For the linear case we find

$$\sum n_k \bar{r}_k = \alpha AC(1 - 4^{p-1}) \frac{5}{3} \sum_{k=0}^{L-1} 4^{k(p-1)}$$

$$= \alpha AC(1 - 4^{p-1}) \frac{5}{3} \sum_{k=0}^{L-1} (4^p)^k$$

$$= \alpha AC(1 - 4^{p-1}) \frac{5}{3} \frac{4^{pL} - 1}{4^p - 1}$$

$$= \alpha AC(1 - 4^{p-1}) \frac{5}{3} \frac{C^p - 1}{4^p - 1} \tag{13}$$

so that $\bar{R}$ is given by (13) divided by (12), and we have

$$\bar{R} = \frac{5}{3} \frac{(1 - 4^{p-1})(C^p - 1)}{(1 - C^{p-1})(4^p - 1)}. \tag{14}$$

For large C this becomes

$$\bar{R} \sim \frac{5}{3} C^p \left(\frac{1 - 4^{p-1}}{4^p - 1} \right).$$

Similarly, we treat the square case

$$\sum_{k<L} n_k \bar{r}_k = \sum_{k=0}^{L-1} \alpha AC(1-4^{p-1})4^{k(p-1)}$$

$$\left(\frac{14}{9}2^k - \frac{2}{9}2^{-k}\right)$$

$$= \sum_{k=0}^{L-1} \alpha AC(1-4^{p-1})4^{k(p-1)}\frac{2}{9}$$

$$\cdot (7.4^{k/2} - 4^{-k/2})$$

$$= \alpha AC(1-4^{p-1}) \sum_{k=0}^{L-1} \frac{2}{9}$$

$$\cdot (7.4^{k(p-(1/2))} - 4^{k(p-(3/2))}).$$

For $p \neq 1/2$, we have

$$= \alpha AC(1-4^{p-1})\frac{2}{9}\left(7 \times \frac{4^{(p-(1/2))L}-1}{4^{p-(1/2)}-1} - \frac{1-4^{(p-(3/2))L}}{1-4^{p-(3/2)}}\right)$$

while for $p = 1/2$ the equation becomes

$$= \alpha AC(1-4^{p-1})\frac{2}{9}\left(7L - \frac{1-4^{(p-(3/2))L}}{1-4^{p-3/2}}\right).$$

Use of (4), (11), and (12) gives us

$$p \neq \frac{1}{2} \quad \bar{R} = \frac{2}{9}\left(7\frac{C^{p-(1/2)}-1}{4^{p-(1/2)}-1} - \frac{1-C^{p-(3/2)}}{1-4^{p-(3/2)}}\right)$$

$$\cdot \frac{1-4^{p-1}}{1-C^{p-1}}$$

$$p = \frac{1}{2} \quad \bar{R} = \frac{2}{9}\left(7\log_4 C - \frac{1-C^{p-(3/2)}}{1-4^{p-(3/2)}}\right)$$

$$\cdot \frac{1-4^{p-1}}{1-C^{p-1}}. \quad (15)$$

We note that for large C and $p > 1/2$, we have asymptotically,

$$\bar{R} \sim C^{p-(1/2)}$$

while for $p < 1/2 \bar{R}$ approaches a finite limit for large C. Clearly, for $p = 1/2$, R is asymptotical as $\log C$.

We need to derive lastly the relationship for $\bar{r}_k$ in (9) and (10); first, the linear case.

We have a set of locations numbered 1 to $4n$, grouped in 4 units of n (see Fig. 1). The distance between locations i and j is given as $|i-j|$, and we wish to know the average length of all pairs of locations *not* belonging to the same group of n units. We can express this as

$$\frac{\sum_{k=2}^{4}\sum_{l=1}^{k-1}\sum_{i=1}^{n}\sum_{j=1}^{n}((k-l)n+i-j)}{6n^2} = \frac{\sum_{k=2}^{4}\sum_{l=1}^{k-1}((k-l)n^3)}{6n^2}$$

because the term $\sum_{i=1}^{n}\sum_{j=1}^{n}i-j$ becomes zero.

We develop further and cancel the term in n^2

$$= \frac{\sum_{k=2}^{4}\sum_{l=1}^{k-1}(k-l)n}{6}$$

and the summation can be developed by hand as

$$\frac{(1+2+1+3+2+1)n}{6} = \frac{5}{3}n.$$

If $n = 4^k$, we have (9).

For the square array, let us consider Fig. 1. We first consider the average distance between points in two neighboring square arrays of size $\omega \times \omega$. This gives

$$\frac{\sum_{i_A=1}^{\omega}\sum_{j_A=1}^{\omega}\sum_{i_B=1}^{\omega}\sum_{j_B=1}^{\omega}[\omega+i_B-i_A+|j_B-j_A|]}{\omega^4}$$

$$= \frac{\sum_{i_A=1}^{\omega}\sum_{i_B=1}^{\omega}\omega^2[\omega+i_B-i_A]+\omega^2\sum_{j_A=1}^{\omega}\sum_{j_B=1}^{\omega}j_B-j_A}{\omega^4}.$$

The first term is simply ω, so we find

$$= \omega + \frac{\sum_{i=1}^{\omega}\sum_{j=1}^{\omega}|j-i|}{\omega^2}.$$

This transforms further to

$$= \omega + \frac{\sum_{i=1}^{\omega}\sum_{j=1}^{i-1}(i-j)+\sum_{i=1}^{\omega}\sum_{j=1}^{\omega}(j-i)}{\omega^2}$$

$$= \omega + \frac{2\sum_{i=1}^{\omega}\sum_{j=1}^{i-1}(i-j)}{\omega^2}$$

$$= \omega + \frac{2\sum_{i=1}^{\omega}\left(i(i-1)-\frac{i(i-1)}{2}\right)}{\omega^2}$$

$$= \omega + \frac{\sum_{i=1}^{\omega}i(i-1)}{\omega^2}$$

$$= \omega + \frac{\omega(\omega+1)(\omega-1)}{3\omega^2}$$

$$= \frac{4}{3}\omega - \frac{1}{3\omega}.$$

The average distance for two squares diagonally opposed is given by

TABLE I
COMPARISON OF EXPERIMENTAL AND THEORETICAL PLACEMENT
DISTANCES

Graph	Gates	p	Array Size	No. of Connections	$\bar{R}$ Exp.	$\bar{R}$ Theor.	$\dfrac{\bar{R} \text{ (Expt)}}{\bar{R} \text{ (Theor.)}}$
A(1)	60	∿0.67	8x8	100	1.29	2.76	0.47
B	528	0.59	24x24	1007	2.15	4.02	0.53
C(2)	576	0.75	24x24	1111	2.85	5.26	0.54
D(3)	671	0.57	26x27	1670	2.63	4.07	0.65
E(4)	1239	0.47	36x36	2687	2.14	3.76	0.57
F(5)	2148	0.75	48x48	7302	3.50	7.37	0.48

(1) This graph was provided by T. R. Raymond of IBM Poughkeepsie.

(2) Graph C is a subportion of graph F.

(3) This result is due to M. Hanan and P. K. Wolff, Sr.; the graph is L1 of ref. 6.

(4) This result was obtained in cooperation with M. Hanan and P. K. Wolff, Sr.

(5) This graph is L2 of ref. 6.

$$\frac{\displaystyle\sum_{i_A=1}^{\omega}\sum_{j_A=1}^{\omega}\sum_{i_B=1}^{\omega}\sum_{j_B=1}^{\omega}(2\omega+i_A+j_A-i_B-j_B)}{\omega^4}$$

and becomes, because of cancellations

$$2\omega.$$

The average distance $\bar{r}$ then is

$$\bar{r}=\frac{(2\omega)2+4\left(\dfrac{4_\omega}{3}-\dfrac{1}{3\omega}\right)}{6}$$

$$\bar{r}=\frac{28\omega-4/\omega}{13}$$

$$=\frac{14\omega-2/\omega}{9}.$$

Noting that $\omega=2^k$ gives us (10).

III. COMPARISON WITH EXPERIMENT

We present in Table I results for some placements of logic complexes on essentially square arrays. The partitioning exponent is not known for graph A; a reasonable value is assumed. Distance is measured on a Manhattan grid and minimal spanning basis. It can be seen that the ratio of empirical $\bar{R}$ to the theoretical upper bound is reasonably constant, lending credence to the theory. We can see that the theory accounts largely for the dependence on C of the empirical $\bar{R}$ observed, since results are given for $C=60$ as well as $C=2000$. The dependence on the partitioning exponent p seems also to be confirmed by these results.

IV. CONCLUSION

We presented here a theoretical upper bound on the expected average placement distance, which would take into account to a first order the effect of placement on distance. This theory also predicts trends in average ideal length of connections with gate count and the partitioning exponent p, which are in agreement with the results given here for some experimental graphs. It is to be noted that we can interpret the exponent p as the degree of parallelism of the logic complex, so that we relate placement distance to machine organization.

We do not include effects on wiring length due to effects of wiring algorithms, the detailed connection of wires to the gates, and other detailed effects of layouts. These effects we would expect to be normally of second order.

ACKNOWLEDGMENT

The author would like to name a number of people for support of this work: H. Freitag, W. R. Heller, and R. L. Russo; furthermore, P. Oden, M. Hanan, A. Weinberger, W. F. Mikhail, and P. K. Wolff, Sr., contributed to clarifying the ideas of this paper in some interesting discussions.

REFERENCES

[1] M. Hanan and J. M. Kurtzberg, "Placement techniques," in *Design Automation of Digital Systems*, M. A. Breuer, Ed. Englewood Cliffs, NJ: Prentice-Hall, 1972, ch. 5, pp. 213–282; this chapter contains extensive literature references in this field.
[2] E. N. Gilbert, "Random minimal trees," *SIAM J. Appl. Math.*, vol. 13, pp. 376–387, 1965.
[3] W. E. Donath, "Statistical properties of the placement of a graph," *SIAM J. Appl. Math.*, vol. 16, no. 2, pp. 376–387, Mar. 1968.
[4] ——, "Equivalence of Memory to Random Logic," *IBM J. Res. Develop.*, vol. 58, no. 5, pp. 401–407, Sept. 1974; "Stochastic model of the computer logic design process," IBM T. J. Watson Res. Cent., Yorktown Heights, NY Rep. RC 3136, Nov. 5, 1970.
[5] I. Sutherland and D. Oestreicher, "How big should a printed circuit board be?" *IEEE Trans. Comput.*, vol. C-22, pp. 537–542, May 1973.
[6] B. S. Landman and R. L. Russo, "On a pin versus block relationship for partitions of logic graphs," *IEEE Trans. Comput.*, vol. C-20, pp. 1469–1479, Dec. 1971.
[7] B. Mandelbrot, "The Parceto-Levy law and the distribution of income," *Int. Econom. Rev.*, vol. I, p. 79, 1960; "Information theory and psycholinquistics: A theory of word frequencies," in *Readings in Mathematical Sciences*, P. F. Lazarsfeld and N. W. Henry, Eds. Cambridge, MA: MIT Press, 1968, p. 350.
[8] R. L. Russo, "On the tradeoff between logic performance and circuit to pin ratio for LSI," *IEEE Trans. Comput.*, vol. C-21, pp. 147–155, Feb. 1972.
[9] W. E. Donath and R. B. Hitchcock, "Path lengths in combinational computer logic graphs," IBM T. J. Watson Res. Center, Yorktown Heights, NY, Rep. RC 3383, June 2, 1971.

Prediction of Wiring Space Requirements for LSI

W. R. HELLER,* W. F. MIKHAIL* and W. E. DONATH†

A stochastic model is developed for estimating wiring space requirements for one-dimensional layouts. This model uses as input the number of devices in the complex to be wired, the average length of a connection, and the average number of connections per device, to compute the probability of successfully wiring the devices as a function of the number of tracks provided.

A heuristic approach is used to extend this model to the two-dimensional case, and tested against experimental studies. Satisfactory agreement is found between a priori calculations of track requirements for the two-dimensional case against global wiring solutions for artificially generated problems, and for some layouts of actual logic complexes.

I. INTRODUCTION

In the design of complex computing systems, the provision of adequate space for the interconnections is of major importance. Without this provision, the physical design of interconnections essential to circuit layout may require excessive effort. An alternative, or worse, simultaneous, difficulty may be that major design changes are required to reduce wiring density. This usually entails using more chips or other gate carriers, which means increased manufacturing costs. Such revisions also imply that schedules may slip and design costs rise. It becomes clear, then, that methods for a priori estimation of wiring space requirements have an attractive if not essential return in providing the proper amount of space for efficient wiring.

On the other hand, it is better not to provide so great a margin of extra wiring space that the total number of wiring carriers becomes unnecessarily high. Whether the carrier be made of silicon, ceramic, or organic material, manufacturing costs must be minimized in conjunction with design costs and other charges against the overall project. The whole problem becomes more urgent in the era of large scale integration of circuits. Both capital costs and extensive and complex processing make the integration of many circuits on one silicon surface a critical procedure. Changes of interconnections or circuits after first manufacture are either practically not possible or may raise questions of overall reliability and yield.

This paper tackles the problem of estimating wiring space requirements for a regular array of gates on a regular wiring grid. The questions which arise in the formulation of the problem can be summarized as follows:

Wiring Demand and Capacity
1) Preparation of the Logic Image
2) Average Length of Interconnections
3) Distribution and Number of Interconnections
4) Capacity
5) Discussion of Approximations

In the literature on this subject, the first problem to be attacked was that of interconnection length. Gilbert [1] developed relationships for average numerical spanning tree lengths of randomly placed nets of points. While this is of interest, the model does not take into account the effect of purposeful prior arrangement of the logic gates and terminals. Referring to the previous outline, it is essential that some technique be used to simulate the essential effects of proper preliminary disposition of the logic. In principle, this implies that, even with a fixed, regular image, the operations of partitioning and placement of gates and assignment of input-output terminals must be carried out skillfully. The results, for our purposes, should also be capable of being characterized by a small number of relevant parameters.

One of us has addressed this problem [2, 3]. A reasonable correspondence was found between the average wire lengths deduced from a theoretical model which simulates a well designed gate placement, and actual mean connection lengths taken from the results of operation of effective placement algorithms. An important aspect of this work was the recognition of the relevance for the model of an empirical regularity sometimes known as Rent's rule [4]. This rule gives the ratio of the total number of input and output terminals to the total of circuits or circuit carriers in an interconnected set of logic circuits or circuit carriers, all housed upon one "super-carrier". This ratio is found to vary inversely as a fractional power $(1 - p)$ of the total gate (or carrier) count. The exponent p is called the "Rent" exponent. In turn, this reflects the designer's effectiveness in partitioning and placement, i.e., in putting close together those gates which belong together. The larger the collection of gates, the more self-contained the package, i.e., the fewer the external input and output terminals required per circuit of the collection, hence, the slower the growth of average interconnection length with increasing circuit count. An alternative theoretical treatment has been given deriving a value for the mean length as a function of the previously mentioned ratio [5]. Results were obtained which were asymptotically similar to Donath's for average

*IBM System Products Division, East Fishkill, Hopewell Junction, N.Y.
†IBM Thomas J. Watson Research Center, Yorktown Heights, N.Y.

Reprinted from the *J. Design Automation and Fault Tolerant Computing*, vol. 2, no. 2, pp. 117–144, May 1978, with the permission of the publisher Computer Science Press, Inc., 1803 Research Blvd., Rockville, MD 20850.

length. In later work, it was possible to deepen the understanding of Rent's rule with a theoretical model of the design process [6].

Regardless of the method eventually used to derive an expression for average length of interconnections as a function of gate count, the essential step for our purpose is that such a relationship is assumed to exist. The present work carries forward the investigation by providing a model not only for the overall wiring requirement, but also for the locally fluctuating demand for wiring space internal to interconnected collections of logic gates. Given some regular geometric arrangement of the gates, one can model the demand for wiring space as a stochastic process. Wires will originate and terminate at particular gate locations according to probabilistic models.

There is an apparent difficulty in handling the question of wiring capacity. As the channels fill up with wires, one has to take account of the fact that actual wiring programs pay attention, or should pay attention, to the amount of space remaining. Taking literal account of this fact in the model creates unnecessary difficulties. Instead, the approach taken here is to model the space-filling as a "final snapshot," assuming that the tendency to uniformize the demand for wiring is taken into account by the stochastic nature of the model itself.

A final point is that of "end effects." Real wiring distributions pay attention to the fact that there are fewer neighbors of a logic gate at the periphery of a logic complex than at its center. It is found best here to ignore this difference among logic gates as an initial approximation. We wish to aim at a slightly conservative solution, not an optimistic one. Hence an estimate of wirability should aim at the wiring problems at the center of the logic complex.

It is convenient, in the context of our stochastic model, to take $\bar{R}$, the connection length, as a specified input parameter. With this explicit separation of the problem into two sub-problems, one can deal directly with the fluctuations which give rise to local wiring congestion. The reader may question the separation on several grounds. An infinite mass of logic, while not yet encountered in real life, would, according to the only idealized model we have, possess infinite $\bar{R}$ [3, 5]. In practical programs, it is important that placement of gates actually take account of local congestion during operation of the placement algorithm and hence pay attention to more than merely wiring length [7].

The answers to these two questions are: To the first, a realistic model of actual computer logic would produce finite $\bar{R}$ for its logic carriers, and our stochastic wiring model, with the sub-problem separation indicated, is in a position to use empirical data as well as the predictions for $\bar{R}$ given by an idealized model of logic. To the second question, examination of empirical data shows that enhancing wirability has strong correlation with (although it does not coincide with) minimizing wire length in practical algorithm

operation. Hence, especially in first approximation, average wire length is our parameter characterizing placement. The overflow count, in contrast, evaluated as it eventually has to be, near the end of the wiring task, is extremely sensitive to the final tuning of algorithms to peculiarities of images and logic.

The only published treatment of the joint problems of wiring demand and capacity that we have been able to find is that of Sutherland and Oestreicher [8]. They deal with track requirements for printed circuit boards and assume random placement of the interconnected sub-units. For arrays exceeding several hundred in size, this assumption leads to a rapidly increasing demand for wiring space and a tendency toward uneconomic design. Using the approach we have outlined, the "reason" for this is best interpreted as the rapidly increasing average interconnection length for a random placement scheme, as circuit count increases.

In section II, we introduce and outline the results for a one-dimensional array of terminals. Statements of the required theorems are given in Appendix A, together with proofs.

In section III, the treatment is sketched of a generalized, two-dimensional model which permits wirability assessment. Details are given in Appendix B.

In section IV, we present a comparison of theoretical and experimental results. The latter are derived in part from computer-programmed wiring with idealized connection distributions, and in part from automatic placement and wiring of actual logic networks. Appendix C describes the global wiring program used for the evaluation.

II. OUTLINE OF THE APPROACH TAKEN FOR WIRABILITY PREDICTION

Estimating wirability of an image to be wired means estimating the required number of tracks (channels) to satisfy projected wiring demand after allowing for fluctuations.

For a given number of ports (a port may be a gate, a chip, etc.), the demand is a function of the average number of connections (wires) per port, and as this number increases, the demand increases and vice versa. Another factor that affects demand is the average connection length and, in general, as it increases, the demand increases. In computer wiring, wires in the same net often connect three or more ports. In spite of this correlation, it has been observed, a posteriori, that the assumption of independent origination and termination of single two-point wires does not introduce serious error into the congestion calculations that follow.

To estimate wirability, we first treat the one-dimensional case. Here we are dealing with a large number of ports that are lined up in one row with wires connecting pairs of ports. The problem of estimating the required number of tracks/port reduces to obtaining the largest number of wires crossing between two ports (or to the left of any port) as illustrated in Figure 1. (The numbers below the row of ports indicate the required number of tracks between any two ports. Since the maximum is 4, the stretch of 7 ports is wirable if 4 tracks are available.)

A more precise way of stating the problem is: Given T tracks (available) per port, what is the probability of successfully wiring a given stretch of ports? This paper deals with obtaining a solution to the above stated problem in the one-dimensional (1–D) case and a heuristic generalization to the two-dimensional case is given. The nature of the solution is illustrated in Figure 2.

The mathematical treatment can be carried through with the aid of the model in Figure 3, representing an infinite stretch of ports. Wires may originate at any port and pass to the right. There is no loss of generality in assuming that wires always pass from "inputs" at the left to "outputs" at the right, never the reverse. The number of wires originating at a given port is assumed to follow a Poisson distribution with parameter y, where y is the average number of wires (two-port connections) per port.* With this assumption, it turns out that the marginal probability that k wires lie to the left of some port, p_k, is given by

$$p_k = [e^{-y\bar{R}}(y\bar{R})^k/k!]$$

Here the distribution of *wire lengths* does not figure in the calculation, only its average, $\bar{R}$. It can then be shown (see Appendix A for statement of theorem) that the *conditional* probability, π_{kl}, that if k wires lie to the left of some port, l wires will lie to its right, is

$$\pi_{kl} = \sum_{j=0}^{\min(k,l)} \binom{k}{j}(1 - 1/\bar{R})^j (1/\bar{R})^{k-j} \cdot \frac{e^{-y}\,y^{l-j}}{(l-j)!}$$

Fig. 1   An idealized one-dimensional wiring image.

*Note that the quantity y is equal to the fan-in of a gate (port) if one can neglect dotting and the (small) proportion of *chip* output pads to used circuits; y equals fan-in plus number of dots per circuit plus number of *chip* output pads per circuit. For calculation of *wiring demand*, it is, of course, the connections per circuit (*not* the used pins per circuit) which is relevant. Overall, the connection count is obtained from the relation: number of used pads and pins equals the number of nets plus the number of connections.

Now consider a segment of M ports from an infinitely long stretch of ports. Let T be the number of tracks available per port and let Π_{00} denote the $(T + 1)$ by $(T + 1)$ transition matrix whose elements, π_{kl}, are given by the equation above, for $0 \leq k, 1 \leq T$. Then the probability of successful wiring, P_s, (i.e., the probability that none of the M ports requires more than T tracks) is given by

$$P_s = P_0 \Pi_{00}^{M-1} Q_0$$

where Q_0 is a $(T + 1)$ column vector whose elements are all unity and P_0 is a $(T + 1)$ row vector whose kth element is given by the expression for p_k given above, where $k = 0, 1, \ldots, T$.

For sufficiently large values of M, P_s can be written as

$$P_s \approx P_0\{\lambda_T^{M-1} \cdot 1\}Q_0$$

Fig. 3   Description of wire entrance and exit from ports.

Fig. 2   Confidence level in wiring vs wiring channels available.

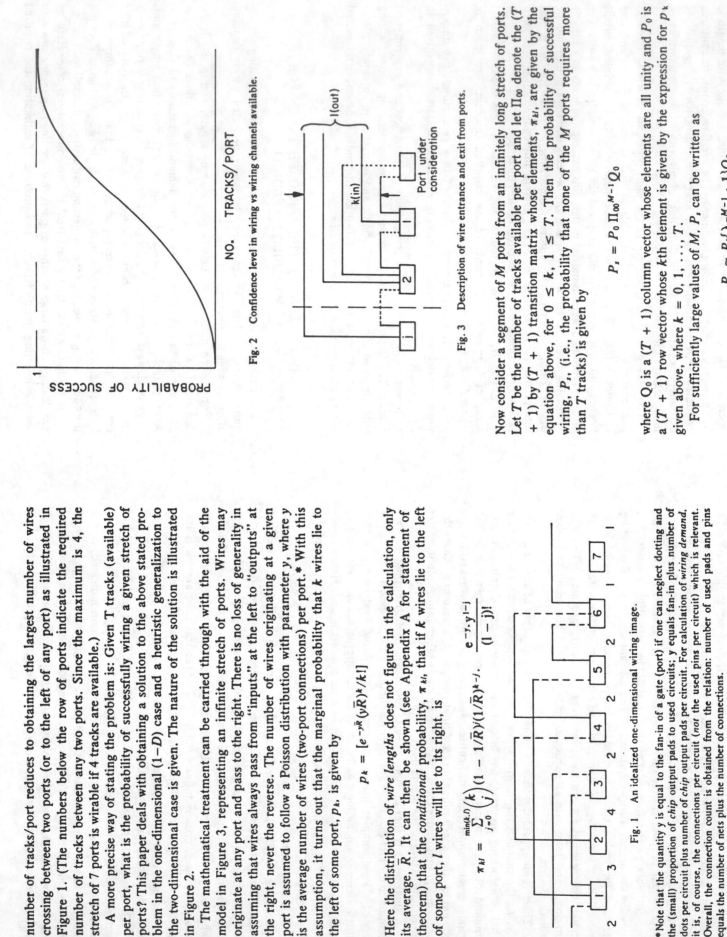

where λ_T is the largest characteristic root of the $(T + 1)$ by $(T + 1)$ matrix Π_∞, is positive, and quite close to unity in value. 1 is the unit matrix of $T + 1$ rows and columns. Now, if we write $\lambda_T = 1 - \alpha_T$, the result can be written $P_s \approx e^{-\alpha_T M}$ and the numerical value of α_T can be closely approximated. Later tables and graphs show the numerical values for a set of interesting values of M, T, R, and y.

III. OUTLINE OF THE TWO-DIMENSIONAL MODEL

A more realistic wiring situation is that encountered in attempting to interconnect a two-dimensional array of cells (or ports). In a common arrangement, wires connect pins on the cells, and may jog from the horizontal to the vertical direction. Wires are routed horizontally on one plane, vertically on another. Changes in direction are made by use of a plane-to-plane access hole, called a "via". It is assumed here that such vias can be created anywhere in either plane without restrictions.

The approach chosen in this work was to use a parameter to describe the degree of flexibility of the wire routing in changing planes. A compromise in the choice of flexibility parameter is singled out by merging into one port, a certain number, W, greater than one, of the ports in a row or a column of the array. The effective number of channels per row or column now includes W of the real channels. In effect, we treat the two-dimensional problem as two one-dimensional problems, one for the horizontal plane, and one for the vertical plane. We refer the reader to Appendix B for the method of estimating W, the effective channel width. We consider a square array of M cells (ports), let N be the total number of connections, and $\bar{R}$, the average connection length, measured in port-to-port units. Figures 4 and 5 illustrate the situation.

Summing up the results of Appendix B, the two-dimensional problem is reduced to two independent one-dimensional problems, each having the following set of input data:

$$M_{eff} = [M(2 - g)/(R + 1 - g)]; y_{eff} = [y(\bar{R} + 1 - g)/2]; \bar{R}_{eff} = \bar{R}/(2 - g)$$

Here g is the expected fraction of cases in which a wire has *either* a vertical component only, *or* a horizontal component only. $y = N/M$ is the number of connections per original port, and

$$W = (\bar{R} + 1 - g)/(2 - g)$$

This situation is equivalent to *one* one-dimensional problem with input data 2 M_{eff}, y_{eff}, $\bar{R}_{eff}$, by the use of theorems stated in Appendix A (see section II). Now one can obtain the effective number of tracks per cell, T_{eff}, required for a given level of probability of successful wiring from the one-dimensional model. The total number of tracks (horizontal and vertical) required per real cell is therefore given by

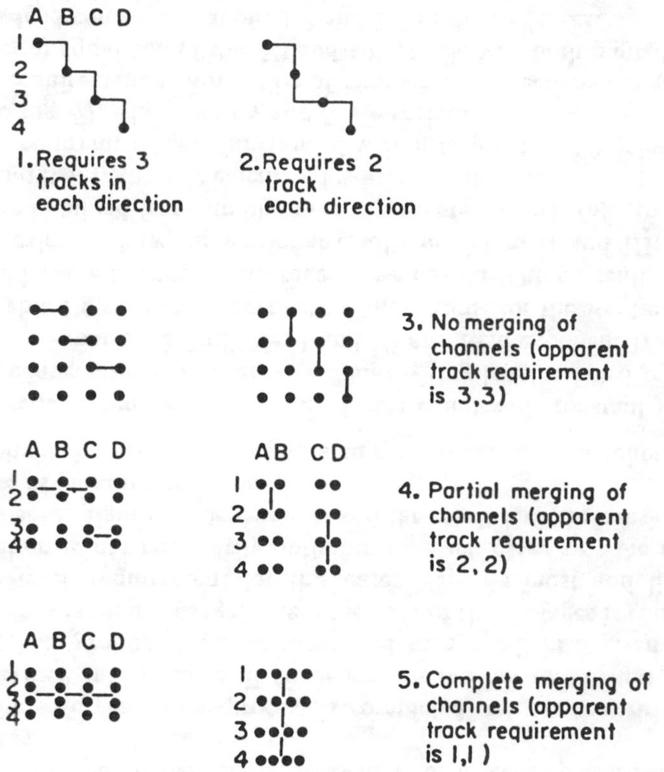

$$T = (2T_{eff}/W)$$

IV. EXPERIMENTAL AND THEORETICAL RESULTS

Simulation

In choosing a method for comparison between theory and experiment, it was necessary to find a compromise. With actual chip design, there is a need to amass, disentangle, and interpret the relevant data from particular chip images. Details of final wiring completion and overflow data are quite dependent upon the final wiring techniques such as "clean-up" algorithms. To avoid too great dependence of our interpretation of experiments on such details, it was clear that results of global wiring afforded the best data base.

A second point needing consideration is the basis for comparison

Fig. 4   Alternate routing of interconnections on 4 × 4 grid.

1. Requires 3 tracks in each direction
2. Requires 2 track each direction
3. No merging of channels (apparent track requirement is 3,3)
4. Partial merging of channels (apparent track requirement is 2,2)
5. Complete merging of channels (apparent track requirement is 1,1)

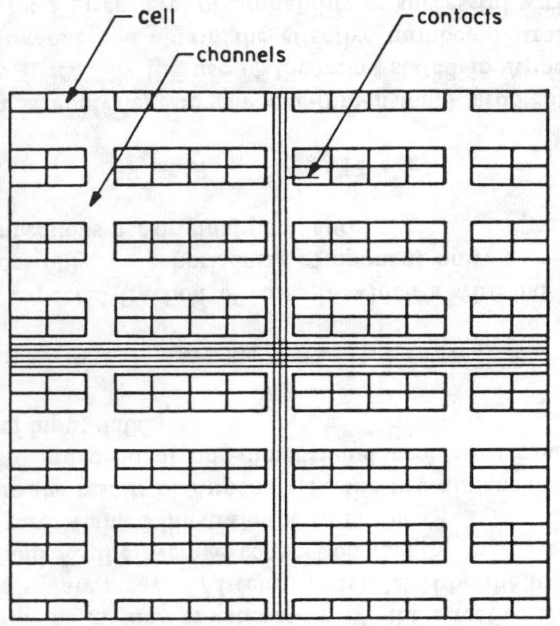

metric is the "Manhattan" distance, $R = |\Delta x| + |\Delta y|$, between pairs of (points) ports with x-separation of Δx and y-separation of Δy.

We shall now describe a procedure for generating wires with some given probability distribution; first, we generate a random number z and determine the length of the wire as being the least value of l such that

$$\sum_{d \leq l} f(d) < z, \quad o \leq z \leq 1 \qquad (4.2)$$

We assign an orientation for the wire by generating a second random number z', and let

$$\Delta x = \langle (2z' - 1)l \rangle$$
$$\Delta y = l - |\Delta x|, \quad o \leq z' \leq 1$$

be the horizontal and vertical extensions of the wire, where $\langle m \rangle$ denotes the least integer greater than or equal to m, and $|m|$ denotes its absolute value. To embed that wire in the $\sqrt{M}$ by $\sqrt{M}$ array we select the upper left corner point of the $|\Delta x|$ by $|\Delta y|$ rectangle spanned by the wire. This is accomplished by selecting a point at random among the $(\sqrt{M} - |\Delta x|)$ by $(\sqrt{M} - |\Delta y|)$ which constitute the possible locations for the upper left hand corner of that rectangle. We repeat this process N times to generate the number of wires we desire. N is the total number of wires (connections), and $\sqrt{M}$ is the square root of the total number of cells (ports) M, in the square array.

After the wire generation process is completed, we use the wiring program described in Appendix B to determine the horizontal and vertical segments (i.e., groups of tracks associated with a cell in a given row or column) with maximum usage, where the sum of these usages is considered to be the track requirement for this case. Here we must add that this determination is not necessarily optimum for each case, i.e., the number obtained here is algorithm dependent, and the result must be considered in some sense as an approximation.

We conducted several sets of simulated runs to examine the following:

1) Track requirement as a function of gate count with constant y (number of connections per gate) = 2.5 and wiring lengths fitted by theory with Rent exponent 2/3 (Tables I and II) and Rent exponent 1/2 (Table III). Table I gives the histograms for those runs, for the Rent exponent 2/3, and Table II gives the average, as well as the 90th percentile values. The calculated two-dimensional results of Tables II and III, as well as those obtained from simulation are presented in Fig. (6). (In Figures 6, 7, and 8 only the 2-D calculated values are plotted.)

2) Track requirement as a function of wiring length for 400 gates (Table IV and Fig. 7), where both γ and $\bar{R}$ are varied.

3) Track requirement as a function of number of connections per gate with gate count equal 100 (Table V) and 400 (Table VI), both plotted in Fig. 8; $\bar{R}$ is held constant at about 1.8 and 2.4 in the two cases.

Fig. 5 Idealized two-dimensional wiring image.

between the experimental wiring results and the wirability theory. Clearly sufficient data would be necessary so that some statistic representing channel requirement in the experiments could be compared to an appropriate level of probability of successful wiring. It was found that a comparison statistic was the average value of simulated channel requirement; this was found to correspond to a 90% level of probability of successful wiring.

In performing the simulated experiments, a global routing program—described in Appendix C and written by one of us (WED)—is used. The wires being generated are assumed to follow the following wiring length distribution:

$$\begin{aligned} f(d) &= A/d^\gamma & \text{for} \quad d < L \\ f(d) &\leq A/d^\gamma & \text{for} \quad d = L \\ f(d) &= 0 & \text{for} \quad d > L \end{aligned}$$

where A. L and f(L) were chosen such that

$$\sum_{d=1}^{L} f(d) = 1 \quad \text{and} \quad \sum_{d=1}^{L} d f(d) = \bar{R}.$$

$\bar{R}$ (the average connection length) and γ are input parameters, and the

TABLE I

Simulated Track Requirement Distribution

No. Tracks Per Cell	Number of Cells/Array						
	(6 × 6)	(8 × 8)	(10 × 10)	(15 × 15)	(20 × 20)	(30 × 30)	(40 × 40)
9	6						
10	88	8					
11	112	20	1				
12	211	88	36				
13	48	59	50	1			
14	33	64	94	46	9		
15	2	8	13	22	11		
16		3	5	31	74	7	
17			1		3	10	
18					3	33	6
19							2
20							2
Average $\bar{T}$ Tracks	11.6	12.7	13.5	14.83	15.8	17.5	18.6
$\bar{R}$	1.39	1.59	1.77	2.12	2.41	2.89	3.28
No. of Runs	500	250	200	100	100	50	10

$y = 2.5 = $ number of connections/cell, $\gamma = 5/3$.

TABLE II

Comparison Between Simulated and Calculated Track Requirement

M	$\sqrt{M}$	$\bar{R}$	No. of Runs	$\bar{T}$ (Sim.)	T (calculated)	
					2-D	1-D
6 × 6	6	1.387	500	11.6 (13)*	11.1	9.0
8 × 8	8	1.59	250	12.7 (14)	12.0	10.5
10 × 10	10	1.771	200	13.5 (14)	13.0	11.7
15 × 15	15	2.117	100	14.83 (16)	14.6	13.9
20 × 20	20	2.41	100	15.8 (16)	15.9	15.6
30 × 30	30	2.889	50	17.5 (18)	17.7	18.4
40 × 40	40	3.276	10	18.6 (20)	19.1	20.6
44 × 44	44	2.469	10	19.3	19.7	21.5

$y = 2.5$, $\gamma = 5/3$.
*90th percentile experimental values.

TABLE III

Comparison Between Simulated and Calculated Track Requirement

M	$\sqrt{M}$	$\bar{R}$	No. of Runs	$\bar{T}$ (Sim.)	$\bar{T}$ (calculated)	
					2-D	1-D
6 × 6	6	1.28	10	11.3	11.0	8.7
8 × 8	8	1.425	10	12.1	11.8	9.7
10 × 10	10	1.544	10	12.7	12.6	10.6
15 × 15	15	1.766	10	13.1	13.9	12.3
20 × 20	20	1.93	10	14.1	14.8	13.5
30 × 30	30	2.16	10	14.9	16.1	15.4

$y = 2.5 = $ number of connections/cell, $\gamma = 2$.

V. COMPARISON BETWEEN SIMULATED AND THEORETICAL RESULTS

The theoretical (calculated) results for both the two-dimensional and one-dimensional models correspond to a 90% level of probability of successful wiring; while the observed ones correspond to the averages of the simulated runs.

A comparison between the simulated and calculated track requirements is given in Tables (II-VI) and presented in Figures 6, 7, and 8—from which we observe the following:

1) The calculated results obtained from the 2-D model track are close to those obtained from simulation. The maximum difference observed between the two results did not exceed 10%.

2) The difference between the 2-D and 1-D results is small when $\bar{R}$ is roughly in the range (2.25-2.75).

3) Generally for $\bar{R} \leq 2.25$, the required number of tracks as estimated from the 2-D model exceeds that from the 1-D model and the situation is reversed for $\bar{R} > 2.75$.

4) For smaller arrays simulated track requirements tend to be slightly larger than theory. While this effect is less than 10%, we believe it is mainly due to the fact that there is an end-effect when the ratio of $\bar{R}$ to array size becomes larger.

VI. COMPARISON OF MODEL PREDICTIONS TO A SELECTION OF REAL LOGIC DESIGNS

Earlier we mentioned some difficulties in comparisons with particular designs. It is important to show, nevertheless, that the predictions of the model are borne out by comparison with actual wiring data.

To carry out this comparison, we have chosen to deal with the wiring completion data. In references (7, 9) about 100 chip designs, correspond-

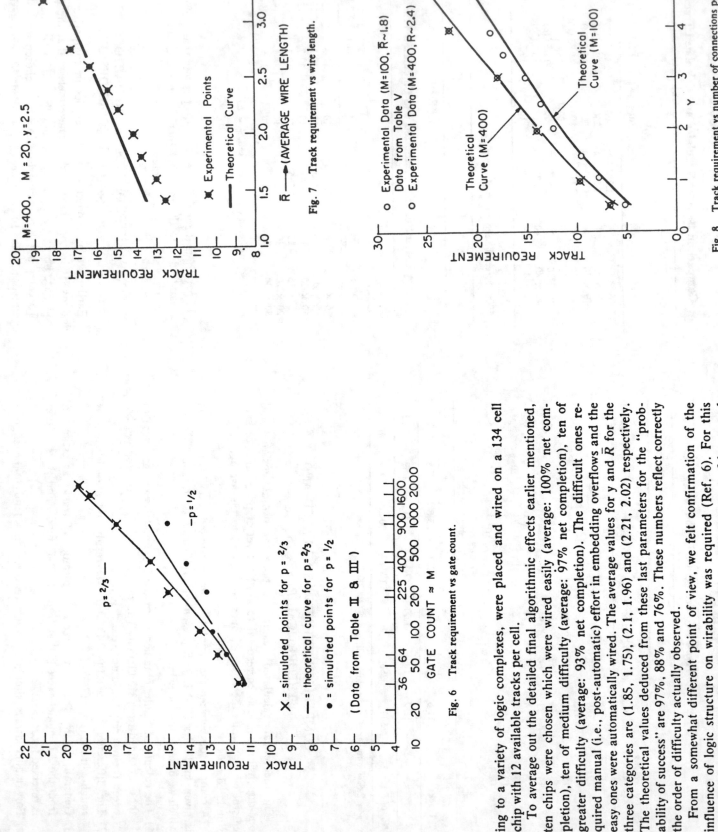

Fig. 6 Track requirement vs gate count.

Fig. 7 Track requirement vs wire length.

Fig. 8 Track requirement vs number of connections per cell.

ing to a variety of logic complexes, were placed and wired on a 134 cell chip with 12 available tracks per cell.

To average out the detailed final algorithmic effects earlier mentioned, ten chips were chosen which were wired easily (average: 100% net completion), ten of medium difficulty (average: 97% net completion), ten of greater difficulty (average: 93% net completion). The difficult ones required manual (i.e., post-automatic) effort in embedding overflows and the easy ones were automatically wired. The average values for y and R̄ for the three categories are (1.85, 1.75), (2.1, 1.96) and (2.21, 2.02) respectively. The theoretical values deduced from these last parameters for the "probability of success" are 97%, 88% and 76%. These numbers reflect correctly the order of difficulty actually observed.

From a somewhat different point of view, we felt confirmation of the influence of logic structure on wirability was required (Ref. 6). For this purpose, designs were chosen from the two known extremes of low speed and high speed logic, corresponding to small and large p. One expects average connection length to increase from the low speed to the high speed designs. This was indeed the case where a reasonably effective placement algorithm was used. Global wiring results are presented in Table VII. We can see that the expected difference in wirability indeed is present.

TABLE VII

Data on Two Real Logic Graphs

	Hand Calculator p = .59	Subportion of High Speed Machine: p = .75
Number of circuits	528	675
Size of array	24 × 24	24 × 24
Number of connections	1007	1111
y	1.91	1.93
$\bar{R}$	2.15	2.85
Theoretical Track Req.	13.0	14.4
Experimental Track Req.	12	14

SUMMARY

Beginning with basic concepts, we have developed a mathematical model for two-dimensional wirability analysis. We have compared track requirements obtained from our model with those obtained from either simulated wire distributions or actual logic graphs. The prediction from our model was always within 10% of the simulated or actual result.

ACKNOWLEDGEMENTS

The authors would like to thank F. J. Worthmann and Dr. R. Russo for their support and encouragement of this joint work. They would also like to thank the authors of the two other papers being presented for numerous valuable discussions of results and techniques [7, 9].

REFERENCES

[1] E. N. Gilbert, "Random Minimal Trees", SIAM J. Appl. Math. 13 (1965) pp. 376-387.
[2] W. E. Donath, "Statistical Properties of the Placement of a Graph", SIAM J. Appl. Math. 16 (1968) pp. 376-387.
[3] W. E. Donath, "Placement and Average Interconnection Lengths of Computer Logic", IBM Research Report 4610, to be published, IEEE Journal of Solid State.
[4] B. S. Landman and R. L. Russo, "On a Pin Versus Block Relationship for Partitions of Logic Graphs", IEEE Trans. on Computers C-20 (1971) pp. 1469-1479.
[5] M. Feuer, "Connectivity of Random Logic", IBM Internal Report, 1974 (to be published).
[6] W. E. Donath, "Equivalence of Memory to Random Logic", IBM Journal of Research

TABLE IV

Track Requirement As a Function of Wire Length

$\bar{R}$	γ	No. of Runs	$\bar{T}$ (Sim.)	$\bar{T}$ (calculated) 2-D	$\bar{T}$ (calculated) 1-D
1.395	2.0	10	12.6	13.6	11.0
1.593	2.0	10	13.0	14.0	12.0
1.78	2.0	10	13.7	14.6	12.9
1.98	2.0	10	14.1	15.0	13.7
2.17	1.7	10	14.8	15.3	14.5
2.37	1.67	10	15.4	15.9	15.6
2.58	1.6	10	16.3	16.3	16.3
2.75	1.55	10	17.3	16.8	17.0
3.19	1.5	10	18.3	17.7	18.7

$M = 400, \sqrt{M} = 20, y = 2.5.$

TABLE V

Track Requirement As a Function of Connections per Cell

y	$\bar{R}$	$\bar{T}$ (Sim.)	$\bar{T}$ (calculated) 2-D	$\bar{T}$ (calculated) 1-D
.49	1.83	5.4	5.0	4.5
.98	1.79	8.0	7.5	6.6
1.45	1.71	9.4	9.2	8.0
2.04	1.77	12.3	11.5	10.1
2.5	1.77	13.4	13.0	11.7
2.96	1.76	15.0	14.6	13.0
3.49	1.72	17.5	16.3	14.2
3.92	1.78	18.4	17.5	15.8
5.13	1.79	22.8	21.0	19.3

$\sqrt{M} = 10, M = 100, \bar{R} \sim 1.8,$ number of runs = 10, $\gamma = 5/3.$

TABLE VI

Track Requirement As a Function of Connections per Cell

y	T (Sim.)	T (cal. 2-D)	T (cal. 1-D)
.5	6.5	6.1	5.9
1.0	9.3	8.9	8.8
2.0	13.5	13.7	13.7
3.0	17.9	17.8	17.7
4.0	22.2	21.7	21.7
5.0	26.5	25.4	25.2

$M = 20 \times 20, \bar{R} = 2.4,$ number of runs = 10, $\gamma = 5/3.$

and Development 18 (1974) pp. 401–407; see also "Stochastic Model of the Computer Logic Design Process", IBM Research Report RC 3136.

[7] K. Khokhani and A. M. Patel, "The Chip Layout Problem: A Placement Procedure for LSI", Proceedings 14th Design Automation Workshop, New Orleans, La, June 22, 1977.

[8] I. Sutherland and D. Oestreicher, "How Big Should a Printed Circuit Board Be?", IEEE Trans. on Comp., May 1973, C-22, pp. 537-542.

[9] K. A. Chen, M. Feuer, K. Khokhani, N. Nan, S. Schmidt, "The Chip Layout Problem: An Automatic Wiring Procedure", Proceedings 14th Design Automation Workshop, New Orleans, La, June 22, 1977.

APPENDIX A: MATHEMATICAL TREATMENT OF THE ONE-DIMENSIONAL MODEL

In this appendix, we give some details and proofs of the work summarized in Section II of the text.

DEFINITIONS

Consider one port out of a very large (infinite) stretch of ports as illustrated in Figure 3. Let:

• The random number $X(L)$ represent the number of wires that terminate at the port under consideration, i.e., lie immediately to its left but not to its right.

• The random variable $X(LR)$ represent the number of wires that cross the port, i.e., lie both to its left and its right.

• The random variable $X(R)$ represent the number of wires that originate at the port itself (and thus move to its right).

ASSUMPTIONS

We shall assume that we are dealing with an infinitely long stretch of ports. Wires may originate at any port and then move to the right. It is clear on grounds of symmetry that one may assume that wires pass only from left to right, since there is nothing in the model which needs to distinguish input from output ports. The number of wires originating at a given port $X(R)$ is assumed to follow the Poisson distribution with parameter y where y is the average number of wires (connections) per port.

LEMMA

If the number of wires originating at a given port $X(R)$ follows the Poisson distribution with parameter y then, for any wiring length distribution whose average is R, the random variables $X(L)$ and $X(LR)$ are independently Poisson distributed as Poisson with parameters y and $y(\bar{R} - 1)$ respectively.

Proof

Referring to Fig. 3, it was assumed that:

• All wires that originate at the port under consideration move to the right.

• We shall assume that a wire will have length l with probability q_l where

$$\sum_{i=1}^{\infty} q_i = 1.$$

Hence a wire that starts at a port j units to the left of the one under consideration belongs to one of the following three categories: (a) It terminates at the port of interest if it is of length j, i.e., with probability q_j. In other words, *it lies immediately to the left of the port under consideration* with probability q_j; (b) It crosses the port and thus lies *both to its left and its right* if its length $> j$, i.e., with probability

$$\sum_{i=j+1}^{\infty} q_i;$$

(c) It does not affect the port under consideration, i.e., terminates at some port to its left, if its length $< j$, i.e., with probability

$$\sum_{i=1}^{j-1} q_i$$

• Denoting by $X_j(L)$ and $X_j(LR)$ the numbers of wires belonging to categories (a) and (b) above, and further assuming that the number of wires originating at any port follows the Poisson distribution with parameter y, we obtain for the joint probability distribution of $[X_j(L), X_j(LR)]$ the following:

$$\text{prob}(X_j(L) = n_1, X_j(LR) = n_2) = e^{-y} \sum_{r=n_1+n_2}^{\infty} \frac{y^r}{r!} \frac{r!}{n_1! n_2! (r - n_1 - n_2)!} \cdot$$

$$q_j^{n_1} \sum_{i=j+1}^{\infty} q_i^{n_2} \sum_{i=1}^{j-1} q_i^{r-n_1-n_2} = \frac{e^{-y q_j}(y q_j)^{n_1}}{n_1!} \left(e^{-y \sum_{i=j+1}^{\infty} q_i} \right).$$

$$\frac{\left(y \sum_{j+1}^{\infty} q_i \right)^{n_2}}{n_2!}$$

A(1)

Equation (1) implies that the number of wires $X_j(L)$ originating at a port j units to the left of the one under consideration and terminating at the latter is independent of the number of wires $X_j(LR)$ that cross the port under consideration (i.e., lie both to its left and its right). Moreover, both $X_j(L)$ and $X_j(LR)$ have Poisson distributions with parameters (yq_j) and

$$y \sum_{i=j+1}^{\infty} q_i \quad \text{respectively.}$$

Assuming that wires originate independently at different ports, let $X(L)$ and $X(LR)$ represent, respectively, the total number of wires to the left at a given port (i.e., lie immediately to its left), and the total number of wires that cross the port (i.e., lie both to its left and its right). Hence

$$X(L) = \sum_{j=1}^{\infty} X_j(L)$$

and

$$X(LR) = \sum_{j=1}^{\infty} X_j(LR) \qquad \text{A(2)}$$

Keeping in mind that the distribution of a sum of random variables, each obeying the Poisson law, is also Poisson with parameter the sum of the individual parameters, it follows that $X(L)$ and $X(LR)$ have Poisson distributions with parameters

$$y \sum_{j=1}^{\infty} q_j = y$$

and

$$y \sum_{j=1}^{\infty} \sum_{i=j+1}^{\infty} q_i = y(\bar{R} - 1)$$

respectively, where

$$\bar{R} = \sum_{j=1}^{\infty} j q_j.$$

Corollary (1)

The joint probability $p(k,l)$ that l wires lie to the right of some port while k wires lie to its left is given by

$$p(k,l) = \sum_{j=0}^{min(k,l)} \left(e^{-y(\bar{R}-1)} \frac{(y(\bar{R}-1))^j}{j!} \right) \cdot$$

$$(e^{-y} y^{k-j}/(k-j)!) \cdot (e^{-y} y^{l-j}/(l-j)!) \qquad \text{A(3)}$$

Proof

The proof follows immediately from the results of the above lemma by observing that the first term implies that j wires lie both to the left and right of the port; the second term implies that $(k - j)$ wires lie to the left (thus making the total number of wires to the left $=k$) and the third term implies that $(l - j)$ wires originated at the port itself in order to obtain a total of l wires to the right.

Corollary (2)

The marginal probability that k wires lie to the left of some port, p_k, is given by

$$p_k = e^{-y\bar{R}}(y\bar{R})^k/k! \qquad \text{A(4)}$$

Proof

The proof follows from observing that the number of wires to the left is simply the sum of $X(L)$ and $X(LR)$.

Corollary (3)

The conditional (transition) probability, π_{kl}, that if k wires are to the left of some port, l wires will lie to its right is

$$\pi_{kl} = \sum_{j=0}^{min(k,l)} \binom{k}{j}(1 - 1/\bar{R})^j(1/\bar{R})^{k-j}[e^{-y} y^{l-j}/(l-j)!] \qquad \text{A(5)}$$

Proof

Follows from equations (3) and (4) by observing that

$$\pi_{kl} = p(k,l)/p_k$$

Theorem

Consider a segment of M ports from an infinitely long stretch of ports. Let T be the number of tracks available per port and let Π_{00} denote the $(T + 1)$ by $(T + 1)$ transition matrix whose elements π_{kl} are given by equation (5) for $0 \leq k, l \leq T$. Then the probability of successful wiring, P_s, (i.e., the probability that none of the M ports require more than T tracks) is given by

$$P_s = P_0 \Pi_{00}^{M-1} Q_0 \qquad \text{A(6)}$$

71

where Q_0 is a $(T + 1)$ column vector whose elements are all one and P_0 is a $(T + 1)$ for vector whose kth element is given by

$$p_k = e^{-yR}(y\bar{R})^k/k! \qquad (k = 0, 1, \ldots T)$$

Proof

The probability $W(k_1, k_2, \ldots, k_M)$, that exactly $k_1, k_2 \ldots k_M$ wires lie to the left of the first, second, ..., Mth port, respectively, is given by

$$W(k_1, k_2, \ldots k_m) = p_{k_1} \pi_{k_1 k_2} \pi_{k_2 k_3} \ldots \pi_{k_{M-1} k_M}$$

Now, the probability of successful wiring P_s, can be expressed as the sum of all $W(k_1, k_2, \ldots k_m)$ such that no k_i exceeds T for $i = 1, 2, \ldots M$, i.e.,

$$P_s = \sum_{k_1=0}^{T} \sum_{k_2=0}^{T} \cdots \sum_{k_M=0}^{T} W(k_1, k_2, \cdots k_M)$$

$$= \sum_{k_1=0}^{T} p_{k_1} \sum_{k_2=0}^{T} \pi_{k_1 k_2} \sum_{k_3=0}^{T} \pi_{k_2 k_3} \cdots \sum_{k_M=0}^{T} \pi_{k_{M-1} k_M}$$

$$= P_0 \Pi_{00}^{M-1} Q_0$$

For a given set of inputs $(M, \bar{R}, y, T)$, the probability of successful wiring is given by P_s of equation (6). Also, for a prescribed level of probability, the required number of tracks can be obtained by using the same equation.

Methodology for Evaluating Equation (6)

Here, we derive an approximation to the formula expressed by equation A(6) that is more appropriate for numerical evaluation.
For large values of M, P_s can be written as

$$P_s \simeq P_0 \lambda_T^{M-1} Q_0 \qquad A(7)$$

where λ_T is the largest characteristic root of the $(T + 1)$ by $(T + 1)$ matrix Π_{00} and is therefore positive and relatively close to 1. Let

$$\lambda_T = 1 - \alpha_T \qquad A(8)$$

Substituting from (A-8) into (A-7) we get

$$P_s \simeq (1 - \alpha_T)^{M-1} P_0 Q_0 \simeq e^{-\alpha_T \cdot M} \qquad A(9)$$

One method of evaluating α_T is as follows:

$$(1 - \alpha_T) = \lambda_T = \lim_{k \to \infty} (P_0 \Pi_{00}^k Q_0 / P_0 \Pi_{00}^{k-1} Q_0) \qquad A(10)$$

or

$$\alpha_T = \lim_{k \to \infty} [1 - (P_0 \Pi_{00}^k Q_0 / P_0 \Pi_{00}^{k-1} Q_0)]$$

$$= \lim_{k \to \infty} [P_0 \Pi_{00}^{k-1}(1 - \Pi_{00}) Q_0 / P \Pi_{00}^{k-1} Q_0] \qquad A(11)$$

Let $\Pi_{01}, \Pi_{10}, \Pi_{11}$ denote the transition matrices with elements $\pi_{k,l}$ given by Eq. A(5) for values of (k,l) given by $(k \leq T, l > T)$; $(k > T, l \leq T)$ and $k > T, l > T)$ respectively.
Since the sum of elements in each row of the transition matrix

$$\begin{pmatrix} \Pi_{00} & \Pi_{01} \\ \Pi_{10} & \Pi_{11} \end{pmatrix}$$

is unity, it follows that

$$\begin{pmatrix} \Pi_{00} & \Pi_{01} \\ \Pi_{10} & \Pi_{11} \end{pmatrix} \begin{pmatrix} Q_0 \\ Q_1 \end{pmatrix} = \begin{pmatrix} Q_0 \\ Q_1 \end{pmatrix}$$

Here Q_0 is a $(T + 1)$ column vector of all 1's and Q_1 is an infinite column vector of all 1's, hence

$$Q_0 = \Pi_{00} Q_0 + \Pi_{01} Q_1 \qquad A(12)$$

or

$$\Pi_{01} Q_1 = (1 - \Pi_{00}) Q_0$$

Substituting from A(12) into A(11) we get

$$\alpha_T = \lim_{k \to \infty} (P_0 \Pi_{00}^{k-1} \Pi_{01} Q_1 / P_0 \Pi_{00}^{k-1} Q_0) \qquad A(13)$$

It is to be noticed that Eq. A(11) as well as Eq. A(13) may be used for expressing α_T. However Eq. A(13) is preferred because, for very small values of α_T, it is less sensitive to rounding-off errors. It also approaches a limit relatively fast. To sum up: The solution to Eq. A(6) is approximately given by Eq. A(9) where, in turn, α_T is given by Eq. A(13).

APPENDIX B: THE TWO-DIMENSIONAL PROBLEM
(An Outline of Results Appeared in Section III)

In this case we have a two-dimensional array of cells (or ports); wires connecting pins on the cells move horizontally and vertically on tracks. There is a considerable degree of freedom in how the individual wires are routed, which leads to possibilities of different algorithms.

Several features of the layout deserve amplification; for one, wires are routed horizontally on one plane and vertically on the other plane (Fig. 5). It is assumed that wires may access to the other plane almost anywhere via a so-called "via hole". (We call this the infinite via capability.)

Our approach here is essentially one of attempting to convert the two-dimensional problem into a set of independent one-dimensional problems. We shall discuss two straightforward extreme approaches and then develop a "compromise" approach, which, we hope, accounts for the flexibility introduced by the two-dimensional aspects of wire routing (Fig. 4).

One extreme approach is as follows: We divide the problem into two problems for each of the 2 planes—one horizontal, one vertical. On the horizontal (vertical) plane we merge all the cells in a vertical (horizontal) line into one port. A wire that has a horizontal and vertical component generates one wire on each plane while wires which connect cells on the same horizontal or vertical line generate wires on only one plane. This approach would be optimistic about wirability, because it assumes that wires can run anywhere on the plane, which is in fact not true.

The second extreme approach—which would yield pessimistic results—is to consider every line of ports as a separate problem. This assumes a certain lack of flexibility in the routing.

The compromise we worked out here is to merge into one port, not all the ports in a column (or row) but just a certain number greater than one—i.e., make effective channels, which include several of the real channels. The extra width of the effective channels is to account for the fact that there is a certain flexibility as to which channel a wire gets assigned.

Put in other words, we treat the two-dimensional problem as two one-dimensional problems; one for the horizontal plane and the other for the vertical plane. On the horizontal (vertical) plane we consider effective ports, effective channels (by "channel" we mean a row or column of ports with track space available to connect these ports), effective connections, and effective distance. An effective port consists of W of the original ports where W is greater than 1 and is less than the number of ports in any column (or row). Moreover W defines the width of the effective channel.

To obtain an unambiguous method for estimating a priori the effective channel width W, we shall use the following heuristic approach: The average vertical (horizontal) length of wires, which have non-zero horizontal (vertical) component, will be used to estimate the *extra* width of the

effective channel that reflects the flexibility of wire routing in a two-dimensional array. (It is to be noticed that this extra width becomes zero if we have a one-dimensional setup where no freedom in wire routing exists and thus W = 1, i.e., the effective channel becomes identical to the original channel.) This seems to us as a natural way of reflecting in the model the degree of freedom inherent in the two-dimensional model. We proceed now to develop the formulas required for computing the probability of success for the two-dimensional model.

Consider a square array of M cells (ports) and let N be the total number of connections and $\bar{R}$ be the average connection length. Any connection belongs to one of the following categories: (a) it has only a horizontal component; (b) it has only a vertical component; and (c) it has both a horizontal and a vertical component. A connection belonging to the last category generates one connection on each plane. Let n_x, n_y, and n_{xy} denote, respectively, the number of wires that belong to the above three categories. Then,

$$N = n_x + n_y + n_{xy} \tag{B.1}$$

Moreover, let N_x and N_y denote the total number of connections on the horizontal and vertical plane respectively. Then

$$N_x = n_x + n_{xy} \quad \text{and} \quad N_y = n_y + n_{xy} \tag{B.2}$$

Let p_k be the probability that a connection (wire) has length k; we assume that all possible orientations of this wire are equally likely—i.e., it may be split into a horizontal component of length $|m|$ and vertical component of length $k - |m|$, where m ranges in value from $1 - k$ to k. The probability that the horizontal (or vertical) component of such a wire is zero is given by $1/2k$. The average length of the horizontal (vertical) component of a wire with non-zero horizontal (vertical) component is given by

$$\bar{R}_{hk} = k^2/(2k - 1) \tag{B.3}$$

and the average length of the vertical component of a wire of length k with non-zero horizontal component is given by

$$\bar{R}_{vk} = k(k - 1)/(2k - 1) \tag{B.4}$$

Also, the expected fraction, g, that a wire has either a horizontal component only or a vertical component only is given by:

$$g = \sum_k p_k \times 2/2k = \sum_k p_k/k \tag{B.5}$$

Hence,

$$n_x + n_y = g N \qquad (B.6)$$

Assuming that $n_x = n_y$ (and thus $= g N/2$), it follows from equations (B.1) and (B.2) that:

$$N_x = N_y = N(1 - g/2) \qquad (B.7)$$

Let $\bar{R}_1$ denote the average horizontal (vertical) component length of a wire that has a non-zero horizontal (vertical) component and let $\bar{R}_2$ denote the average vertical (horizontal) component length of a wire with a non-zero horizontal (vertical) component. Then from (B.3), (B.4) we get

$$\bar{R}_1 = \sum_k p_k \left(1 - \frac{1}{2k}\right)\frac{k^2}{(2k-1)} \Big/ \sum_k p_k \left(1 - \frac{1}{2k}\right) = \bar{R}/(2 - g) \qquad (B.8)$$

where, g is given by (B.5) and

$$\bar{R} = \sum_k k p_k. \qquad (B.9)$$

Also

$$\bar{R}_2 = \sum_k p_k \left(1 - \frac{1}{2k}\right)\frac{k(k-1)}{2k} \Big/ \sum_k p_k \left(1 - \frac{1}{2k}\right) = (\bar{R} - 1)/(2 - g) \qquad (B.10)$$

We let the effective width W of a channel be given by

$$W = 1 + \bar{R}_2$$
$$= (\bar{R} + 1 - g)/(2 - g). \qquad (B.11)$$

Let $M_{eff}(x)$ and $M_{eff}(y)$ denote the number of effective ports on the horizontal and vertical planes. Then

$$M_{eff}(x) = M_{eff}(y) = M/W = M(2 - g)/(\bar{R} + 1 - g) \qquad (B.12)$$

Also, let $y_{eff}(x)$ and $y_{eff}(y)$ denote the effective number of connections per effective port on the two planes respectively. Hence

$$y_{eff}(x) = \frac{N_x}{M_{eff}(x)} = \frac{N(1 - g/2)}{M/W} = \frac{N(2 - g)W}{2M}$$
$$= \frac{y(2 - g)W}{2} = \frac{y(\bar{R} + 1 - g)}{2} \qquad (B.13)$$

where, $y = N/M =$ number of connections per original port. Similarly,

$$y_{eff}(y) = \frac{N_y}{M_{eff}(y)} = y(2 - g)w/2 = y_{eff}(x) \qquad (B.14)$$

To sum up, the two-dimensional problem is reduced to two independent one-dimensional problems each having the following sets of inputs:

$$M_{eff} = M(2 - g)/(\bar{R} + 1 - g)$$

$$y_{eff} = y(\bar{R} + 1 - g)/2$$

This is equivalent to one one-dimensional problem with the inputs; $2M_{eff}$; y_{eff} and $\bar{R}$ whose solution has been obtained in the previous section. In other words, for a given level of probability we obtain the effective number of tracks/cell, T_{eff}, from the one-dimensional model with inputs $2M_{eff}$; y_{eff} and $\bar{R}_{eff}$. The total number of tracks (horizontal & vertical) required per cell is given by

$$T = 2T_{eff}/W.$$

APPENDIX C: A GLOBAL WIRING PROBLEM

Figure 9 shows an arrangement of cells and wiring segments; each cell has associated with it a horizontal and vertical wiring segment. The wiring segments are considered to carry a load of wires—i.e., any wire that uses a cell segment adds a load of 1 to the cell segment load.

Wires start at some cell, where they can access either the horizontal or vertical wiring segment; accessing a horizontal segment permits access to either of the horizontal segments of the cells horizontally adjacent, while access of the vertical segment of a cell permits access to the vertical segments of the cells just above and below. Access to a horizontal (vertical) segment of a cell permits also access to the vertical (horizontal) segment of the same cell. A "path" is a sequence of segments such that any two consecutive segments on the path have access to each other—i.e., they must either be horizontal segments of cells which are horizontal neighbors, or vertical segments of cells that are directly above each other, or a horizontal and vertical segment of the same cell. A path connects cells i and j if its end points are segments on cell i and j. A wire connection requires a path connecting the two cells which specify the wire; a net consists of a set of cells, and the interconnection pattern of a net consists of a set of segments such that for each pair of cells in the net, there exists also a sequence of segments in the interconnection pattern connecting these two cells. If a set

the program can be restricted in the number of steps it can use above the minimum required.

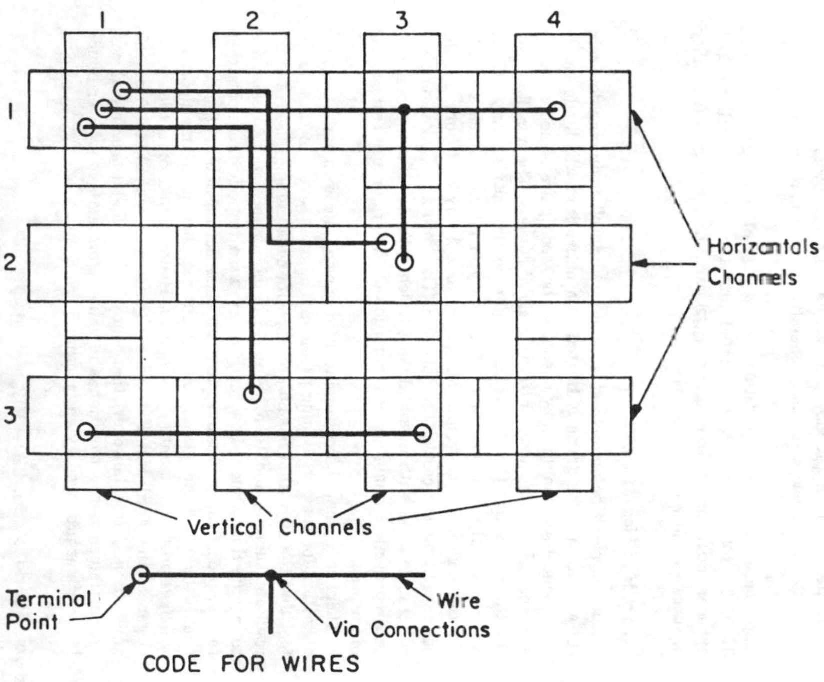

Prior to the maze run, a net is split into a set of pairs of connections using a minimum spanning tree program. The cost criterion used in this program is the minimum cost of a shortest path between pairs of points, the determination of which is done by the maze running program. The wiring program works in this order:

1. Find minimal spanning tree for a net.
2. Determine the actual routing of the set of pairwise connections generated.
3. Generate a list of the segments used by the net and update the load count.

The program has been written in such a way that the nets could be rerouted; in the first pass, all nets are routed. In subsequent passes, each net is individually removed and routed with the new cost functions. The number of backward steps in the maze runner can be varied from pass to pass; the first pass is set to 0, the second to 1, and the third to 2 backward stages; experimentation showed that no significant improvement could be made by increasing the number of passes and backward steps in the maze runner beyond these values.

Fig. 9 Global wiring model used for experimental studies.

of segments is assigned to a net or wire, the "load" on each of the segments is incremented by 1; alternatively, if the connections for a net are to be removed, then the "loads" of the segments used to connect the net are decremented by 1.

In connecting a net, or routing the wires connecting a net, a cost function is assigned as follows: let p denote a path going through cell segments $s_1, s_2, \ldots, s_i$; then the cost $C(p)$ of the path p is

$$C(p) = \sum_i f(l(s_i)).$$ (C.1)

where $l(s_i)$ is the load on the ith segment and $f(l)$ is a non-decreasing function, which can be inputted to the program. The function chosen here is steeply increasing (3^l for most of the range). A maze runner is used to determine the optimal routing; however, this maze runner could recognize what segments have been used by connections, which are part of the same net and have already been routed, and assign a cost of 0 to these segments, so that multiple use of a segment can be made by a net. The maze runner is furthermore restricted to a rectangle, which exceeds the minimum distance rectangle by a pre-assigned number of steps. As a matter of fact,

An Efficient Heuristic Procedure for Partitioning Graphs

By B. W. KERNIGHAN and S. LIN

We consider the problem of partitioning the nodes of a graph with costs on its edges into subsets of given sizes so as to minimize the sum of the costs on all edges cut. This problem arises in several physical situations—for example, in assigning the components of electronic circuits to circuit boards to minimize the number of connections between boards.

This paper presents a heuristic method for partitioning arbitrary graphs which is both effective in finding optimal partitions, and fast enough to be practical in solving large problems.

I. INTRODUCTION

1.1 *Definition of the Problem*

This paper deals with the following combinatorial problem: given a graph G with costs on its edges, partition the nodes of G into subsets no larger than a *given maximum size*, so as to minimize the total cost of the edges cut.

One important practical example of this problem is placing the components of an electronic circuit onto printed circuit cards or substrates, so as to minimize the number of connections between cards. The components are the nodes of the graph, and the circuit connections are the edges. There is some maximum number of components which may be placed on any card. Since connections between cards have high cost compared to connections within a board, the object is to minimize the number of interconnections between cards.

This partitioning problem also arises naturally in an attempt to improve the paging properties of programs for use in computers with paged memory organization. A program (at least statically) can be thought of as a set of connected entities. The entities might be subroutines, or procedure blocks, or single instruction and data items, depending on viewpoint and the level of detail required. The connections between the entities might represent possible flow or transfer of control, or references from one entity to another. The problem is to assign the objects to "pages" of a given size so as to minimize the number of references between objects which lie on different pages.

To pose the partitioning problem mathematically, we shall need the following definitions. Let G be a graph of n nodes, of sizes (weights) $w_i > 0$, $i = 1, \cdots, n$. Let p be a positive number, such that $0 < w_i \leq p$ for all i. Let $C = (c_{ij})$, $i, j = 1, \cdots, n$ be a weighted connectivity matrix describing the edges of G.

Let k be a positive integer. A *k-way partition* of G is a set of nonempty, pairwise disjoint subsets of G, $v_1, \cdots, v_k$ such that $\bigcup_{i=1}^{k} v_i = G$. A partition is *admissible* if

$$|v_i| \leq p \quad \text{for all} \quad i,$$

where the symbol $|x|$ stands for the *size* of a set x, and equals the sum of the sizes of all the elements of x. The *cost* of a partition is the summation of c_{ij} over all i and j such that i and j are in different subsets. The cost is thus the sum of all external costs in the partition.

The partitioning problem we consider here is to find a minimal-cost admissible partition of G.

There are three other problems which are equivalent to this one. First, minimizing external cost is equivalent to maximizing internal cost because the total cost of all edges is constant. Further, by changing the signs of all c_{ij}'s, we can maximize external cost, or minimize internal cost.

1.2 *Exact Solutions*

A strictly exhaustive procedure for finding the minimal cost partition is often out of the question. To see this suppose that G has n nodes of size 1 to be partitioned into k subsets of size p, where $kp = n$. Then there are $\binom{n}{p}$ ways of choosing the first subset, $\binom{n-p}{p}$ ways for the second, and so on. Since the ordering of the subsets is immaterial, the number of cases is

$$\frac{1}{k!} \binom{n}{p} \binom{n-p}{p} \cdots \binom{2p}{p} \binom{p}{p}.$$

For most values of n, k, and p, this expression yields a very large number; for example, for $n = 40$ and $p = 10$ $(k = 4)$, it is greater than 10^{20}.

Formally the problem could also be solved as an integer linear programming problem, with a large number of constraint equations necessary to express the uniformity of the partition.

Because it seems likely that any direct approach to finding an optimal solution will require an inordinate amount of computation, we turn to an examination of heuristics. Heuristic methods can produce good solutions (possibly even an optimal solution) quickly. Often in practical applications, several good solutions are of more value than one optimal one.

The first and foremost consideration in developing heuristics for combinatorial problems of this type is finding a procedure that is powerful and yet sufficiently fast to be practical. A process whose running time grows exponentially or factorially with the number of vertices of the graph is not likely to be practical. In most cases, a growth rate of more than the square of the number of vertices is still not too practical. (If the running time of a procedure grows as $f(n)$, where n is the number of vertices involved, we shall refer to it as an $f(n)$-procedure.)

1.3 *False Starts*

To point out a few pitfalls, we mention some unsuccessful attempts at heuristic solutions to the partitioning problem.

1.3.1 *Random Solutions*

One tactic is simply to generate random solutions, keeping the best seen to date, and terminating after some predetermined time or value is reached. This is quite fast, although actually an n^2-procedure. Unfortunately, this approach is unsatisfactory for problems of even moderate size, since there are generally few optimal or near-optimal solutions, which thus appear randomly with very low probabilities. Experience with 2-way partitions for a class of 0–1 matrices of size 32×32, for example, has indicated that there are typically 3 to 5 optimal partitions, out of a total of $\frac{1}{2} \binom{32}{16}$ partitions, giving a probability of success on any trial of less than 10^{-7}.

1.3.2 *Max Flow-Min Cut*

Another partitioning method is the Ford and Fulkerson max flow-min cut algorithm[1]. The graph is treated as a network in which edge costs correspond to maximum flow capacities between pairs of nodes. A cut is a separation of the nodes into two disjoint subsets. The max flow-min cut theorem states that the maximal flow values between any pair of nodes is equal to the minimal cut capacity of all cuts which separate the two nodes. In our terminology, a cut is a 2-way partition, and the cut capacity is the cost of the partition. The Ford and Fulkerson algorithm finds a cut with maximal flow, which is thus a minimal cost cut; this represents a minimum cost partition of the graph into two subsets of unspecified sizes.

There are several difficulties involved in using the Ford and Fulkerson algorithm for our partitioning problem. The most severe of these is the fact that the algorithm has no provision for constraining the sizes of the resultant subsets, and there seems to be no obvious way to extend it to include this. Thus if flow methods are used to perform a split, then further processing is necessary to make the resulting subsets the correct size. If the subsets are greatly different in size, then use of this algorithm will have produced essentially no benefit. Hence in spite of its theoretical elegance, the Ford and Fulkerson algorithm is not suitable for this application. (Note however, that since it does find the minimal cost unconstrained 2-way partition, the value it produces is a lower bound for solutions produced by any method.)

Reprinted with permission from *Bell Syst. Tech. J.*, vol. 49, no. 2, pp. 291–307, Feb. 1970.

1.3.3 Clustering

A class of much more intuitive methods is based on identifying "natural clusters" in the given cost matrix—that is, groups of nodes which are strongly connected in some sense. For example, one can use very simple heuristics for building up clusters, based on collecting together elements corresponding to large values in the cost matrix. But again these methods do not in general include much provision for satisfying constraints on the sizes of the subsets, nor do they provide for systematic assignment of "stragglers" (nodes which do not obviously belong to any particular subset).

1.3.4 λ-Opting

Lin, working on the Traveling Salesman Problem, [See Ref. 2] categorized a set of methods of improving given solutions by rearranging single links, double links, triplets, and in general, λ links. He referred to a change involving the movement of λ links as a λ-change. If a configuration of the system is reached in which no λ-change can be made which results in a decrease in cost, the configuration is said to be "λ-opt."

For the partitioning problem, an analogous operation is the interchange of groups of λ points between a pair of sets. Thus a 1-change is the exchange of a single point in one set with a single point in another set. A configuration is then said to be "1-opt" if there exists no interchange of two points which decreases the cost of the partition. Experiments to evaluate 1-opting for 2-way partitions of 0-1 matrices (32×32) within which about one-half of the elements were nonzero, show that apparently optimal values can be achieved in about 10 percent of the trials; values within 1 or 2 of the optimal can be achieved in about 75 percent of cases.

It appears fruitless to extend λ beyond 1 (1-opting is already an n^2-procedure), or to extend 1-opting experiments to partitions into more than two subsets, since more powerful methods have been developed. These methods are the topic of the next sections.

II. TWO-WAY UNIFORM PARTITIONS

2.1 Introduction

The simplest partitioning problem which still contains all the significant features of larger problems is that of finding a minimal-cost partition of a given graph of $2n$ vertices (of equal size) into two subsets of n vertices each. The solution of the 2-way partitioning problem is the subject of this section. The solution provides the basis for solving more general partitioning problems. In Section 2.6, we discuss 2-way partitions into sets of unequal size.

Let S be a set of $2n$ points, with an associated cost matrix $C = (c_{ij})$, $i, j = 1, \cdots, 2n$. We assume without loss of generality that C is a symmetric matrix, and that $c_{ii} = 0$ for all i. There is no assumption about nonnegativity of the c_{ij}'s. We wish to partition S into two sets A and B, each with n points, such that the "external cost" $T = \sum_{A \times B} c_{ab}$ is minimized.

In essence, the method is this: starting with any arbitrary partition A, B of S, try to decrease the initial external cost T by a series of interchanges of subsets of A and B; the subsets are chosen by an algorithm to be described. When no further improvement is possible, the resulting partition A', B' is locally minimum with respect to the algorithm. We shall indicate that the resulting partition has a fairly high probability of being a globally minimum partition.

This process can then be repeated with the generation of another arbitrary starting partition A, B, and so on, to obtain as many locally minimum partitions as we desire.

Given S and (c_{ij}), suppose A^*, B^* is a minimum cost 2-way partition. Let A, B be any arbitrary 2-way partition. Then clearly there are subsets $X \subset A$, $Y \subset B$ with $|X| = |Y| \leq n/2$ such that interchanging X and Y produces A^* and B^* as shown below.

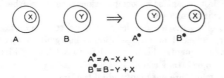

$$A^* = A - X + Y$$
$$B^* = B - Y + X$$

The problem is to identify X and Y from A and B, without considering all possible choices. The process we describe finds X and Y approximately, by sequentially identifying their elements.

Let us define for each $a \in A$, an external cost E_a by

$$E_a = \sum_{y \in B} c_{ay}$$

and an internal cost I_a by

$$I_a = \sum_{z \in A} c_{az} .$$

Similarly, define E_b, I_b for each $b \in B$. Let $D_z = E_z - I_z$, for all $z \in S$; D_z is the difference between external and internal costs.

Lemma 1: Consider any $a \in A$, $b \in B$. If a and b are interchanged, the gain (that is, the reduction in cost) is precisely $D_a + D_b - 2c_{ab}$.

Proof: Let z be the total cost due to all connections between A and B that do not involve a or b. Then

$$T = z + E_a + E_b - c_{ab} .$$

Exchange a and b; let T' be the new cost. We obtain

$$T' = z + I_a + I_b + c_{ab}$$

and so

$$\text{gain} = \text{old cost} - \text{new cost} = T - T'$$
$$= D_a + D_b - 2c_{ab} .$$

2.2 Phase 1 Optimization Algorithm

In this subsection we present the algorithm for 2-way partitioning.

First, compute the D values for all elements of S. Second, choose $a_i \in A$, $b_i \in B$ such that

$$g_1 = D_{a_i} + D_{b_i} - 2c_{a_i b_i}$$

is maximum; a_i and b_i correspond to the largest possible gain from a single interchange. (We will return shortly to a discussion of how to select a_i and b_i quickly.) Set a_i and b_i aside temporarily, and call them a_1' and b_1', respectively.

Third, recalculate the D values for the elements of $A - \{a_i\}$ and for $B - \{b_i\}$, by

$$D_x' = D_x + 2c_{x a_i} - 2c_{x b_i}, \qquad x \in A - \{a_i\},$$
$$D_y' = D_y + 2c_{y b_i} - 2c_{y a_i}, \qquad y \in B - \{b_i\}.$$

The correctness of these expressions is easily verified: the edge (x, a_i) is counted as internal in D_x, and it is to be external in D_x', so $c_{x a_i}$ must be added twice to make this correct. Similarly, $c_{x b_i}$ must be subtracted twice to convert (x, b_i) from external to internal.

Now repeat the second step, choosing a pair a_2', b_2' from $A - \{a_1'\}$ and $B - \{b_1'\}$ such that $g_2 = D_{a_2'} + D_{b_2'} - 2c_{a_2' b_2'}$ is maximum (a_1' and b_1' are *not* considered in this choice). Thus g_2 is the additional gain when the points a_2' and b_2' are exchanged as well as a_1' and b_1'; this additional gain is maximum, given the previous choices. Set a_2' and b_2' aside also.

Continue until all nodes have been exhausted, identifying $(a_3', b_3'), \cdots$, (a_n', b_n'), and the corresponding maximum gains $g_3, \cdots, g_n$. As each (a', b') pair is identified, it is removed from contention for further choices so the size of the sets being considered decreases by 1 each time an (a', b') is selected.

If $X = a_1', a_2', \cdots, a_k'$, $Y = b_1', b_2', \cdots, b_k'$, then the decrease in cost when the sets X and Y are interchanged is precisely $g_1 + g_2 + \cdots + g_k$. Of course $\sum_1^n g_i = 0$. Note that some of the g_i's are negative, unless all are zero.

Choose k to maximize the partial sum $\sum_{i=1}^k g_i = G$. Now if $G > 0$, a reduction in cost of value G can be made by interchanging X and Y. After this is done, the resulting partition is treated as the initial partition, and the procedure is repeated from the first step.

If $G = 0$, we have arrived at a locally optimum partition, which we shall call a *phase 1 optimal partition*. We now have the choice of repeating with another starting partition, or of trying to improve the phase 1 optimal partition. We shall discuss the latter option shortly. Figure 1 is a flowchart for the phase 1 optimization procedure.

2.3 Effectiveness of the Procedure

One general approach to solving problems such as this one is to find the *best* exchange involving say λ pairs of points, for some λ specified in advance[2]. The difficulty encountered is that use of a small value of λ is not sufficient to identify good exchanges, but the computational effort required grows rapidly as λ increases.

The procedure we have described *sequentially* finds an approximation to the best exchange of λ pairs. λ is not specified in advance, but rather is chosen to make the improvement as large as possible. This technique sacrifices a certain amount of power for a considerable gain in speed. Since we construct a sequence of gains g_i, $i = 1, \cdots, n$, and find the

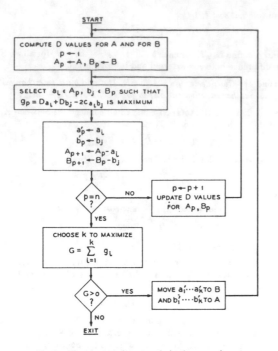

Fig. 1 — Flowchart of phase 1 optimization procedure.

maximum partial sum, *the process does not terminate immediately when some g_i is negative.* This means that the process can sequentially identify sets for which the exchange of only a few elements would actually increase the cost, while the interchange of the entire sets produces a net gain.

Numerous experiments have been performed to evaluate the procedure on different types of cost matrices. The matrices used have included (*i*) 0-1 matrices, with density of nonzero elements ranging from 5 percent to 50 percent, (*ii*) integer matrices with elements uniformly distributed on $[0, k]$, $k = 2, \cdots, 10$, (*iii*) matrices with clusters of known sizes and binding strength. Results on all of these matrices have been similar, so we shall only summarize them here. A more extended discussion may be found in Ref. 3.

A useful measure of the power of a heuristic procedure is the probability that it finds an optimal solution in a single trial. Suppose that p is the probability that a phase 1 optimal solution found using a random starting partition is globally optimal. We have examined the behavior of this probability as the size of the matrices involved is varied. Experiments show p is around 0.5 for matrices of size 30×30, 0.2 to 0.3 for 60×60, and 0.05 to 0.1 for 120×120. The functional behavior of p is approximately $p(n) = 2^{-n/30}$.

These values are derived primarily from 0-1 matrices having about 50 percent 1's (randomly placed). Experiments on matrices with lower densities of 1's yield larger variances, but substantially identical mean values for p.

2.4 *Running Time of the Procedure*

Let us define a *pass* to be the operations involved in making one cycle of identification of $(a_1', b_1'), \cdots, (a_n', b_n')$, and selection of sets X and Y to be exchanged. The total time for a pass can be estimated this way. First, the computation of the D values initially is an n^2-procedure, since for each element of S, all the other elements of S must be considered. The time required for updating the D values is proportional to the number of values to be updated, so the total updating time in one pass grows as

$$(n - 1) + (n - 2) + \cdots + 2 + 1$$

which is proportional to n^2.

The dominant time factor is the selection of the next pair a_i', b_i' to be exchanged. The method we have used to perform this searching is to sort the D values so that

$$D_{a_1} \geqq D_{a_2} \geqq \cdots \geqq D_{a_n}$$

and

$$D_{b_1} \geqq D_{b_2} \geqq \cdots \geqq D_{b_n}.$$

When sorting is used, only a few likely contenders for a maximum gain

need be considered. This is because when scanning down the set of D_a's and D_b's, if a pair D_{a_i}, D_{b_j} is found whose sum does not exceed the maximum improvement seen so far in this pass, then there cannot be another pair a_k, b_l with $k \geqq i$, $l \geqq j$, with a greater gain, (assuming $c_{ij} \geqq 0$) and so the scanning can be terminated. Thus the next pair for interchange is found rapidly. Sorting is an $n \log n$ operation, so in this method, the total time required to sort D values in a pass will be approximately

$$n \log n + (n - 1) \log (n - 1) + \cdots + 2 \log 2$$

which grows as $n^2 \log n$.

To reduce the time for selection of an (a, b) pair, it is possible to use techniques which are faster than sorting, but which do not necessarily always give the maximum gain at each stage. For example, one method is to scan for the largest D_a and the largest D_b, and use the corresponding a and b as the next interchange. This method is essentially linear-time and would probably be implemented as part of the recomputation of the D values. It is best suited for sparse matrices, where the probability that $c_{ab} > 0$ is small. A slight extension, involving negligible extra cost, is to save the largest two or three D_a's and D_b's, so that if the largest pair does not give the maximum gain (because c_{ab} is too large), then another can be tried. Experience indicates that three values are sufficient in virtually all cases, even for matrices with a relatively high percentage of nonzero entries. Use of this method reduces running time by about 30 percent in the present implementation, with very small degradation of power.

The number of passes required before a phase 1 optimal partition is achieved is small. On all matrix sizes tested at the time of writing (up to 360 points), it has been almost always from 2 to 4 passes. On the basis of this experimental evidence, the number of passes is not strongly dependent on the value of n.

From the foregoing observations, it is possible to estimate the total running time of the procedure. If we use a method which sorts the D values at each stage (time proportional to $n^2 \log n$), then the running time should grow as $n^2 \log n$. If a fast-scan method is used, and the number of passes is constant, the running time should have an n^2 growth rate; this is a lower bound.

For comparison, examination of all pairs of sets X and Y, and evaluation of the costs would require time proportional to

$$n^2 \sum_{k=1}^{n/2} \binom{n}{k}^2 \sim \frac{n^2}{2} \sum_{k=0}^{n} \binom{n}{k}^2$$
$$= \frac{n^2}{2} \binom{2n}{n}$$
$$\sim \frac{n^2}{2} 4^n \left(\frac{1}{\pi n}\right)^{\frac{1}{2}}$$

for large n. This function grows as $n^{3/2} 4^n$.

Running times have been plotted in Fig. 2. The observed times have an apparent growth rate of about $n^{2.4}$, which is reasonably close to n^2. Although on the logarithmic plot this curve is close to linear over the range $n = 20$ to $n = 130$, it may actually be $n^2 \log n$; insufficient data is available to check this. All times are based on an implementation in FORTRAN G on an IBM System 360 Model 65.

2.5 *Improving the Phase 1 Optimal Partition*

In this section, we discuss a method which might be used to improve the partition produced by the phase 1 procedure, which may not be globally optimum. The method suggested in this section is based heavily on experimental evidence, although there are quite plausible reasons for performing the particular set of operations. The basic idea is to perturb the locally optimal solution in what we hope is an enlightened manner, so that an iteration of the process on the perturbed solution will yield a further reduction in the total cost. If this tactic fails, nothing has been lost except some computation time, since the best solution seen so far is always saved.

Computer results for problems with up to 64 points suggest that whenever a phase 1 optimal solution is not globally optimal, $|X| = |Y| \approx n/2$. Roughly, this implies that if $|X|$ and $|Y|$ had been small compared to $n/2$, they would have been found by the process; it is only larger sets which are not identified all the time.

A successful heuristic to find the correct X and Y in this case is to find a phase 1 optimal partition for each of the sets A and B, say $A \rightarrow [A_1, A_2]$ and $B \rightarrow [B_1, B_2]$. (That is, find near-optimal partitions of A and of B separately.) Recombine the 4 sets into 2, say $A_0 = A_1 \cup B_1$ and $B_0 = A_2 \cup B_2$, and continue with phase 1 optimization. If our expectation is correct, the new X and Y will be small, and thus readily identified by the phase 1 process.

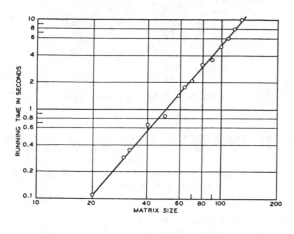

Fig. 2 — Running time.

When A is split into A_1, A_2 and B into B_1, B_2 there are two ways in which the smaller sets can be recombined. A series of tests was made on matrices of moderate size (up to 64 × 64), in which both possible recombinations were done, generating three phase 1 optimal values for each starting partition. For matrices of size 32 × 32, the apparent optimal value was observed at least once in each triple of values, for a large number of cases. With matrices of size 64 × 64, there were occasional failures.

It might be noted that the extra time involved for the recombination approach is three times that required to do a completely new partition from a random start, assuming an n^2-procedure.

It is possible to estimate whether a particular improvement tactic is profitable or not in the following way. Suppose that some method increases the probability of finding an optimal partition from p to p', while it increases the running time from t to t'. Then in a fixed amount of time, it is possible to do k trials of the basic procedure, and kt/t' trials of the improved method. The corresponding probabilities of achieving an optimal solution are $1 - (1 - p)^k$ and $1 - (1 - p')^{kt/t'}$ respectively. The improved method is then desirable if the second expression is greater than the first; by simple manipulation, this condition becomes

$$1 - p' < (1 - p)^{t'/t}.$$

On the basis of the numerical values in this section, it may be useful to try the recombination method.

2.6 Partitioning into Unequal-Sized Sets

It is simple to modify the procedure to partition a set S with n elements into two sets of specified sizes n_1 and $n_2(n_1 + n_2 = n)$. Assume $n_1 < n_2$. Then restrict the maximum number of pairs that can be exchanged in one pass of the procedure to n_1. All other operations are performed on all elements of each set. (The starting partition is into two sets, of n_1 and n_2 elements respectively.)

Suppose we wish to partition S into two sets, such that there are at least n_1 elements and at most n_2 elements in each subset; $n_1 + n_2 = n$, but they are not specified further.

The procedure is easily modified to handle this sort of constraint by the addition of "dummy" elements. These are elements which have no connections whatsoever; that is, they have zero entries in the cost matrix wherever they appear. Add $2n_2 - n$ dummies so S has $2n_2$ elements, and perform the procedure on it. The resulting partition will assign the dummy elements to the two subsets so as to minimize the external cost; at this point the dummies are discarded, leaving a partition into two subsets that satisfy the size constraints given.

2.7 Elements of Unequal Sizes

We have made the assumption so far that the elements (vertices) of the graph are all of the same size. This requirement may be relaxed to a large extent by converting any node of size $k > 1$ to a cluster of k nodes of size 1, bound together by edges of appropriately high cost. The size of the problem will obviously increase proportionally to the value of k, so it may be necessary to sacrifice some accuracy to keep the number of generated nodes within reasonable bounds.

III. MULTIPLE-WAY PARTITIONS

3.1 Reduction to 2-Way Partitioning Problem

So far, the discussion has been concerned exclusively with the basic problem of performing a 2-way partition on a set of $2n$ objects. In this section we extend the technique to perform k-way partitions on a set of kn objects, using the 2-way procedure as a tool.

The essential idea is to start with some partition into k sets of size n and by repeated application of the 2-way partitioning procedure to pairs of subsets, make the partition as close as possible to being pairwise optimal. (Section 3.2 treats the question of what starting sets to use.) Of course pairwise optimality is only a necessary condition for global optimality. There may be situations where some complex interchange of three or more items from three or more subsets is required to reduce a pairwise optimal solution to globally optimum; at the moment, no reasonable method for identifying such sets is known.

There are $\binom{k}{2}$ pairs of subsets to consider, so the time for one pass through all pairs is (assuming an n^2-procedure) $\binom{k}{2}n^2 \approx (kn)^2/2 = $ (number of points)$^2/2$. In general, more passes than this will actually be required, since when two sets are made optimal, this may change their optimality with respect to other sets.

Experience indicates that the number of passes is small and the process converges quickly. For example, our algorithm selects (i, j) as the next pair of sets to be optimized, where either i or j has been changed since the last time the pair (i, j) was selected. Using this selection process, the average number of passes through each pair of sets is a slowly growing function of both k and n. For matrices of size 100 or less and $k < 6$, the number of passes has been less than 5. [The average number of passes is computed as the average number of pairs considered to reach pairwise optimality, normalized by $\binom{k}{2}$.]

In any particular trial, there is a correlation between the number of pairs selected and the quality of the final partition. To get a better solution requires more work.

Convergence is rapid: two passes account for more than 95 percent of the improvement in most cases; the remaining passes contribute only small further reductions. Let $p(n, k)$ be the proportion of minimum cost solutions found for a particular n and k. For k fixed and small compared to n, the functional behavior of $p(n, k)$ is similar to the case $k = 2$, but the actual values are lower. Roughly, we observe $p(n, k + 1) \approx \frac{1}{2}p(n, k)$ for k in the range 2–4, and n up to 100, with considerable variation depending on the matrix being tested. For instance, for matrices of size about 40, $p(40, 2) \approx 0.4$, $p(42, 3) \approx 0.2$, and $p(40, 4) \approx 0.1$.

Another interesting question is measurement of how close to optimum the partitions found are. The solutions obtained by pairwise optimization have values concentrated in a narrow range. In almost all cases, the largest value found by the procedure is within 4–5 percent of the smallest. As another measure, if c is the mean cost of random partitions and b is the cost of the best partition observed, then virtually all partitions found have values v such that

$$v - b \leqq 0.1(c - b).$$

For instance, one test case was a series of 4-way partitions of a 0–1 matrix of size 80. This matrix had 1278 nonzero entries (a density of 0.2), corresponding to 639 edges in the graph. The mean value of randomly chosen partitions was 480.6. Twenty-four partitions of this matrix were found using the method described above. The lowest value encountered was 352 (1 time), the highest 365 (1 time); the mean value was 359.5, the median 360.

3.2 Starting Partition

In this subsection we discuss various methods of generating good starting partitions, based on modifications of the basic procedure.

The primary reason for choosing good starting partitions is that this particular form of preprocessing reduces the amount of work required to make the system pairwise optimal. It may also make the probability of an optimal solution higher, although this tendency is very difficult to evaluate.

Several methods for finding good multi-way starting partitions which are based on repeated application of the procedure itself have been investigated. The essential idea is to generate a k-way starting partition by first forming an r-way partition, then an s-way partition on each of the resulting subsets, and so on, up to t-way. (Here $k = rs \cdots t$.) The partitions found this way will in general be better than those which are completely arbitrary. A pairwise optimization stage is applied to the final set of subsets.

For example, if k is a power of 2, then perform a 2-way split, then a

2-way split on each of these subsets, and so on until the desired size of subsets is found.

This general approach is prone to the following difficulty: the first split divides the original set into r subsets by trying to make the internal connections in each subset as large as possible. Obviously this may conflict directly with the next stage, which is to try to divide each subset further. Carried to several levels, it can lead to a relatively poor overall solution. In experiments with 4-way partitions of matrices of sizes up to 64×64, this method yields optimal solutions approximately as often as does starting with a 4-way partition in the first place. In addition, this method will be effective if the matrix happens to have natural clusters of approximately the correct size (that is, equal to the final subset size).

A second method which can be used is to partition the set of kn elements into a set of n and a set of $(k - 1)n$, using the slightly modified version of the basic procedure discussed in the first part of Section 2.6. The set of n elements is set aside, and the next n elements from the remaining $(k - 1)n$ are identified. This continues until k subsets have been formed; again the pairwise optimization technique is used to improve on this partition.

This method can make an error in the identification of the first set which will bias the choice of the second, and so on; the effect is most severe for the case where k is large, so each set is small.

The method of breaking off subsets sequentially has another potential flaw: regardless of the starting configuration, it will identify approximately the same set each time it is used on a particular problem, and hence little is gained by using it twice on one cost matrix. However variations in the order of performing pairwise optimizations can still produce different final partitions in general.

Limited computational experience with sequential break-off followed by pairwise optimization suggests that it yields solutions which are on the average at least as good as (and sometimes slightly better than) those provided by pairwise optimization applied to an arbitrary k-way starting partition. Pairwise optimization yields the optimum with a higher probability, however, because it is less susceptible to error caused by a bad choice made early. For instance, in tests on the 80 point matrix mentioned previously, sequential break off yielded 4-way solutions with a mean value of 358.6, but the lowest value found was 355. (The highest was 363.) These may be compared to 359.5, 352 and 365 for the standard partitioning method.

Running time for the sequential break-off method is lower than for straight pairwise optimization.

Insufficient data is available for a direct comparison between sequential break off and the method of repeated subdivision.

In all cases, the original process, be it a completely random generation of some initial configuration, or the production of a good starting partition, is followed by a pairwise optimizing phase. It is unlikely that using better starting partitions will lead to worse results than random starts, on the average. Whether the possible improvement in results and running times will justify the extra computational effort required to generate the starting partition depends on the characteristics of the particular class of matrices being studied.

Fig. 3 — Cost reduction by expansion.

Some limited experiments were performed to compare the present procedure with a multi-dimensional scaling technique[4], on a Boolean matrix of 316 points, with about 1400 nonzero entries. The results indicated that the procedure identifies clusters well, even when no attempt is made to provide a good starting partition.

3.3 *Expansion Factor*

The introduction of dummy elements was mentioned in Section 2.6 as a method of handling partitioning into subsets of unequal sizes. This can be viewed equally well as a means of introducing "slack" into a solution, in an attempt to get a lower overall cost by allowing "expansion." That is, so far we have treated the problem of finding a partition with a constraint on the sizes of the subsets, *and* on the number of subsets, since given kn points, we have tried to find the best partitions into exactly k subsets of n points each. Suppose we now relax this second constraint by permitting the addition of dummy elements to increase the size of the problem, and attempt to find the best solution involving *any number* (greater than or equal to k) of subsets, with *at most* n points in each. This solution with k or greater subsets will in general have a lower cost than the constrained solution.

Figure 3 shows an example in which introducing slack permits a lower overall cost. Assume n is 3 and all nodes are size 1. The vertical edges have cost 1 and the horizontal ones cost 2. Any partition into 2 equal subsets has a cost of at least 3, but there is an obvious partition into 3 subsets with cost 2. Any nontrivial partition into 4 or more subsets has a cost greater than 2, so 3 subsets represents the optimal expansion. It is possible to find the minimal cost solution and the corresponding optimal amount of expansion as follows. Suppose the problem has kn points to be partitioned into k sets of n points each. Starting with no slack (kn points), the optimal assignment is found. Then n dummies, enough to create one extra subset, are added, making a $(k + 1)n$ problem, and so on. Eventually, one subset is produced which consists entirely of dummies. When this occurs, we take the partition with this set of dummies removed as our optimum solution.

REFERENCES

1. Ford, L. R., and Fulkerson, D. R., *Flows in Networks*, Princeton, New Jersey: Princeton University Press, 1962, p. 11.
2. Lin, S., "Computer Solutions of the Traveling Salesman Problem," B.S.T.J., 44, No. 10 (December 1965), pp. 2245–2269.
3. Kernighan, B. W., "Some Graph Partitioning Problems Related to Program Segmentation," Ph.D. Thesis, Princeton University, January 1969, pp. 74–126.
4. Kruskal, J. B., Multi-Dimensional Scaling by Optimizing Goodness of Fit to a Non-Metric Hypothesis," *Psychometrika, 29*, No. 1 (March 1964), pp. 1–27, and No. 2 (June 1964), pp. 115–129.

A PROPER MODEL FOR THE PARTITIONING OF ELECTRICAL CIRCUITS

D. G. Schweikert and B. W. Kernighan
Bell Telephone Laboratories, Incorporated
Murray Hill, New Jersey   07974

ABSTRACT

Partitioning algorithms for electrical circuits are often based on the heuristic manipulation of a simple element-to-element interconnection matrix. However, the element-to-element interconnection matrix does not properly represent an <u>electrical</u> interconnection, or "net", among more than two elements. This paper expands on several aspects of the discrepancy: 1) its source, 2) the circumstances under which it is likely to be significant, and its magnitude for typical circuits, and 3) the comparative difficulty and expense of using a more appropriate representation.

A physically correct "net-cut" model is presented. This model is computationally straightforward and is easily adapted to the typical heuristic solution strategies. The "net-cut" model is coupled with the Kernighan-Lin partitioning algorithm [3]; using the same algorithm, comparisons with the "edge-cut" model demonstrate that the correct model reduces net-cuts by 19 to 50% for four digital logic circuits.

1.   INTRODUCTION

Partitioning and placement algorithms for electrical circuits are often based on the heuristic manipulation of a simple element-to-element interconnection matrix. From a graph theoretical viewpoint, the partitioning problem is usually represented as finding a partition that cuts the minimum number of edges of the graph corresponding to this matrix. The placement problem is typically represented as minimizing a total interconnection "length" computed as the term-by-term multiplication of the interconnection matrix by an element-to-element distance matrix. (This representation is, in fact, the quadratic assignment problem.)

In both cases the element-to-element interconnection matrix does not properly represent an <u>electrical</u> interconnection, or "net", among more than two elements.

The discrepancy has been noted by several authors (Hanan [1], Rutman [2]) in the case of the placement problem.

This paper expands on several aspects of the discrepancy: 1) its source, 2) the circumstances under which it is likely to be significant, and its magnitude for typical circuits, and 3) the comparative difficulty and expense of using a more appropriate representation. Since the basis of the discrepancy is the same in both partitioning and placement, we will restrict discussion to the partitioning problem.

A physically correct "net-cut" model is presented. This model is computationally straightforward and is easily adapted to the typical heuristic solution strategies. This "net-cut" model has been coupled with the Kernighan-Lin partitioning algorithm [3]; using the same partitioning algorithm, comparisons with the "edge-cut" model demonstrate that the correct model reduces net-cuts by 19 to 50% for four digital logic circuits.

2.   REPRESENTATION ERROR IN THE GRAPH PARTITIONING MODEL

The physical problem of dividing an electrical circuit into two parts, with a minimum number of interconnecting wires, is typically cast as the following mathematical problem:
Let each electrical device be a vertex of a graph.
Let every electrical connection between any two devices be represented by an edge between the corresponding vertices of the graph. Find the minimum number of edges which, when removed, divide the vertices into two disjoint subgraphs. (In other words, divide the graph into two parts by "cutting" the minimum number of edges - hence, it is termed the minimum edge-cut problem.)

In practice there may be additional constraints (e.g., the two parts must be equal) or other variations (e.g., divide into n parts, or, minimize <u>weighted</u>

edge-cuts). However, none of these variations mitigate the fact that the minimum edge-cut problem does not properly represent the physical problem for an electrical circuit in which some of the nets have three or more connected devices.

For example, consider the circuit shown in Fig. 1.

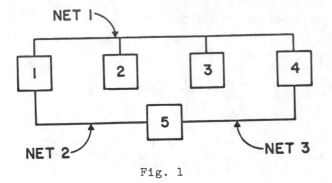

Fig. 1

Net 1 connects elements 1, 2, 3, 4; Net 2 connects elements 1, 5; Net 3 connects elements 4,5. The usual representation of the circuit as a graph is shown in Fig. 2. This graph has a two-way minimum edge-cut partition

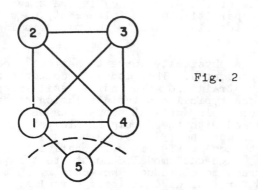

Fig. 2

P_E: $\{(1,2,3,4)(5)\}$ with two edge cuts (see dashed line).

However, it is clear from Fig. 3 that physically this is not the best partition. For example, P_N: $\{(2,3)(1,4,5)\}$ requires only one wire between the two parts, rather than the two required for P_E.

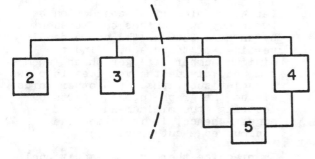

Fig. 3

The error arises because the edge-cut model treats Net 1, a 4-element net, as six 2-element nets created from all the pair-wise combinations of its four terminals. Since P_N cuts four edges of the graph it is not even close to being a minimum edge-cut partition.

Generally, the edge-cut model treats a k-element net as (k-1) + (k-2) + ... +1 two-element nets - a physically realizable interconnection pattern but hardly the one involving the minimum number of wires.

Thus the effect of the edge-cut model is to exaggerate the importance of any net with more than two elements. The exaggeration grows rapidly as the number of elements increases; for example, the cost of dividing an 11-element net ranges between 10 and 30 edge cuts, when physically only one wire need be cut. Such a grossly disproportionate weighting of an 11-element net, compared with two-element nets, almost insures that all its attached elements will be in one package of the partition - unfortunately "dragging their tentacles behind them."

3. AN EXAMPLE FROM AN ACTUAL CIRCUIT

To illustrate this effect in more detail, let us examine an actual example circuit (Circuit 1). Circuit 1 is a digital logic circuit contained on a single silicon chip. For partitioning, certain nonessential net information was deleted - such as external connections. Some elements were omitted or clustered where the effect on partition cost was obvious: elements (logic gates) having only one terminal and connected to a two-element net were deleted; any element in the middle of a "string" of connected elements was clustered with one of its neighbors. The resulting circuit is smaller but retains the essential complication of the original circuit. The resulting distribution of elements per net is shown in Table 1.

Table 1.

Circuit 1: 39 elements - 41 nets - 116 element-net connections

Elements per Net	Number of Nets	Nets Cut P_N	Nets Cut P_E
2	25	0	8
3	9	3	6
4	3	0	1
5	2	2	0
7	1	1	0
10	1	1	0
Totals	41	7	15

There exists a two-way partition with only seven nets connecting the two parts; this partition is labeled P_N and is the best known equal size partition (actually 19 vs. 20 gates). Taking a naive, statistical approach, one might expect that the chances of a net being cut by a partition increases with the number of elements attached to it. P_N confirms that speculation: it divides all nets with five or more elements while dividing none of 25 two-element nets (see column 3 in Table 1).

The results for the best minimum edge-cut partition P_E are shown in the fourth column of Table 1. As expected, none of the nets with high element counts (above 4, in this case) were divided, while significantly more two-element nets were divided. The count of net-cuts is increased from 7 (for P_N) to 15. Clearly the minimum edge-cut representation is not the appropriate approximation to this physical problem.

These experiments also show that a good net-cut solution such as P_N with seven net-cuts is a poor edge-cut solution: P_N has 57 edge-cuts, while P_E has 23 edge-cuts.

4. A "NET-CUT" MODEL FOR PARTITIONING ELECTRICAL CIRCUITS

Consider a two-way partition of an electrical circuit into blocks A and B. Let the circuit elements (devices) a_i be in block A and elements b_i be in block B. The electrical interconnections of these devices are called nets. A net connects two or more circuit elements. (An element may be connected to as many different nets as it has different terminals. In the physical layout of the circuit it is usually important that a specific terminal of a device is connected to a specific net; for partitioning, however, we need to know only that a net is connected to an element not the specific terminal involved.)

The wire (or other conductor) which physically implements a net may have many different forms; for example the wire may be either tree-like or a single line, contacting elements in any order. For partitioning, however, these forms are generally considered equivalent.

Given a particular partition, or assignment of elements to blocks A and B, we need to know the minimum number of wires required to connect A and B. The following statement is the essential part of this model: A net which is divided by the partition (i.e., interconnects elements in both A and B) requires one and only one wire to connect its elements in A to its elements in B.

Now, the "net-cut" model itself can be stated very simply: For a given partition the number of wires needed to connect A and B is equal to the number of divided, or "cut", nets. This very simple and seemingly obvious model is at variance with the interconnection matrix ("edge-cut") model which counts, for each divided net, all the pairwise combinations of its elements in A with its elements in B.

5. SOLUTION OF THE MINIMUM NET-CUT PROBLEM

Heuristic algorithms for solving the minimum edge-cut problem typically modify a starting partition by pairwise exchanges of elements which will reduce the "cost" of the partition. The "cost" is determined from the element-to-element interconnection (incidence) matrix. In its simplest form, the strategy is as follows: An element in A which is more strongly connected to elements in B than in A is considered a good candidate for a move to B. Similarly, elements in B are inspected for A candidates. The candidates from A and B are paired and the (hopefully positive) improvement from a simultaneous exchange is calculated for each pair, then the best pair is exchanged and the algorithm repeats until no further improvement can be found.

Adapting such algorithms to solve the minimum net-cut problem presents the following difficulty: there is no equivalent of the element-to-element "cost" matrix representation. It is no longer possible to say that since a and b are connected, separating them will cost an additional net-cut. This depends instead on the locations of all elements on the net joining a and b. For instance, if a net is already cut, all elements but the last one move across the partition gap at no cost - but any element could be the "last" one.

Nevertheless, minimum edge-cut algorithms which use a "differential cost" strategy, such as Kernighan-Lin [3], can be easily modified to solve the net-cut partitioning problem. For illustration we use the Kernighan-Lin algorithm and notation.

The basic approach of [3] is to find, not just single pairs of elements to exchange, but entire groups. The groups are still identified a pair at a time, but the algorithm does not give up when a single pair yields a negative improvement: it only stops when no group that gives an improvement can be identified in this sequence of pairs. This makes the Kernighan-Lin algorithm much more powerful than simple interchange algorithms.

Assume an initial partition into blocks A and B has been given. We want to improve it (i.e., reduce the number of nets which have elements in both A and

B) by exchanging pairs of elements between A and B.

Suppose that for each element a of A, we define D_a as the decrease in the number of nets cut if we move a from A to B. D_b is defined similarly for elements of B. The larger the D value of an element, the more "out of place" it is likely to be in its current position.

Since D_a is the gain involved in moving a to B, and D_b that from moving b to A, the gain in a simultaneous exchange of a and b is $D_a + D_b$ minus a correction term, which allows for the possibility that a and b are connected. (If, on the ith net, a is the sole element of i in A, and if b is also on i, D_a contains a +1 associated with moving a to B which clearly will not be realized if b is simultaneously moved to A. The correction simply decreases the gain $D_a + D_b$ by one for each such situation.)

The next step is to select that a and b which maximize

$$D_a + D_b - \text{correction}(a,b)$$

and tentatively exchange them.

The D values of previously unexchanged elements are now updated to reflect the tentative exchange of a and b, and the whole process repeats, selecting a new a and b from previously unexchanged elements.

After all elements have been considered for tentative exchange, we have a sequence of possible exchanges and the total improvement resulting at each step. The actual exchange selected is just that part of the sequence that gives the best total improvement. If the best total improvement is positive, the exchange is made, and the process repeats with the new sets; otherwise, it halts.

We stress again that the process does not terminate just because the exchange of a single pair a,b results in a loss; it only stops when no part of the sequence yields a profit. This lets us find groups for which exchanging any subset results in a loss, but the whole group gives a profit.

The D values may be straightforwardly computed in the net-cut model from the following observations. (All D values are initialized to zero.)

1. If all elements of a particular net are in one set, then moving any one of these elements to the other set cuts the net and thus the D value for each element of this net is decreased by one.

2. If all elements but one of a particular net are in one set, moving the isolated element to the opposite set reconnects a previously cut net, and is thus a profit of one. The D value of this one element is increased by one.

3. If a net has less than two elements in the two sets under consideration, or there are two or more elements of the net in each set, this net contributes nothing to the D values.

6. COMPUTATIONAL EXPERIENCE

The cost model described above has been embedded in the basic Kernighan-Lin solution algorithm [3], and experiments have been performed on partitioning a spectrum of representative circuits* (in addition to Circuit 1, discussed above). We will summarize computational experience from some of them. Computational data for all four circuits is tabulated in Table 2.

Circuit 2
This is a conventional printed circuit of modest size: 117 nets and 40 elements. The distribution of net sizes is 87 two-element nets, 20 three-element, 5 four-element, 4 five-element, and 1 six-element, with a mean of 2.4 elements per net. This heavy bias of two-element nets suggests that the edge-cut model is not seriously in error. (In fact, this circuit showed the smallest discrepancy).

The requirement here is a two-way partition into 20-element sets. Random initial partitions have a mean of about 75 net-cuts. The minimum cost obtained was 22 net-cuts which corresponded to 50 edge-cuts. The best edge-cut partition found had 46 edge-cuts, corresponding to 27 net-cuts. As before, this illustrates the discrepancy between edge and net models.

It is interesting to note that for this particular circuit, a constructive algorithm under development [4] produced a partition of 34 net-cuts; when that solution was used as a starting partition for this program, reduction to 32 was achieved, a substantially poorer result than that achieved from random starts. The opposite result (discussed below) occurred for circuit 3.

---

*For the convenience of persons wishing to compare the results of other partitioning programs with those presented here, an Appendix containing the net lists for all four circuits is available from the authors in either printed or machine readable form.

Run times averaged about 1.5 seconds per case (Honeywell 6070).

Circuit 3

This circuit is substantially larger than circuits 1 and 2, having 117 nets and 162 elements. The distribution of net sizes is 42 two-element, 29 three-element, 12 four-element, 14 five-element, 10 six-element, 2 seven-element, 5 eight-element, and 3 nine-element nets, for a mean of 3.8 elements per net. As might be expected, this preponderance of more than two-element nets makes the edge-cut model quite inaccurate.

A series of four-way partitions into sets of 41 elements from random initial configurations produced a range of minimum net-cut partitions with costs of 52 to 61 net-cuts. Mean initial cost for these random starts was about 160 net-cuts.

The best edge-cut partition had a cost of 198 edge-cuts, representing 70 net-cuts, and in fact this partition could be improved to 58 net-cuts by using the net-cut model. The final net-cut solutions typically corresponded to more than 300 edge-cuts, substantially above the best edge-cut solution, again demonsteating that the best edge-cut partition is unrelated to the best net-cut, and vice versa.

We also experimented with less constrained four-way partitions, i.e., into sets of potentially more than 41 elements, and with partitions into more than four sets. Of course, costs for the less constrained partitions were lower, because of additional freedom of arrangement.

Run time for four-way partitions of this circuit averaged about 25 seconds per case.

As an (as yet unexplained) aside, the constructive procedure alluded to above was also used to partition this circuit; it produced a 61 net-cut partition. But using that partition as the initial start of our iterative improvement procedure, we obtained a 50 net-cut partition which was of lower cost than that obtained from any of the random starts. This result is in sharp contrast with that of Circuit 2, where a constructive starting partition led to a quite poor final answer. We are still investigating this effect.

Circuit 4

This is a very large digital logic circuit, representing the controller of a small switching computer. It has 1048 nets and 402 elements. Because of sheer size, experimental results are somewhat scanty for the edge-cut model, but extrapolating from the fact that the mean number of elements per net in this circuit is 6.7, the edge-cut model is likely to be substantially in error. For instance, for a 20-way partition, the best edge-cut partition involves 3584 edges, and has 999 net-cuts. This may be compared to the best net-cut partition, which has 757 net-cuts.

This circuit is being experimentally partitioned into several partition sizes, to evaluate different proposed technologies. Since one of these must be selected for actual construction, use of the appropriate model here is more than an academic exercise.

7. GENERAL OBSERVATIONS

For the current program implementations, run times for the edge-cut model are roughly one-fourth to one-half of those for the net-cut model. (The two implementations differ only in a few subroutines and data representation.) This is not unexpected, since many of the computations in the net-cut model are somewhat more complex than their edge-cut analogs. However, the improved results from the net-cut model are so much better than for the edge-cut model that the extra time is immaterial.

Storage requirements for the two models may be compared on the basis of the basis of the number of non-zero elements of the respective data matrices: in either case, the matrices are extremely sparse for typical circuits, usually less than 1% non-zero entries. The storage required is usually lower for the net-cut model, in direct proportion to the number of elements per net. This may not be immediately apparent: since the edge-cut cost matrix is symmetric, it would appear to need half as much storage; furthermore, there are often many more nets than elements. The reason is this: in the edge-cut model, a k-terminal device is represented by up to $k(k-1)/2$ non-zero entries; the same data appears in the net model as only k entries. Thus the net model is more economic of storage, especially for larger values of k. This expectation is borne out in most of these circuits. A summary appears in Table 3.

TABLE 3

Circuit	Storage for Edge Model	Storage for Net Model
1	147	116
2	230	280
3	770	430
4	7000	3540

8. SUMMARY

The analysis and examples convincingly demonstrate that the interconnection matrix, or graph partitioning, model is an inappropriate representation for the partitioning of typical electrical circuits. The edge-cut model results in a substantial increase in the number of nets needed to interconnect the packages. This, in turn, increases the difficulty of routing and the area required for these interconnections. The increase in the number of nets usually means an increase in the number of external connections (or pins) on each package. For designs which are pin-limited, this increase may require the use of substantially more packages than are actually necessary.

We have demonstrated that a partitioning algorithm for the proper, net-cut model requires a moderate increase in run-time and reduced storage requirements.

REFERENCES

1. M. Hanan and J. M. Kurtzberg, A review placement and quadratic assignment problems. IBM Report RC-3046, 20 April 1970. (This report was the basis of a tutorial given by Hanan at the 8th Design Automation Workshop, June 1971, but did not appear in the Proceedings.)

2. R. A. Rutman, An algorithm for placement of interconnected elements based on minimum wire length. Proc. SJCC (1964), 477-491.

3. B. W. Kernighan and S. Lin, An efficient heuristic procedure for partitioning graphs. Bell System Tech. J. Feb. 1970, pp. 291-308. U. S. Patent 3,617,714 (Nov. 1971).

4. A. H. Scheinman, Private communication.

Multiterminal Flows in a Hypergraph

T. C. HU AND K. MOERDER

I. INTRODUCTION

THE circuit layout problem, in its simplest form, is that of placing a set of modules on a chip in a nonoverlapping manner and then connecting the terminals on those modules by noninterfering wires according to a wiring list. An optimal solution to the circuit layout problem minimizes the area of the bounding rectangle of the chip. Due to the problem's complexity, it is usually divided into subproblems, with intermediate goals established to provide a basis for optimizing these subproblems.

Placement Problem: Place the modules on the chip in a nonoverlapping manner, thereby maximizing the routability of the chip and minimizing the area required.

Routing Problem: Given a placement of modules on a chip, connect the terminals of those modules by noninterfering wires according to the wiring list.

A. Network Flow

In this paper, we consider only the placement problem. We approach this problem by formulating a network flow model of the circuit and exploring algorithms for partitioning the circuit recursively. Using the minimum cut criteria, we can partition and place the modules recursively. First compute a minimum cut dividing the chip into two regions, then divide each region into two subregions, and repeat the process until each small region contains only one module. Kernighan and Lin [1], and Breurer [2] have suggested this criteria of minimum cut placement.

Let W be a wiring list consisting of a set of nets. Each net is a list of terminals on modules that must be connected. Let M be a set of modules, where each module has a weight that represents the chip area required by that module. The minimum cut placement problem is that of placing the modules on the chip in a nonoverlapping manner, such that the number of wires crossing dividing lines on the chip is minimized. This placement tends to minimize the area required to connect the modules according to the wiring list.

B. Edge Cut Model

Previous authors have used the edge cut model to represent the structure of the circuit. In the edge cut model, a graph represents the circuit. Modules are represented by vertices, and wires are represented by edges. Two vertices M_a and M_b are connected by an edge if there is a net connecting terminals on M_a with terminals on M_b. Figure 1 provides an example problem. The circuit consists of five modules and three nets.

Manuscript received December 18, 1984.

T. C. Hu is with the Dept. of Elec. and Computer Science, University of California, San Diego, La Jolla, CA 92093.

K. Moerder is with Linkabit, 3033 Science Park Road, San Diego, CA 92121.

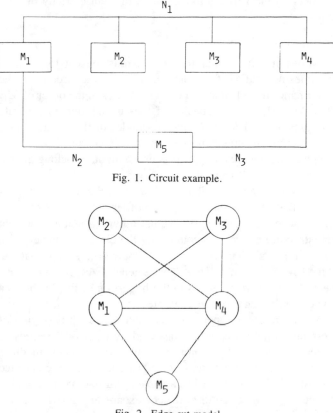

Fig. 1. Circuit example.

Fig. 2. Edge cut model.

Net N_1 connects modules M_1, M_2, M_3, and M_4. Net N_2 connects modules M_1 and M_5. Net N_3 connects modules M_4 and M_5. The edge cut model for this circuit is shown in Fig. 2, where net N_1 is represented by a clique, a complete subgraph, of the four vertices M_1, M_2, M_3, and M_4.

As an example, consider the chip as a rectangle partitioned into the left and right regions by an imaginary vertical line L. After we have placed the modules into the two regions, we must connect all the nets in the wiring list. If a net N connects modules in both regions, then we say net N spans the vertical line L. Each net spanning L requires at least one wire across the vertical line to make the connection. If there are more nets spanning L than there are available tracks on the chip across L, it is impossible to connect all the nets. Ideally, we would place the modules in such a way that a minimum number of nets span the vertical line L.

We can consider the graph in Fig. 2 as a flow network and find the minimum cut that partitions the network into two regions. The *minimum cut* separating two sets of vertices A and B is the smallest set of edges whose removal disconnects A and B. The minimum cut separating vertices M_1 and M_2 from the remaining vertices of the network has capacity five because it contains five edges. However, the five modules could be connected as shown in Fig. 1, where only two nets span the vertical line separating M_1 and M_2 from the remaining modules of the circuit. The reason for this discrepancy is the representation of net N_1 by a clique of size four. Any partition of the vertices connected by this clique requires a cut capacity of three or four. In reality, we could connect all the vertices in a chain with a cut capacity of only one.

C. Net Cut Model

Notice in Fig. 1 that four wires must be cut to separate modules M_1 and M_3 from the remainder of the circuit. On the other hand, if we had arranged the modules in the order of M_1, M_3, M_2, M_4, a cut capacity of two is required to separate modules M_1 and M_3 from the remainder of the circuit. Since the optimum arrangement is not known in advance, we need a graph model where any partition of the modules belonging to a net requires a cut capacity of one.

Schweikert and Kernighan [3] propose the use of a net cut model that reflects the ability to reorder the vertices of a net and reduce the cut capacity. We use a hypergraph to represent the structure of the nets connecting the circuit. A hypergraph $H(V, E)$ is similar to an ordinary network, except that the edged set E is generalized. These generalized edges connect sets of two *or more* nodes in the hypergraph. Fig. 3 shows a hypergraph model of the example.

We represent a hypergraph as a network with two types of vertices, which we call nodes and stars. Nodes represent modules, and stars represent nets. Any set of modules connected by a net N is represented as a set of nodes connected to a star S. Every star S has a capacity of one, corresponding to the single wire required if net N spans any line L. This is shown in Fig. 3. A cut separating two sets of nodes A and B is now a set of stars, whose removal from the network disconnects A and B. The capacity of the cut is the sum of the capacities of the stars composing the cut.

D. Organization of the Paper

The remainder of this paper is divided into five sections. Section two demonstrates that the value of any maximum flow in a hypergraph from M_s and M_t equals the capacity of a minimum cut separating M_s and M_t. The next section shows noncrossing minimum cuts separating the nodes of a hypergraph always exist. These noncrossing cuts mean that $n - 1$ flow computations are sufficient to find the minimum cuts separating every pair of nodes in a hypergraph. Section four then provides a necessary and sufficient condition for a set of numbers to be the flow values between pairs of nodes of a hypergraph. Following this section comes an algorithm for computing a flow-equivalent cut tree corresponding to these $n - 1$ flow computations. The concluding section discusses limitations and further applications of the hypergraph model for circuit layout.

II. MAXIMUM FLOW MINIMUM CUT

A *cut* in a hypergraph is the collection of all stars from a subset of nodes to its complement. The cut is denoted by $(X, \bar{X})$, where X is the subset of nodes and $\bar{X}$ is its complement. Thus, a cut $(X, \bar{X})$ is a set of all stars S connected to both a

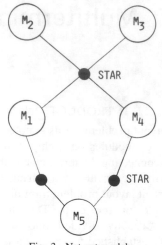

Fig. 3. Net cut model.

node in X and a node in $\bar{X}$. The capacity of this cut is represented by $c(X, \bar{X})$.

We shall prove first that in a hypergraph the maximum flow between any two nodes M_s and M_t equals the capacity of a minimum cut separating M_s and M_t. The proof shows a flow with value equal to the capacity of a minimum cut always exists. Because any flow is always less than or equal to the capacity of any cut that separates the source from the sink, the theorem is proved.

Theorem 1: Maximum flow minimum cut theorem. For any hypergraph with positive integer star capacities, the maximum flow value from the source to the sink equals the capacity of a minimum cut separating the source and the sink.

Proof: Figure 4 shows two nodes, M_i and M_j, connected by edges e_{iq} and e_{qj} to a star S_q. The capacity of S_q is b_q. Equation (1) states that x_q—the flow through

$$\sum_i x_{iq} = \sum_j x_{qj} = x_q \qquad (1)$$

$$\sum_q x_{qi} = \sum_p x_{ip} M_i \notin \{M_s, M_t\} \qquad (2)$$

$$0 \le x_q \le b_q \qquad (3)$$

S_q—equals the total flow into S_q and the total flow out of S_q. By (2), the total flow into any node M_i except M_s and M_t equals the total flow out of M_i. M_s has only outgoing flow, while M_t has only incoming flow. Equation (3) bounds the nonnegative flow x_q through a star by the capacity b_q of the star.

We start with initial conditions satisfying (1), (2), and (3), where $M_s \in X$, and all x_q equal zero. Based on the current flow in the hypergraph, we define a subset X of the nodes recursively by the following rules:

1) $M_s \in X$.
2) If $M_i \in X$, $x_q < b_q$ then $M_j \in X$.
3) If $M_i \in X$, $x_{qi} > 0$ then $M_j \in X$.
4) If $M_i \in X$, $x_{jq} > 0$ then $M_j \in X$.

Any nodes not in X belong to $\bar{X}$. Using these rules to define the set X, we have two cases to consider.

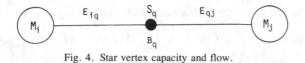

Fig. 4. Star vertex capacity and flow.

Case 1: $M_t \in \bar{X}$. By rule 2), $x_q = b_q$ for all stars separating X from $\bar{X}$. From rules 3) and 4) there is no flow x_{ji} from $\bar{X}$ to X. Thus we have a maximum flow from M_s to M_t with value equal to $c(X, \bar{X})$.

Case 2: $M_t \in X$. Then there exists a chain from M_s to M_t composed of edges, each satisfying rule 2), 3), or 4). Let the chain be $M_s \cdots M_{i_1} e_{i_1 q_1} S_{q_1} e_{q_1 i_2} M_{i_2} \cdots M_t$, where every star in this chain satisfies rule 2), 3), or 4).

If the star S_{q_j} satisfies rule 2), we can send additional flow through S_{q_j} from M_{i_j} to $M_{i_{j+1}}$. S_{q_j} is called an unsaturated star because $x_{q_j} < b_{q_j}$.

If S_{q_j} satisfies rule 3) or 4) we can also send additional flow from M_{i_j} to $M_{i_{j+1}}$ through S_{q_j}, effectively canceling the existing flow in an edge connected to this star. This path is called a *flow-augmenting path* with respect to the current flow.

Let e be the minimum additional flow through every star S_q on this chain, then increase the flow along this chain by e. The new x_{ij} satisfies (1), (2), and (3).

Now we can redefine the set X based on the new flow and again consider Case 1 and Case 2. Since the star capacities b_q are integers, we increase the flow through the network by an integer amount each time we apply Case 2. But the network has finite flow; therefore, this process eventually terminates. □

Theorem 1 motivates our interest in flows through hypergraphs. The minimum cut provides the minimum cost partition of the circuit into two components.

III. NONCROSSING CUTS

Let $(X, \bar{X})$ and $(Y, \bar{Y})$ be two cuts separating M_s and M_t in a hypergraph. We say these two cuts *cross* each other if and only if each of the following sets are nonempty:

$$\bar{X} \cap \bar{Y}, \quad \bar{X} \cap Y, \quad X \cap \bar{Y}, \quad X \cap Y.$$

Theorem 2 states that if two minimum cuts separating M_s and M_t cross each other, there are two other minimum cuts, also separating M_s and M_t, that do not cross each other. In order to prove our theorem, we must define some additional notation.

Let $N(A, B, C, D)$ represent the set of all stars connected to nodes in precisely the disjoint sets A, B, C, and D. For example, if the union of A, B, C, and D contains all of the nodes in the hypergraph, the capacity of $(A \cup B, C \cup D)$ equals

$$N(A, C) + N(A, D) + N(B, C) + N(B, D)$$
$$+ N(A, B, C) + N(A, B, D) + N(A, C, D)$$
$$+ N(B, C, D) + N(A, B, C, D).$$

Theorem 2: Let $(X, \bar{X})$, and $(Y, \bar{Y})$ be two minimum cuts. Then $(X \cup Y, \overline{X \cup Y})$ and $(X \cap Y, \overline{X \cap Y})$ are also minimum cuts.

Proof: Case 1: If $X \subset Y$, then $X \cup Y = Y$ and $X \cap Y$

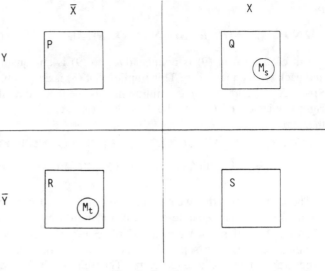

Fig. 5. Crossing cuts.

$= X$, so $(X \cup Y, \overline{X \cup Y}) = (Y, \bar{Y})$ and $(X \cap Y, \overline{X \cap Y}) = (X, \bar{X})$.

Case 2: If $Y \subset X$, then $Y \cup X = X$ and $Y \cap X = Y$, so $(Y \cup X, \overline{Y \cup X}) = (X, \bar{X})$ and $(Y \cap X, \overline{Y \cap X}) = (Y, \bar{Y})$.

Case 3: We now consider the case in which $(X, \bar{X})$ and $(Y, \bar{Y})$ cross, as shown in Fig. 5, where

$$\begin{aligned}
P &= \bar{X} \cap Y, \\
Q &= X \cap Y, \\
R &= \bar{X} \cap \bar{Y}, \\
S &= X \cap \bar{Y}.
\end{aligned}$$

Equations (4), (5), (6), and (7) express the cut capacities of in terms of the stars connecting P, Q, R, and S.

$$\begin{aligned}
c(X, \bar{X}) = {} & N(P, Q) + N(P, S) + N(Q, R) + N(R, S) \\
& + N(P, Q, R) + N(P, Q, S) + N(P, R, S) \\
& + N(Q, R, S) + N(P, Q, R, S)
\end{aligned} \tag{4}$$

$$\begin{aligned}
c(Y, \bar{Y}) = {} & N(P, R) + N(P, S) + N(Q, R) + N(Q, S) \\
& + N(P, Q, R) + N(P, Q, S) + N(P, R, S) \\
& + N(Q, R, S) + N(P, Q, R, S)
\end{aligned} \tag{5}$$

$$\begin{aligned}
c(Q, P \cup R \cup S) = {} & N(P, Q) + N(Q, R) + N(Q, S) \\
& + N(P, Q, R) + N(P, Q, S) \\
& + N(Q, R, S) + N(P, Q, R, S)
\end{aligned} \tag{6}$$

$$\begin{aligned}
c(R, P \cup Q \cup S) = {} & N(P, R) + N(Q, R) + N(R, S) \\
& + N(P, Q, R) + N(P, R, S) \\
& + N(Q, R, S) + N(P, Q, R, S)
\end{aligned} \tag{7}$$

Since $(X, \bar{X})$ and $(Y, \bar{Y})$ are minimum cuts,

$$\begin{aligned}
c(X, \bar{X}) + c(Y, \bar{Y}) \leq {} & c(P \cup Q \cup S, R) \\
& + c(Q, P \cup R \cup S).
\end{aligned} \tag{8}$$

Substituting (4), (5), (6), and (7) into (8) and canceling yields

89

(9):

$$2N(P, S) + N(P, R, S) + N(P, Q, S) \leq 0. \qquad (9)$$

But each term in (9) is nonnegative, so (9) is an equality, and each term equals zero. This implies that (8) is an equality. Since $(X, \bar{X})$ and $(Y, \bar{Y})$ are minimum cuts, neither cut on the right-hand side of (8) can be less than $(X, \bar{X})$ or $(Y, \bar{Y})$, therefore,

$$c(X, \bar{X}) = c(Y, \bar{Y}) = c(P \cup Q \cup S, R) = c(X \cup Y, X \cup Y)$$

$$= c(Q, P \cup R \cup S) = c(X \cap Y, X \cap Y). \qquad (10)$$

$\square$

Theorem 2 states that we can find a set of $n - 1$ noncrossing minimum cuts that separate any two nodes in a hypergraph. Theorem 2 implies that we need calculate only $n - 1$ flows to obtain minimum cuts separating each pair of nodes in the hypergraph. The next two theorems, Theorem 3 and Theorem 4, are also proved by summing flows across cuts. Their proofs are similar and are omitted.

Theorem 3: Let $(X, \bar{X})$ be a minimum cut separating $M_i \in X$ from some other node. Let M_e and M_k be any two nodes contained in $\bar{X}$. Then there exists a minimum cut $(Z, \bar{Z})$ separating M_e from M_k, such that $(Z, \bar{Z})$ and $(X, \bar{X})$ do not cross each other.

Theorem 4: Let $(X, \bar{X})$ be a minimum cut separating $M_i \in X$ from some node $M_j \in \bar{X}$. Let M_k be any other node contained in $\bar{X}$. Then there exists a minimum cut $(Z, \bar{Z})$ separating M_i from M_k, such that $(Z, \bar{Z})$ and $(X, \bar{X})$ do not cross each other.

Theorems 3 and 4 simplify the computation necessary to find the $n - 1$ noncrossing minimum cuts. Theorem 3 means that if $(X, \bar{X})$ is a minimum cut, we can regard the flow between all nodes within X as infinite when calculating the maximum flow values between any two nodes in $\bar{X}$. Therefore, we can regard X as a single node or condense X into a single node when calculating the maximum flow values between any two nodes in $\bar{X}$. Similarly, we can condense $\bar{X}$ into a single node when calculating the maximum flow values between any two nodes in X. Theorem 4 shows that if $(X, \bar{X})$ is a minimum cut separating M_i from some other node, and if we wish to find the maximum flow from M_i to any other node in $\bar{X}$, we can condense X into a single node.

IV. TRIANGLE INEQUALITY

The following theorem, Theorem 5, provides realizability constraints for the flows in a hypergraph. These constraints are the same as for the flows in a network. Given any three nodes in a hypergraph M_i, M_j, and M_k, consider the flows f_{ij}, f_{jk}, and f_{ik}. The flow $f_{ik} \geq \min (f_{ij}, f_{jk})$. Therefore, at least two of these flows must be equal, because if the three flows were distinct, the smallest would be less than the minimum of the other two. Further, if one flow value is distinct, it is the largest of the three flow values.

Theorem 5: A necessary and sufficient condition for a set of nonnegative numbers

$$f_{ij} = f_{ji} \qquad i, j \epsilon \{1, \cdots, n\}$$

to be the maximum flow values of a hypergraph is

$$f_{ik} \geq \min (f_{ij}, f_{jk}) \qquad i, j, k \epsilon \{1, \cdots, n\}. \qquad (11)$$

Proof: NECESSITY: Let $(X, \bar{X})$ be a minimum cut separating M_i from M_k, where $M_i \in X$ and $M_k \in \bar{X}$. If $M_j \in X$, then $(X, \bar{X})$ is a cut separating M_j and M_k; hence,

$$f_{jk} \leq c(X, \bar{X}) = f_{ik}. \qquad (12)$$

If $M_j \in \bar{X}$, then $(X, \bar{X})$ is a cut separating M_i and M_j; hence,

$$f_{ij} \leq c(X, \bar{X}) = f_{ik}. \qquad (13)$$

Notice that whether $M_j \in X$ or $M_j \in \bar{X}$, if M_j is contained in a component of the hypergraph disconnected by $(X, \bar{X})$,

$$f_{ij} \leq f_{ik} \text{ and } f_{jk} \leq f_{ik}. \qquad (14)$$

Either (12) or (13) holds; therefore, we have

$$f_{ik} \geq \min (f_{ij}, f_{jk}). \qquad (15)$$

Further, by induction we have

$$f_{ik} \geq \min (f_{ia}, f_{ab}, \cdots, f_{dk}), \qquad (16)$$

where $M_i, M_a, M_b, \cdots, M_d, M_k$ is any sequence of nodes in the hypergraph.

SUFFICIENCY is proven by the construction of a network with flow values satisfying (11). Let T be the maximum spanning tree of the specified flow values, where the edges of T are labeled b_{ij}.

If edge e_{ik} is in T, then $b_{ik} = f_{ik}$. If edge e_{ik} is not in T, the flow v_{ik} in T is

$$v_{ik} = \min (b_{ia}, b_{ab}, \cdots, b_{dk})$$

$$= \min (f_{ia}, f_{ab}, \cdots, f_{dk}),$$

where $b_{ia}, b_{ab}, \cdots, b_{dk}$ are the capacities of the edges forming the unique path from M_i to M_k in T. But from (16), and since edge e_{ik} is not in T,

$$f_{ik} = v_{ik}.$$

Therefore, we can always construct a flow-equivalent tree for any set of flow values satisfying (11). $\square$

V. FLOW EQUIVALENT CUT TREE

We say that two networks are *flow-equivalent* if and only if they have the same maximum flow values between each pair of nodes. Because we can partition the n nodes of a hypergraph into n regions with the $n - 1$ noncrossing minimum cuts, we can construct a tree that is flow-equivalent to the hypergraph. This tree, similar to a Gomory–Hu cut tree [4], contains an edge corresponding to each of the $n - 1$ noncrossing minimum cuts, as Fig. 6 illustrates. We first give an informal description of the algorithm that computes this tree, and then a detailed description of the algorithm, followed by a proof that the tree is flow-equivalent to the hypergraph.

Select any two nodes M_i and M_j in the hypergraph H. Perform a flow computation from M_i to M_j in order to find a

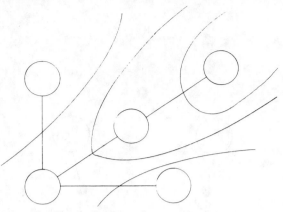

Fig. 6. Cut tree and noncrossing cuts.

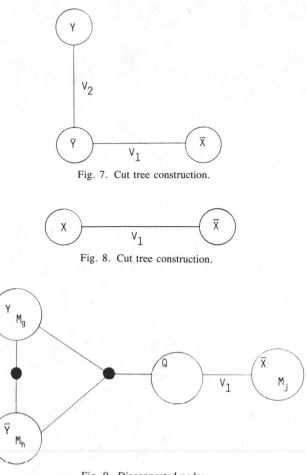

Fig. 7. Cut tree construction.

Fig. 8. Cut tree construction.

Fig. 9. Disconnected nodes.

minimum cut $(X, \bar{X})$ separating M_i and M_j. Now select two nodes M_k and M_l in X, and find the minimum cut $(Y, \bar{Y})$ separating them. All of the nodes in $\bar{X}$ are condensed into a single node for this second flow computation. Figure 7 illustrates the situation, where v_1 is the capacity of $(X, \bar{X})$ and v_2 is $c(Y, \bar{Y})$.

In Fig. 7, $\bar{X}$ is connected to $\bar{Y}$ because $\bar{X}$ and $\bar{Y}$ lie on the same side of the cut $(Y, \bar{Y})$. We repeat this process until there are only single nodes in each set. While performing a flow computation in one set, all components that become disconnected when the set is removed are condensed into single nodes.

We now present a detailed description of the algorithm.

1) Select two terminal nodes arbitrarily and perform a maximum flow computation on the original hypergraph. This gives a minimum cut $(X, \bar{X})$, which is represented by two circles connected by a link in Fig. 8. Label the link with v_1, the capacity of $(X, \bar{X})$. The left circle lists all of the nodes in X; the right circle lists all of the nodes in $\bar{X}$.

2) From the tree diagram, select two nodes—M_g and M_h—from any circle containing two or more nodes. Remove this circle from the tree diagram, condense each disconnected component of the tree into a single node, and replace the circle. Perform a flow computation in order to find the minimum cut $(Y, \bar{Y})$ separating M_g from M_h. Suppose M_g and M_h are in X: if X is removed from the tree, each disconnected component can be condensed into a single node for the flow computation. In Fig. 7, $\bar{X}$ is connected to $\bar{Y}$ because $\bar{X}$ and $\bar{Y}$ are on the same side of the cut with capacity v_2. In general, when a cut partitions a circle A into A' and A'', the adjacent circles are each connected to A' or A'' on the side of the cut containing them.

3) Fig. 9 illustrates a case where a cut in a hypergraph can create disconnected components. The minimum cut separating M_g and M_h disconnects $\bar{X}$. In this situation, if M_i is in Y, we include Q in the circle containing Y and connect $\bar{X}$ to Y. If M_i is in either $\bar{Y}$ or Q, then we include Q in the circle containing $\bar{Y}$ and connect $\bar{X}$ to $\bar{Y}$.

4) The process of computing flows and partitioning circles is repeated. After $n - 1$ times, a tree diagram is produced in which each circle contains exactly one node.

Theorem 6 proves that the tree produced by this algorithm is flow-equivalent to the original hypergraph.

Theorem 6. The maximum flow value between any two nodes M_i and M_j in the hypergraph equals

$$f_{ij} = \min \ (v_{ia}, \ v_{ab}, \ \cdots, \ v_{dj}), \tag{17}$$

where $v_{ia}, v_{ab}, \cdots, v_{dj}$ are the values associated with the links of the tree that form the unique path from M_i to M_j.

Proof: Because every link $v_{ia}, v_{ab}, \cdots, v_{dj}$ in the tree represents a cut that separates M_i from M_j, we have

$$f_{ij} \leq \min \ (v_{ia}, \ v_{ab}, \ \cdots, \ v_{dj}).$$

It is, therefore, sufficient to prove that

$$f_{ij} \geq \min \ (v_{ia}, \ v_{ab}, \ \cdots, \ v_{dj}). \tag{18}$$

Consider any stage in the construction of the tree, where a link with value v connects two circles X and Y. From (16) of Theorem 5, it is sufficient to prove the claim that there is a node M_i in X and a node M_j in Y such that $f_{ij} = v$. Consider the completed tree: we have

$$f_{ij} \geq \min \ (f_{ia}, \ f_{ab}, \ \cdots, \ f_{dj}) = \min \ (v_{ia}, \ v_{ab}, \ \cdots, \ v_{dj})$$

because

$$f_{ia} = v_{ia}, \ f_{ab} = v_{ab}, \ \cdots, \ f_{dj} = v_{dj}.$$

91

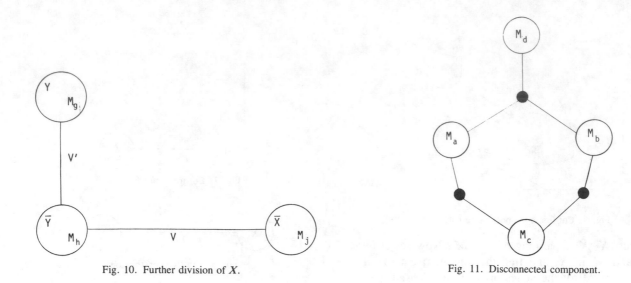

Fig. 10. Further division of X.

Fig. 11. Disconnected component.

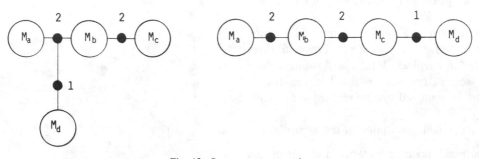

Fig. 12. Star tree representation.

This claim is clearly true when the tree consists of two circles X and $\bar{X}$, because this cut was obtained from a flow computation between M_i and M_j. We must now prove that the claim holds when the tree is further divided. Figure 10 illustrates a later step where X is further partitioned by performing a flow computation from M_g to M_h.

There are now two cases to consider. First, if the cut separating M_g from M_h disconnects $\bar{X}$, then, by construction, M_i is in the circle connected to $\bar{X}$, so $f_{gh} = v'$ and $f_{ij} = v$. Second, if the cut separating M_g from M_h causes no disconnected components, the location of M_i is important.

If M_i is in $\bar{Y}$ (the circle connected to $\bar{X}$), then again $f_{gh} = v'$ and $f_{ij} = v$. If M_i is in Y (the circle not connected to $\bar{X}$), we must show that $f_{hj} = v$.

In the tree diagram we have

$$f_{gh} = v' \text{ and } f_{ij} = v.$$

From Theorem 4, if the nodes in Y are condensed into a single node, f_{gh} and f_{ij} are unaffected; by Theorem 3, f_{hj} is also unaffected. From Theorem 5 and the condensed tree diagram we have

$$f_{hj} \geq \min (f_{ji}, f_{ig}, f_{gh}) = \min (v, \infty, v').$$

However $v' \geq v$ because $(Y, \bar{Y} \cup \bar{X})$ is a cut separating M_i

from M_j. Therefore,

$$f_{hj} = v. \qquad \square$$

VI. CONCLUSION

One feature of the hypergraph model is that a cut $(X, \bar{X})$ separating two nodes M_i and M_j can disconnect other components of the hypergraph. This situation occurs when the set of stars connecting the component to the remainder of the hypergraph is a subset of $(X, \bar{X})$. A natural question is: Can the cut tree representation be generalized to display some additional information about the interactions between the cuts? Figure 11 is an example hypergraph with disconnected components, and Fig. 12 shows two generalized cut trees for the hypergraph.

Notice that the first tree correctly represents the disconnected components of the hypergraph. The minimum cut separating M_a from M_b creates a disconnected component M_d. This feature requires the star connecting M_a, M_b, and M_d correspond to an edge in the tree, not a path in the tree. Figure 13 is an example hypergraph where one of the disconnected components cannot be shown in the tree. In general, a tree cannot represent the structure of the disconnected components of a hypergraph.

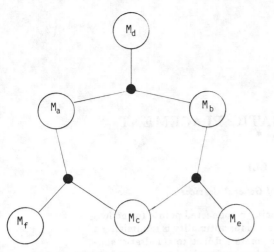

Fig. 13. Multiple disconnected components.

The hypergraph model for circuit layout extends network flow techniques to the placement of modules on a chip. The model captures the ability to choose the routes for the wires after the modules have been placed. This ability reduces the coupling between the placement and routing stages of the design process. The hypergraph model provides a computational tool for minimum *net cut* placement.

REFERENCES

[1] B. W. Kernighan and S. Lin, "An efficient procedure for partitioning graphs," *Bell Syst. Tech. J.*, Feb. 1970.
[2] M. A. Breuer, "Minimum cut placement," *J. Design Automation and Fault Tolerant Comput.*, vol. 1, no. 4, pp. 343–362, Oct. 1977.
[3] D. G. Schweikert and B. W. Kernighan, "A proper model for the partitioning of electrical circuits," in *IEEE Proc. 9th Design Automation Workshop*, pp. 57–62, 1972.
[4] R. E. Gomory and T. C. Hu, "Multi-terminal network flows," *J. Soc. of Industrial and Appl. Math.*, vol. 9, no. 41961, pp. 551–570, 1961.

AN r-DIMENSIONAL QUADRATIC PLACEMENT ALGORITHM*

KENNETH M. HALL†

State of California, Department of General Services

In this paper the solution to the problem of placing n connected points (or nodes) in r-dimensional Euclidean space is given. The criterion for optimality is minimizing a weighted sum of squared distances between the points subject to quadratic constraints of the form $X'X = 1$, for each of the r unknown coordinate vectors. It is proved that the problem reduces to the minimization of a sum or r positive semidefinite quadratic forms which, under the quadratic constraints, reduces to the problem of finding r eigenvectors of a special "disconnection" matrix. It is shown, by example, how this can serve as a basis for cluster identification.

1. Introduction

Many sequencing and placement problems can be characterized as follows: Given n points (or nodes) and an $n \times n$ symmetric connection matrix, $C = (c_{ij})$, where $c_{ii} = 0$, and $c_{ij} \geqq 0$, $i \neq j$, $i = 1, 2, \cdots, n$, is the "connection" between point i and point j, find locations for the n points which minimizes the weighted sum of squared distances between the points (i.e., weighted by c_{ij}).

If x_i denotes the X-coordinate of point i and z denotes the weighted sum of squared distances between the points, then the 1-dimensional problem is to find the row vector $X' = (x_1, x_2, \cdots, x_n)$ which minimizes

$$(1.1) \qquad z = \tfrac{1}{2} \sum_{i=1}^{n} \sum_{j=1}^{n} (x_i - x_j)^2 c_{ij}$$

where the prime denotes vector transposition. To avoid the trivial solution $x_i = 0$, for all i, the following quadratic constraint is imposed:

$$(1.2) \qquad X'X = 1.$$

The solution to (1.1) and (1.2) is given in the next section. It is assumed that the noninteresting solution $x_i = x_j$, for all i and j, is to be avoided. Extensions to higher dimensions are given in §§3 and 4.

2. Optimum Solution in 1-Dimension (placement on a line)

Let $c_i.$ and $c_{.j}$ be the ith row sum and the jth column sum, respectively, of the (symmetric) matrix C. Define a diagonal matrix $D = (d_{ij})$ as follows:

$$d_{ij} = 0, \qquad i \neq j,$$
$$= c_{i.}, \qquad i = j.$$

Now, define the following matrix:

$$(2.1) \qquad B = D - C.$$

* Received July 1968; revised May 1969, December 1969.

† The author wishes to acknowledge C. H. Mays for his valuable contributions in the original formulation of this model and the characterization of its solution. He has also pointed out some applications of this model the author was not aware of. The author also wishes to acknowledge one of the referees for pointing out Reference 5 and its relationship to the present work.

In words, the ith diagonal entry b_{ii} of B is the ith row (or column) sum of the connection matrix C and the off diagonal element b_{ij} is the negative of the corresponding entry in C. The matrix B plays a very fundamental role in this problem as we shall soon see. For brevity, B will be called the *disconnection* matrix.

Let $X' = (x_1, x_2, \cdots, x_n)$ be a row vector of X-coordinates, where the prime denotes vector transposition. Then (1.1) can be rewritten as $z = X'BX$, i.e.

$$(2.2) \qquad z = \tfrac{1}{2} \sum_i \sum_j (x_i - x_j)^2 c_{ij}$$

$$(2.3) \qquad = \tfrac{1}{2} \sum_i \sum_j (x_i^2 - 2x_i x_j + x_j^2) c_{ij}$$

$$(2.4) \qquad = \tfrac{1}{2} (\sum_i x_i^2 c_{i\cdot} - 2 \sum_i \sum_j x_i x_j c_{ij} + \sum_j x_j^2 c_{\cdot j})$$

$$(2.5) \qquad = \sum_i x_i^2 c_{i\cdot} - \sum_j \sum_{i \neq j} x_i x_j c_{ij}$$

$$(2.6) \qquad = X'BX.$$

Equation (2.5) follows because C is symmetrical (i.e., $c_{i\cdot} = c_{\cdot j}$). Equation (2.6) is immediate since (2.5) has yielded a quadratic form. Now we prove the following:

THEOREM. *Let G denote the underlying graph of the connection matrix C (i.e., an arc in G exists between node i and node j if and only if $c_{ij} > 0$). Then the following is true about the disconnection matrix, B:*

(i) *B is positive semi-definite ($B \geqq 0$), and*

(ii) *whenever G is connected, B is of rank $n - 1$.*

PROOF. To prove (i), we simply note from equations (2.6) and (2.2) that $X'BX$ can be written as a sum of nonnegative terms. Thus $B \geqq 0$. That the bound of zero can be reached can be seen from (2.2) by letting $x_i = x_j$ for all i and j.

Before proving (ii), we first note from (2.1) that the row sums of B are zero, so B has an eigenvector which is proportional to the unit vector, $U' = (1, 1, \cdots, 1)$. The associated eigenvalue is zero. If B is to have rank $n - 1$ the remaining $n - 1$ eigenvalues of B must necessarily be positive (a direct result from (i) above). We will prove that the required eigenvalues are, in fact, positive.

Let $0 = \lambda_1 \leqq \lambda_2 \leqq \lambda_3 \leqq \cdots \leqq \lambda_n$ be the eigenvalues of matrix B, with corresponding eigenvectors $E_1, E_2, \cdots, E_n$. E_1 is proportional to the unit vector, U, because the row sums of B are zero. The remaining eigenvectors, $E_2, E_3, \cdots, E_n$, being orthogonal to U (or E_1) must each have the sum of its components equal to zero. Therefore, some components will be negative and some will be positive and hence, not all components will be equal. Therefore, if we can prove: For connected G with $x_i \neq x_j$ for all i and j, that $X'BX$ is positive, our proof would be complete. We will prove this by contradiction, i.e., assume $X'BX = 0$ and show that it contradicts the hypothesis that $x_i \neq x_j$ for all i and j.

Rewrite (2.5) and (2.2) as

$$X'BX = \sum_{i=1}^{n-1} \sum_{i<j} (x_i - x_j)^2 c_{ij} + \sum_{i=1}^{n-1} (x_i - x_n)^2 c_{in}.$$

Whenever $X'BX = 0$, both of the above terms on the right-hand side must be zero. Refer to these as RHSL and RHSR, respectively. Since G is connected, one or more of the coefficients c_{in}, $i \neq n$, must be positive. In all these cases x_i must equal x_n if RHSR is to be zero. Now form two sets of subscripts: $S_1 = \{i : c_{in} = 0\}$, $S_2 = \{i : x_i = x_n\}$. Note that S_2 contains all the subscripts with $c_{in} > 0$. If we can show that S_1 is a subset of S_2, our proof will be complete because then x_i would equal x_j for all i and j (providing us with the desired contradiction). Also, the proof is immediate if S_1 is empty. Therefore, assume S_1 contains $m > 0$ elements. Choose an element from S_1, say i_1; then $c_{i_1 i_2} > 0$ for some $i_2 \neq i_1$ or else G would not be connected. This implies that $x_{i_2} = x_{i_1}$ or else RHSL would not be zero. Two cases must now be considered for i_2; either (1) $c_{i_2 n} > 0$, or (2) $c_{i_2 n} = 0$. If (1) holds then $x_{i_2} = x_n$, (or else RHSR would not be zero) which would imply that i_2 is in the set S_2. Since $x_{i_2} = x_{i_1}$ then i_1 would also be in S_2. If (2) holds then $c_{i_1 i_3} + c_{i_2 i_3} > 0$ for some $i_3 \neq i_2 \neq i_1$ or else G would not be connected. In any case, this implies $x_{i_3} = x_{i_2} = x_{i_1}$ or else RHSL would not be zero. As in (1) above, if $c_{i_3 n} > 0$, then i_3 and, consequently, i_1 and i_2 are in S_2. On the other hand, as in (2) above, if $c_{i_3 n} = 0$ we continue building up (from S_1) a subset of $r - 1 \leq m$ elements, $\{i_j\}_{j=1}^{r-1}$, with $c_{i_1 i_2} > 0$, $c_{i_1 i_3} + c_{i_2 i_3} > 0$, $\cdots$, $\sum_{j=1}^{r-1} c_{i_j i_r} > 0$; $x_{i_1} = x_{i_2} = \cdots = x_{i_{r-1}}$; $c_{i_j n} = 0$, $j = 1, r - 1$, and $c_{i_r n} > 0$. The element i_r will eventually be reached if G is connected. When it is reached then i_r and, consequently, the subset $\{i_j\}_{j=1}^{r-1}$ will be in S_2. If $r = m + 1$, the proof would be finished since S_1 would be a subset of S_2. If $r \leq m$, repeat the above process by building up a new subset of connected elements from S_1 (having equal coordinates if RHSL is to be zero) which eventually become "connected" to element n in S_2. When this happens, the entire subset will be in S_2. Only a finite number (at most, m) of such subsets need to be constructed to account for all the elements of S_1. Then S_1 will be a subset of S_2 since x_i will equal x_j for all i and j. This provides the desired contradiction and completes our proof.

Now the problem has been reduced to the following form. Minimize

$$(2.7) \qquad z = X'BX, \qquad B \geq 0$$

subject to the quadratic constraint

$$(2.8) \qquad X'X = 1.$$

To minimize (2.7) subject to the constraint (2.8) introduce the Lagrange multiplier λ and form the Lagrangian $L = X'BX - \lambda(X'X - 1)$. Taking the first partial derivative of L with respect to the vector X and setting the result equal to zero yields $2BX - 2\lambda X = 0$. If I is the identity matrix, the above can be rewritten as

$$(2.9) \qquad (B - \lambda I)X = 0$$

which yields a nontrivial solution, X, if and only if λ is an eigenvalue of the matrix B and X is the corresponding eigenvector. If (2.9) is premultiplied by X' and the constraint (2.8) is imposed we obtain

$$(2.10) \qquad \lambda = X'BX.$$

Thus, the formal solution to (1.1) and (1.2) is simply that X is the eigenvector of B which minimizes z and λ ($=z$) is the corresponding eigenvalue. The minimum eigenvalue, zero, yields the noninteresting solution $X' = (1, 1, \cdots, 1)/\sqrt{n}$. Therefore the second smallest eigenvalue and the associated eigenvector yields the optimum solution. It is important to note that if the original problem is changed to a maximizing problem, then the maximum eigenvalue of B and the associated eigenvector will be the desired solution.

3. Extension to 2-Dimensions

Let $Y' = (y_1, y_2, \cdots, y_n)$ be a row vector of Y-coordinate of the n points. Then the problem is to determine X and Y which minimizes

$$(3.1) \qquad z = X'BX + Y'BY, \qquad B \geq 0$$

subject to the following constraints

$$(3.2) \qquad X'X = 1,$$

$$(3.3) \qquad Y'Y = 1.$$

To solve (3.1)–(3.3), introduce the Lagrange multipliers α and β and form the Lagrangian $L = X'BX + Y'BY - \alpha(X'X - 1) - \beta(Y'Y - 1)$. Taking the first partial derivative of L with respect to the vector X and also with respect to the vector Y and setting the results equal to zero yields the two systems of equations

$$(3.4) \qquad 2BX - 2\alpha X = 0,$$

$$(3.5) \qquad 2BY - 2\beta Y = 0.$$

These yield nontrivial solutions X and Y if and only if X and Y are eigenvectors of B, associated with the eigenvalues α and β, respectively.

If $0 = \lambda_1 < \lambda_2 \leqq \lambda_3 \leqq \cdots \leqq \lambda_n$ denote the n eigenvalues of matrix B, then (3.1)–(3.3) are solved by taking $\alpha = \beta = \lambda_1$. If it is desired that X not be proportional to Y, then take $\alpha = \lambda_1, \beta = \lambda_2$. If, further, it is desired that not all x_i are equal, and not all y_i are equal, then take $\alpha = \lambda_2, \beta = \lambda_3$. The vectors X and Y will be the eigenvectors associated with the eigenvalues α and β in any case. Sometimes, it is desirable to have X and Y orthogonal. This will be true whenever $\alpha \neq \beta$. If (3.4) and (3.5) are premultiplied by X' and Y', respectively, and the constraints (3.2) and (3.3) are imposed, then we see that $z = \alpha + \beta$. Thus, the sum of the relevant eigenvalues used will yield the final value of z.

4. Extension to r-dimensions

For the r-dimensional problem z is simply the sum of r quadratic forms, one for each dimension. If each of the coordinate vectors is constrained to have inner product equal to 1, then setting each coordinate vector equal to the eigenvector associated with the eigenvalue λ_1 would solve the problem. If in each dimension it is required that not all components of the solution vector be equal, then taking the eigenvector associated with λ_2 would solve the problem. If the coordinate vectors must not be proportional to each other, take the eigenvector associated with the eigenvalues $\lambda_2, \lambda_3, \cdots, \lambda_{r+1}$. Thus, after finding the X-vector (from a knowledge of λ_2) and the Y-vector (from a knowledge of λ_3) additional coordinate vectors are found from a knowledge of successively larger eigenvalues. The final value of z will be the sum of the eigenvalues used.

5. Applications

Let c_{ij} be the flow between work center i and work center j in a job shop. Choosing X and Y to be the eigenvectors associated with λ_2 and λ_3, respectively, results in optimum global placement of the work centers in the plane.

Let c_{ij} denote the "distance" (or "dissimilarity") between animal i and animal j based on a set of measurements. Choosing X to be associated with the maximum eigenvalue, λ_n, results in an optimum sequencing of the animals on a line (numerical taxonomy).

Let c_{ij} represent the number of wires interconnecting a pair of electronic components i and j. Choosing X and Y to be the eigenvectors associated with λ_2 and λ_3, respectively, results in optimum placement of the electronic components in the plane in the sense of minimum squared wirelength.

Let c_{ij} be the flow between economic facility i and economic facility j. Choosing X

and Y to be the eigenvectors associated with λ_2, and λ_3, respectively, results in optimum global placement of the economic facilities in the plane.

6. Examples

Example 1. A 4-node graph, its connection matrix C and disconnection matrix B are illustrated in Figure 6.1. Arcs denote direct connections between nodes, with corresponding values given in matrix C.

The 4 eigenvalues of B and their associated eigenvectors E_1, E_2, E_3 and E_4 are shown in Figure 6.2. A plot of the 4 nodes is also shown in Figure 6.2, where E_2 and E_3 has been used as the X and Y-coordinate vectors, respectively. It should be noted that E_2 has "unraveled" the graph, whereas E_4 (the maximization problem) would have made it worse. E_4 would have yielded the sequence 4, 3, 1, 2 rather than 1, 4, 2, 3.

Example 2. Consider the 4-node graph, its connection matrix C and disconnection matrix B illustrated in Figure 6.3. The eigenvalues of B with their associated eigen vectors are shown in Figure 6.4. A plot of the 4 nodes, using E_2 and E_3 is also shown

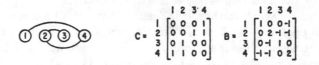

$$C = \begin{array}{c} \\ 1 \\ 2 \\ 3 \\ 4 \end{array} \begin{bmatrix} 0 & 0 & 0 & 1 \\ 0 & 0 & 1 & 1 \\ 0 & 1 & 0 & 0 \\ 1 & 1 & 0 & 0 \end{bmatrix} \qquad B = \begin{array}{c} \\ 1 \\ 2 \\ 3 \\ 4 \end{array} \begin{bmatrix} 1 & 0 & 0 & -1 \\ 0 & 2 & -1 & -1 \\ 0 & -1 & 1 & 0 \\ -1 & -1 & 0 & 2 \end{bmatrix}$$

FIGURE 6.1. A 4-Stage Shift Register

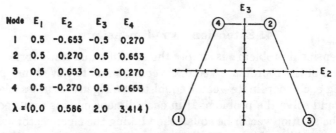

Node	E_1	E_2	E_3	E_4
1	0.5	−0.653	−0.5	0.270
2	0.5	0.270	0.5	0.653
3	0.5	0.653	−0.5	−0.270
4	0.5	−0.270	0.5	−0.653

$\lambda = (0.0 \quad 0.586 \quad 2.0 \quad 3.414)$

FIGURE 6.2. Plot of Figure 6.1 Using Eigenvectors E_2 and E_3

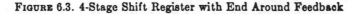

$$C = \begin{array}{c} \\ 1 \\ 2 \\ 3 \\ 4 \end{array} \begin{bmatrix} 0 & 1 & 0 & 1 \\ 1 & 0 & 1 & 0 \\ 0 & 1 & 0 & 1 \\ 1 & 0 & 1 & 0 \end{bmatrix} \qquad B = \begin{array}{c} \\ 1 \\ 2 \\ 3 \\ 4 \end{array} \begin{bmatrix} 2 & -1 & 0 & -1 \\ -1 & 2 & -1 & 0 \\ 0 & -1 & 2 & -1 \\ -1 & 0 & -1 & 2 \end{bmatrix}$$

FIGURE 6.3. 4-Stage Shift Register with End Around Feedback

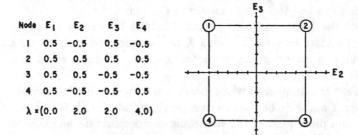

Node	E_1	E_2	E_3	E_4
1	0.5	−0.5	0.5	−0.5
2	0.5	0.5	0.5	0.5
3	0.5	0.5	−0.5	−0.5
4	0.5	−0.5	−0.5	0.5

$\lambda = (0.0 \quad 2.0 \quad 2.0 \quad 4.0)$

FIGURE 6.4. Plot of Figure 6.3 Using E_2 and E_3

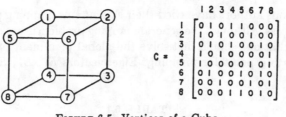

$$C = \begin{bmatrix} & 1 & 2 & 3 & 4 & 5 & 6 & 7 & 8 \\ 1 & 0 & 1 & 0 & 1 & 1 & 0 & 0 & 0 \\ 2 & 1 & 0 & 1 & 0 & 0 & 1 & 0 & 0 \\ 3 & 0 & 1 & 0 & 1 & 0 & 0 & 1 & 0 \\ 4 & 1 & 0 & 1 & 0 & 0 & 0 & 0 & 1 \\ 5 & 1 & 0 & 0 & 0 & 0 & 1 & 0 & 1 \\ 6 & 0 & 1 & 0 & 0 & 1 & 0 & 1 & 0 \\ 7 & 0 & 0 & 1 & 0 & 0 & 1 & 0 & 1 \\ 8 & 0 & 0 & 0 & 1 & 1 & 0 & 1 & 0 \end{bmatrix}$$

FIGURE 6.5. Vertices of a Cube

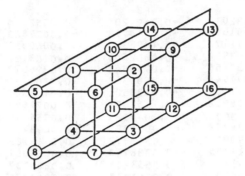

FIGURE 6.6. 4-Dimensional Hypercube

It should be noted that the 2 eigenvalues $\lambda_2 = \lambda_3 = 2$ used in Figure 6.4 are equal. Thus, the "Square" shown in Figure 4 can be rotated any amount about the origin without changing the value of $z = X'BX + Y'BY$ where $X = E_2$ and $Y = E_3$. For example, a 45° rotation clockwise rotation would yield the 2 new eigenvectors.

$$E'_2 = (0, 1, 0, -1)/\sqrt{2}, \qquad E'_3 = (1, 0, -1, 0)/\sqrt{2}$$

which also have eigenvalues $\lambda_2 = \lambda_3 = 2$.

Example 3. Consider the 8-node graph and its connection matrix shown in Figure 6.5. The disconnection matrix B is not shown since it can easily be constructed from C by inspection. The first 4 eigenvectors of B are given by

$$E'_1 = (1, 1, 1, 1, 1, 1, 1, 1)/\sqrt{8}$$

$$E'_2 = (-1, 1, 1, -1, -1, 1, 1, -1)/\sqrt{8}$$

$$E'_3 = (1, 1, -1, -1, 1, 1, -1, -1)/\sqrt{8}$$

$$E'_4 = (1, 1, 1, 1, -1, -1, -1, -1)/\sqrt{8}$$

which have eigenvalues $\lambda_1 = 0$, $\lambda_2 = \lambda_3 = \lambda_4 = 2$. If E_2, E_3 and E_4 are used to reposition the 8 nodes of Figure 6.5 in the X, Y and Z-directions, respectively, we find that the 8 nodes form the vertices of a cube. In this case there are 3 tied eigenvalues. Because of this spherical symmetry the cube can be rotated about the origin without changing the value of the loss function $z = X'BX + Y'BY + Z'BZ$, where $X = E_2$, $Y = E_3$ and $Z = E_4$. If only E_2 and E_3 are used, a 2-dimensional projection of the cube will result.

It is a simple matter to construct graphs with 4 and higher order ties e.g., the 4-dimensional hypercube of Figure 6.6. These situations do not arise much in practice.

Example 4 (Steinberg). Steinberg (1961) has described a 34 node problem in which the objective was to map 34 electronic components into a 4 × 9 rectangular grid (the backboard). Various authors have tried their skill at this mapping problem with varying degrees of success. Although the general mapping problem is not solved in this paper, this author feels that one should first solve the global placement problem and use this solution as a starting point for mapping. Eigenvectors E_2, E_3, E_4 and E_5 are given below:

TABLE 6.1

Eigenvectors for Steinberg Problem

NODE	E2	E3	E4	E5
1	-0.0432012	-0.0694208	0.0052049	-0.0164071
2	-0.0651679	-0.1543718	-0.0077823	0.0296335
3	-0.0571173	-0.1095359	0.0060353	-0.0267673
4	-0.0463958	-0.0802566	0.0103127	-0.0174345
5	-0.0432456	-0.0754842	0.0296073	-0.0188960
6	-0.0371948	-0.0580382	0.0206740	-0.0207142
7	-0.0571307	-0.0633301	-0.0116307	-0.0549437
8	-0.0558912	-0.1151081	-0.0073685	0.0200939
9	-0.0553631	-0.1048123	0.0055148	-0.0274094
10	-0.0524252	-0.0916281	-0.0048700	-0.0069470
11	-0.0093916	-0.0343772	0.0114102	-0.0177897
12	0.0011335	-0.0388740	0.0087087	-0.0193481
13	-0.0241184	-0.0553201	0.0200537	-0.0186262
14	-0.0265103	0.0261925	-0.0049970	-0.0086094
15	-0.0733625	-0.0900708	-0.0422770	-0.2019506
16	-0.1241581	-0.2826598	-0.1615442	-0.7278175
17	-0.1273314	-0.4699478	-0.1974118	0.6141847
18	-0.0796975	-0.2166614	-0.0690472	0.1618230
19	-0.0628896	0.1268366	0.0193869	0.0066971
20	-0.0344391	0.0012820	0.0197461	0.0020348
21	0.0998517	0.0275425	0.4928094	0.0244346
22	0.1002615	0.0190828	0.7292042	0.0356757
23	0.0685555	0.0084281	0.0612788	0.0002766
24	0.6397462	0.0519768	-0.2774723	0.0089584
25	0.4273481	0.0241581	-0.0866789	0.0017937
26	0.4485591	0.0298216	-0.0816017	0.0031537
27	0.0962194	-0.0171636	0.0647895	0.0004566
28	-0.0838568	0.1101782	-0.0325383	-0.0036700
29	-0.1066123	0.2023506	-0.0590732	0.0091841
30	-0.1125110	0.2456635	-0.0769316	0.0328500
31	-0.1507772	0.3917571	-0.1376127	0.0687224
32	-0.0958222	0.2118927	-0.0466467	0.0381619
33	-0.1295471	0.3330358	-0.1028930	0.0697682
34	-0.1274837	0.3169311	-0.0963279	0.0594337
LAMBDA=(	14.9619904	21.5561523	26.0068207	29.4585571)

On examination of the above coordinates, it becomes an easy matter to identify the "top" of the circuit, the "bottom" of the circuit, the "left" and "right", etc. The proximity of elements (e.g., one element lies near another) is perhaps more important than the actual value of their coordinates. In Figure 6.7 Steinberg's data is graphed using E_2 and E_3 for the X and Y-coordinates, respectively. It seems as though, because of the way in which E_2 and E_3 has separated subsets of nodes in this problem (e.g., the nodes 19, 28–34 have been separated from the rest of the circuit), mapping of nodes into discrete locations might be facilitated if algorithms are constructed which incorporate these spacial relationships.

The author could not resist the temptation to try his hand at mapping the 34 components into the required 4×9 grid. In doing so, the graph of Figure 6.7 was followed

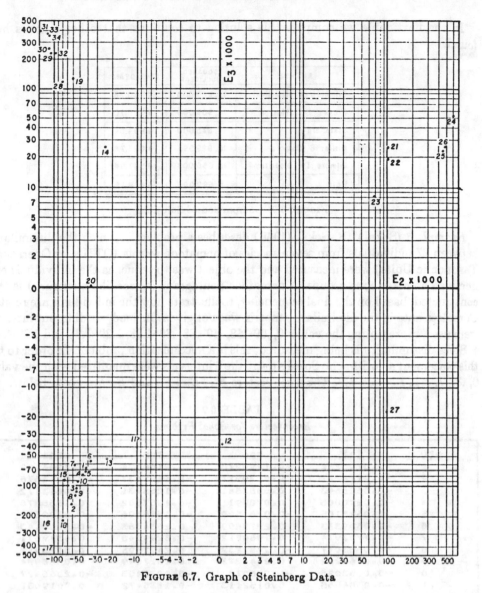

FIGURE 6.7. Graph of Steinberg Data

very closely, along with an occasional look at E_4 to resolve local ties. The grid was turned on end so the mapping was actually done on a 9 × 4 grid. After a little trial and error the result obtained is as follows:

34	33	23	26
31	22	21	25
30	32	27	24
29	14	11	12
19	20	6	13
28	7	1	5
16	15	10	4
17	18	8	9
		2	3

DISTANCES:

EUCLIDEAN = 4419.13

MANHATTAN (Dog-leg) = 5139.00

SQUARED EUCLIDEAN = 9699.00

FIGURE 6.8. Resulting Map from Analog Solution

In the next table we compare this solution to solutions that other authors have found.

AUTHOR	SQUARED EUCLIDEAN	EUCLIDEAN
Gilmore (n^5 algorithm)	11,929.000	4680.36
Graves & Whinston	11,909.000	4490.70
Steinberg	11,875.000	4894.54
Hillier & Connors	10,929.993	4821.78
Gilmore (n^4 algorithm)	10,656.000	4547.54
Hall	9,699.000	4419.13

FIGURE 6.9. Comparison of Solutions

Example 5 (Sokal). R. Sokal (1966) describes a problem in which the dissimilarity between 27 individuals from seven species of nematode worms (OTU's, or Operational Taxonomic Units) were measured and the object was to sequence the individuals on a line into homogeneous groups. A nematode pentagram (a tree-like structure) is then constructed, based on the final sequencing, to illustrate how the individuals are related. Complete data in the dissimilarity matrix was not given. Instead, six different intervals were used, representing the values 0, .09–.48, .49–.88, .89–1.28, 1.29–1.68 and 1.69–2.08.

Since dissimilarity is being measured, this is a maximization problem. In order to test this placement algorithm on Sokal's data, the 6 intervals were quantized with the values 0, .3, .7, 1.1, 1.5, and 1.9. The eigenvectors E_{27}, E_{26}, E_{25} and E_{24} are given below.

TABLE 6.2
Eigenvectors for Sokal Problem

OTU	E27	E26	E25	E24
1	−0.0196284	−0.1299278	−0.2100324	0.2194112
2	−0.0174247	0.0079917	−0.1291837	−0.0143678
3	−0.0526643	0.0083841	−0.1268731	0.0013172
4	−0.0270066	0.0780911	−0.1957602	0.1855876
5	0.0615105	−0.0781528	−0.1544196	−0.3392692
6	−0.0609313	0.2313899	0.2634283	−0.0609143
7	−0.1752974	0.1088812	−0.3299048	0.4549611
8	0.0683077	−0.3481232	0.2492595	0.0473912
9	−0.0163202	0.2641273	0.2327649	0.0386984
10	−0.0153026	−0.0260781	−0.0978453	−0.2366427
11	−0.0384865	0.0152113	−0.1135472	−0.0019631
12	−0.0251980	0.1255451	−0.2313861	0.1865790
13	0.9100240	−0.0595143	−0.1439505	−0.2789607
14	0.0738811	−0.3426431	−0.0258946	0.0992212
15	−0.0153034	0.2720411	0.2676228	0.0000965
16	−0.0695177	0.0152115	−0.1135479	−0.0019636
17	−0.0376632	−0.0838335	−0.1691312	−0.5065462
18	0.0067246	0.1262516	−0.2265009	0.1684259
19	−0.1714458	0.1936861	0.1841411	−0.0356510
20	−0.0473349	−0.3091024	0.1082395	0.1780271
21	0.0769080	0.0807679	−0.1657397	0.1395029
22	−0.1861492	0.2762499	0.2731388	0.0037467
23	−0.1275033	−0.3696333	0.1841339	0.1973473
24	−0.0152448	−0.2019032	0.2215934	0.0420966
25	−0.0149359	0.0154212	−0.1130940	−0.0021011
26	0.0659477	−0.0307920	−0.0886206	−0.2716177
27	−0.1299278	0.2405394	0.2216874	−0.0482839
LAMBDA=(	45.7771912	44.3530121	42.9558411	40.8111115)

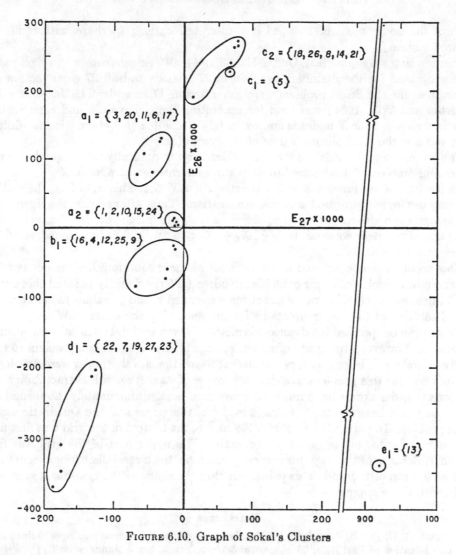

FIGURE 6.10. Graph of Sokal's Clusters

When the above data is plotted in 2 dimensions, 6 distinct clusters appear. Sokal originally defined 7 clusters, which are shown in Figure 6.10. This data indicates that c_1 and c_2 should be combined into one cluster; however, this data is not really accurate since the original data was quantized.

7. Summary

An algorithm has been described for solving quadratic placement problems in r-dimensions. Sums of quadratic forms are either maximized or minimized, depending on the nature of the problem, to yield an optimum solution in any number of dimensions. The r solution (or coordinate) vectors are simple to obtain because they are eigenvectors of a positive semi-definite disconnection matrix B, which is easily constructed from a basic connection matrix C. The n nodes of the graph (i.e., the items which must be positioned) can then be placed at the derived locations.

The solution vectors seem to do a good job separating nodes into local clusters. Therefore, this algorithm may serve as a basis for cluster identification (or separation) problems. Also, it may serve as a basis for mapping problems, where the analog positions of the nodes must be mapped into discrete locations (perhaps with minimum squared motion).

Solution times are quite fast. On the IBM 360/44, all 34 eigenvectors and eigenvalues were generated for the Steinberg problem in 27 seconds and all 27 eigenvectors and eigenvalues for the Sokal problem were generated in 17 seconds. The Jacobi method (Ralston and Wilf, 1964) was used for generating the eigenvalues and eigenvectors. The time required for N nodes is approximately $10(2v + u)N^3$ where v is the addition time and u is the multiplication time of the computer.

The 1-dimensional quadratic form $X'BX$ leads quite naturally into a quadratic programming problem if it is desired to drop the quadratic constraint $X'X = 1$ and replace it with a set of linear inequality constraints, $AX \leq b$. This is, in fact, the method one may use for mapping nodes to discrete locations. Thus, after solving the eigenvector problem, some nodes can be assigned positions (e.g., around the border of a rectangular grid) and then linear constraints (e.g., $X_1 - X_2 \leq 5$, $X_3 \geq 6$) can be imposed to find the remaining positions.

This problem can be reduced to solving a set of linear equations if all the constraints are given in the form of linear equalities. Kodres (1959) originally pointed this out for the 2-dimensional case. He made use of some specified x_i and y_i values to insure a nontrivial solution of the linear equations for the remaining coordinates. When some coordinates can be specified in advance, his method is very useful. It has been the author's experience, however, that quite often not enough is known about the problem to force such constraints. It is for this very reason that the methods of this paper were developed.

The norm for this problem was chosen because of its mathematical tractability. Extensions to other norms have not been considered here. Unfortunately, no formal link has been found between the Quadratic model of this paper and the Quadratic assignment problem. In particular, to obtain Figure 6.8, the bottom of the grid was first filled out with nodes having negative y-coordinates. The upper part of the grid was filled out next. Because of the way this mapping was done, the largest distortions seem to appear at the top of the grid. This points out that the shape of the grid has a great deal to do with the mapping.

References

1. ARMOUR, G. C. AND BUFFA, E. S., "A Heuristic Algorithm and Simulation Approach to Relative Location of Facilities," *Management Science*, Vol. 9, No. 2 (January 1963), pp. 294–309.
2. GILMORE, P. C., "Optimal and Suboptimal Algorithms for the Quadratic Assignment Problem," *Journal of the Society for Industrial and Applied Mathematics*, Vol. 10, No. 2 (June 1962), pp. 305–313.
3. GRAVES, G. W. AND WHINSTON, A. B., "An Algorithm for the Quadratic Assignment Problem," Working Paper No. 110, Western Management Science Institute, University of California, Los Angeles, November 1966.
4. HILLIER, F. S. AND CONNERS, M. M., "Quadratic Assignment Algorithms and the Location of Indivisible Facilities," *Management Science*, Vol. 13, No. 1 (September 1966), pp. 42–57.
5. KODRES, U. R., "Geometrical Positioning of Circuit Elements in a Computer," Conference Paper 1172, AIEE Fall General Meeting, October 1959.
6. LAWLER, E. L., "The Quadratic Assignment Problem," *Management Science*, Vol. 9, No. 4 (July 1963), pp. 586–599.
7. RALSTON, A. AND WILF, S., *Mathematical Methods for Digital Computers*, Vol. 1, John Wiley & Sons, New York, 1964, pp. 84–91.
8. SOKAL, R. R., "Numerical Taxonomy," *Scientific American*, Vol. 215, No. 6 (December 1966), pp. 106–116.
9. STEINBERG, L., "The Backboard Wiring Problem: A Placement Algorithm," *Society for Industrial and Applied Mathematics Review*, Vol. 3, No. 1 (January 1961), pp. 37–50.

Min-Cut Placement

MELVIN A. BREUER

Abstract — In this paper we introduce the concept of min-cut placement algorithms for solving several types of placement (assignment) problems related to the physical implementation of electrical circuits. We discuss the need for abandoning classical objective functions based upon distance, and introduce new objective functions based upon "signals-cut." The number of signals cut by a line c is a lower bound on the number of routing tracks which must cross c in routing the circuit. Three specific objective functions are discussed in detail, and the relationship between one of these and a classical distance measure based upon half-perimeter is derived.

Three min-cut placement algorithms are presented. They are referred to as Quadrature, Bisection, and Slice/Bisection. The concepts of a block and cut line are introduced. These two entities are the major constructs in developing any new min-cut placement algorithm.

Some of the results presented have been implemented, and experimental results are given.

Index terms: placement, PC card design, layout, minimal-cut

I. INTRODUCTION

A. Semi-formal description of placement problem

This paper deals with a classical problem encountered in the physical implementation of circuit cards or chips, referred to as the *placement problem*. The problem is defined, semiformally, as follows. Given a set of elements $\xi = \{e_1, e_2, \ldots, e_n\}$ and a set of signals $\mathcal{S} = \{s_1, s_2, \ldots, s_m\}$. We associate with each element $e \in \xi$ a set of signals $\mathcal{S}_e$, where $\mathcal{S}_e \in \delta$. Similarly, with each signal $s \in \xi$ we associate a set of elements ξ_s, where $\xi_s = \{e \mid s \in \delta_e\}$. ξ_s is said to be a *signal net*. We are also given a set of slots or locations $\mathcal{L} = \{L_1, L_2, \ldots, L_p\}$, where $p \geq n$. The placement problem is to assign each $e_i \in \xi$ to a unique location L_j such that some objective is optimized. The assignment mapping is 1-1 into. Normally each element is considered to be a point, and if e_i is assigned to location L_j then its position is defined by the location of L_j, namely by the coordinate values (x_j, y_j). Usually a subset of the elements in ξ are fixed, i.e., pre-assigned to locations, and only the remaining elements can be assigned to the remaining unassigned locations. Those elements not pre-assigned to locations are called *moveable elements* and those slots not pre-assigned elements are called *open slots*.

There are several different placement type problems which fit the model being described, three of which will be mentioned below.

B. Description of three placement problems

1. Polycell LSI Chips [1,2,3].

In polycell LSI chip technology, the basic element type to be placed is called a cell and consists of units of logic such as gates and flip-flops. These cells are lined

up in rows and are interconnected electrically in the area between rows, which is called an *interconnection channel*. In this model the locations consist of the rows which are available for containing cells. These locations may have to be defined dynamically as cells are assigned to a row, since specific cell types may differ in length.

2. IC Placement on Printed Circuit Cards.

In this model the elements are IC chips and the locations are pre-defined slots on a printed circuit (PC) card.

3. Function Assignment on Printed Circuit Cards.

Many SSI IC chips contain more than one identical function, and there are often many chips of the same type on a card. Hence there is the additional problem of assigning logical functions to specific chips. For example, a specific IC type may contain three 3-input NAND gates. Assume a PC card contains N such chips. If the circuit to be implemented contains N' ($N' \leq 3N$) 3-input NAND gates, then we have the problem of assigning logical NAND gates in the circuit to physical IC chips. Therefore, the placement problem associated with PC cards is typically carried out as follows:

1. Assign functions in circuit to IC chips.
2. Place IC chips on card.
3. Re-assign IC functions to chips.
4. Repeat steps 2 and 3 until some objective is satisfied.

Note that step 3 is a placement problem where the slots are the current locations on the IC's which can realize the type of function being considered. The *function assignment problem* is then to reassign the functions to the IC in some optimal manner.

C. Some distance measures used in placement

For generality we will refer to the area (such as a chip, card or board) containing the slots or locations as the *carrier*.

After the elements are assigned to locations the resulting circuit configuration is routed, i.e., all pins associated with a given signal s must be made electrically common. In this paper we will assume that routing is carried out using the classical two layer etched interconnect technology. Here vertical connections are placed predominantly on one layer and horizontal connections on the second. There are numerous systems and algorithms for carrying out this interconnection process, either automatically or semi-automatically [3,4]. As the density of elements on the carrier increases, the probability of successfully routing all signals decreases. The automatic routing of 100% of the signals in $\mathcal{S}$ is one of the main objectives of the physical implementation portion of a design automation system.

Returning now to the placement problem we see that our actual goal is to assign each element to a location in such a manner as to maximize the routability of the resulting layout. However, this is a very intangible goal, since it is highly

This work was supported in part by the National Science Foundation under Grant ENG74-18647.
The author is with the Departments of Electrical Engineering and Computer Science, University of Southern California, Los Angeles, California.

Reprinted from the *J. Design Automation and Fault Tolerant Computing*, vol. 1, no. 4, pp. 343–362, Oct. 1977, with permission of the publisher Computer Science Press, Inc., 1803 Research Blvd., Rockville, MD 20850.

dependent on a number of factors, one being the basic algorithmic method employed by the router.

Therefore, in an effort to circumvent this problem, researchers have used distance measures to approximate the actual desired goals. The intent here is that if some function of interconnection distance could be minimized, then this would imply less etched interconnect length on the carrier and hence tend to maximize the number of signals successfully routed. The problem here is that during placement it is not known how a signal net is to be interconnected, and hence an accurate distance measure is not easily achieved.

Let d_s be an estimate of the total distance of route required to interconnect net s, and set $N_d = \sum_{s} d_s$. Then the classical placement algorithms attempt to assign elements to locations such that N_d is minimized. Usually d_s is either the length of a minimal spanning tree (MST) for ξ_s, one-half the perimeter of the minimal enclosing rectangle, denoted by ½P, or the Steiner distance. The corresponding total distance using the MST, ½P or Steiner lengths are denoted by $N_d(MST)$, $N_d(P)$, and $N_d(S)$, respectively.

D. Complexity of placement problem

Kurtzberg and Hanan [6] have shown that the placement problem as formulated here is more complex than the quadratic assignment problem, which is of complexity $O(n^3)$, and no known polynomial bounded algorithm has been found to solve this problem. The problem can be solved exhaustively in time proportional to $O(n!)$. Hence numerous heuristic procedures have been proposed to solve this placement problem [6].

E. Motivation and general definition of new objective function

In the algorithm to be proposed in this paper we suggest a new objective function. This objective was motivated by two observations, namely (1) successfully routing of a carrier is dependent on the density of interconnections on the carrier, and (2) some areas of a carrier are more dense than others.

Let c be a horizontal line crossing the surface of a carrier. For an arbitrary signal s, if one or more elements in ξ_s are above c and one or more elements in ξ_s are below c, then when routing signal net ξ_s at least one connection must cross line c. We say that line c is a cut line, and c is said to cut signal s. For a given placement, the value of c, denoted by $v(c)$, is the total number of signals cut by c.

Consider a carrier of width W. If interconnections can be routed at a density of w/inch in the vertical direction, then the carrier is said to have W/w vertical tracks. Assume, in addition, that along some cut line c there are q pads (e.g., IC pins) used to interconnect signals to element pins, and that each pad blocks r vertical tracks. Hence the total available tracks crossing c is $C = (W/w) - qr$. Neglecting signals associated with pads situated along line c, if $v(c) > C$ then the carrier is unroutable. That is, since $v(c)$ is a lower bound on the number of tracks required to route the carrier, C must equal or exceed $v(c)$ if the carrier is to be routable.

Sutherland and Oestreicher [7] have derived expressions for estimating the expected value of $v(c)$ as a function of carrier geometry (slot locations), average signal net size $|\bar{\xi}_s|$, and the location of c.

In general, $v(c)$ is highly dependent on the location of c as well as the geometry of the carrier. For some carriers having a large number of I/O pins located on an edge connector, a cut c near to, and parallel to the edge connector has the maximum value. For other geometries, cuts near the center of the carrier have the highest value.

In the placement procedure to be proposed in this paper we employ a family $\mathcal{C}_V$ and $\mathcal{C}_H$ of vertical and horizontal cut lines. Our objective function is to minimize some function f of the values of each cut line in $\mathcal{C}_V$ and $\mathcal{C}_H$. Placement procedures which minimize the value of such a function are called min-cut placement algorithms.

F. Previous work related to min-cut placement

Druffel and Schmidt [8,9] and Hope [10] have referred to placement procedures that fall into our category of min-cut placement. It appears that Bell Telephone Laboratories [11] has also studied these concepts for placement, but we are not aware of any material they may have published on this subject. Finally, a brochure [12] describing a proprietory DA system also makes reference to using this concept for placement.

II. DEVELOPMENT OF OBJECTIVE FUNCTIONS FOR MIN-CUT PLACEMENT ALGORITHMS

A. Total number of signals cut

The first objective function one might consider is the function

$$N_c(\sigma) = \sum v(c) \qquad (1)$$

where the sum is over all $c \in \mathcal{C}_V \cup \mathcal{C}_H$. To evaluate the properties of this function one must first define the location of the cuts in $\mathcal{C}_V$ and $\mathcal{C}_H$.

Consider a regular carrier geometry where elements are laid out in columns and rows, and where the distance between each column and row is one unit (normalized). We define a canonical set of cut lines as the collection of cut lines between each row and each column.

Theorem 1: Using a canonical set of cut lines, minimizing the objective function $N_c(\sigma) \triangleq \sum v(c)$, where the sum is taken over all cut lines, is equivalent to minimizing the objective function $N_d(P) = \sum d_s$, where the sum is taken over all signal nets and the measure taken for d_s is ½P.

Proof: Consider the signal net shown in Figure 1. Here we see that those elements furthest to the left, right, top, and bottom define the minimal enclosing rectangle. For the situation shown, $P/2 = (i+1)-i+(j+p)-j = p+q$. Note that the number of vertical cut lines between column i and $i+q$ is q, and signal s contributes exactly one unit to each of the cut values associated with these lines. Similarly signal s contributes one unit to each of the p cut lines between row j and $j+p$. Signal s contributes zero to all other cut lines. Hence the total contribution to $\sum v(c)$ due to signal s is $p+q$ which equals $P/2$. ∎

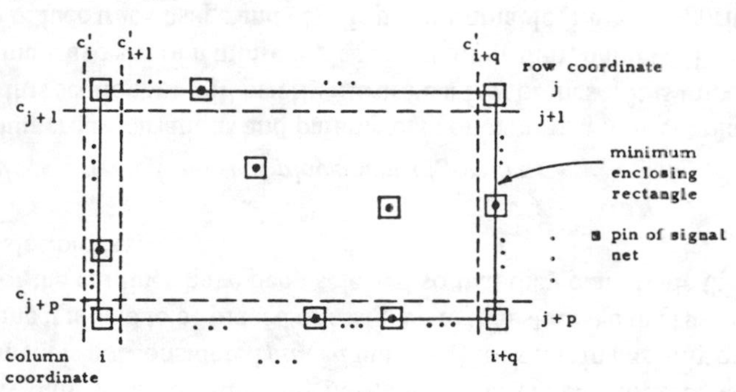

Figure 1. Example signal net.

As one might suspect, minimizing the function $N_c(\sigma)$ is not equivalent to minimizing the distance $N_d(MST)$ nor $N_d(S)$.

Since minimizing the function $N_c(\sigma)$ is equivalent to minimizing the function $N_d(P)$, using the function $N_c(\sigma)$ would not lead to any new results which could not be obtained by existing techniques.

A useful extension of this objective function would be one which tends to minimize some function of both the sum (average) of the cut values as well as the variance of these values. At present we have not found an efficient procedure for achieving such an assignment.

B. *Min-max cut value objective function*

For some carrier technologies a suitable objective function is

$$N_c(mM) = \min(\text{Max}\{v(c)\,|\,c \in \mathscr{C}\}) \tag{2}$$

where $\mathscr{C}$ is a set of cut lines.

Consider, for example, the LSI polycell carrier configuration shown in Figure 2 [2,13]. For simplicity we assume the cell pads are uniformly spaced and that two layers of interconnect are used. For this technology it is important to minimize the

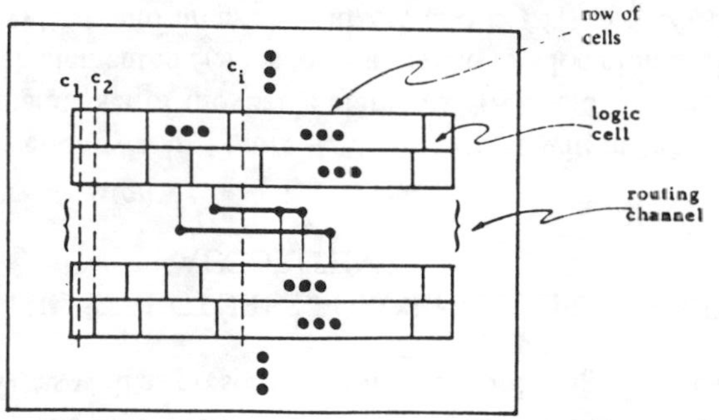

Figure 2. Portion of a polycell LSI chip and a routing channel.

number of horizontal routing tracks used, since by so doing the chip area can be minimized, i.e., the rows of cells can be moved closer to one another. Again, cut lines can be defined, and we recall that a lower bound on the number of tracks required to route a specific routing channel is the value of the maximum cut. Hence, for this type of technology it is desirable to employ the objective function given by (2) rather than minimize N_d or $N_c(\sigma)$. Unfortunately, such an objective function is difficult to satisfy.

In the following example we illustrate the use of the norm $N_c(nM)$.

Consider the polycell row shown in Figure 3, where the distance, for all j, between cells L_j and L_{j+1} is one unit. Assume we are given four vacant locations, L_5, L_7, L_i and L_{i+1}, four unassigned elements A, B, C, D, and three associated signals $\delta = \{1,2,3\}$, where $\xi_1 = \{A, B\}$, and $\xi_2 = \xi_3 = \{C, D\}$. Assume the current cut values, due to previously placed elements, are as shown in the figure, where cut lines occur between cells. If we place C in L_5, D in L_7, A in L_i and B in L_{i+1}, then the total horizontal interconnect-distance will increase by 5, and $v(c_i) = 15$. However, if we place A in L_5, B in L_7, C in L_i and D in L_{i+1}, then the distance will increase by only 3, but $v(c_i) = 16$. If $v(c_i) > v(c_j)$ for all $j \neq i$, then this latter assignment is probably not desirable even though it may correspond to a minimal distance placement.

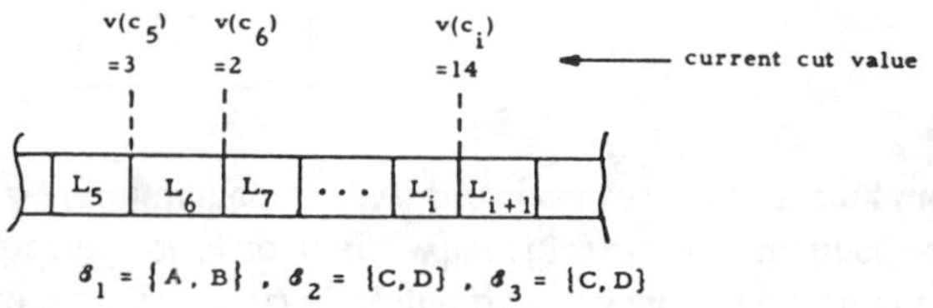

$\delta_1 = \{A, B\}$, $\delta_2 = \{C, D\}$, $\delta_3 = \{C, D\}$

Figure 3. Example where distance measure is inappropriate.

C. *Sequential cut line objective function*

In the preceding discussion we have presented various different min-cut objective functions. Unfortunately, the minimization of these objective functions is difficult to achieve. We will next present an objective function, called a sequential objective function, denoted by $N_c(S)$, whose near minimal value is relatively easy to achieve and which also leads to good placements.

The form of this class of objective function is

$$N_c(S) = \min v(c_{i_r}) \,|\, \min v(c_{i_{r-1}}) \,|\, \cdots \,|\, \min v(c_{i_1})$$

where $c_1, c_2, \ldots, c_r$ are a given set of cut lines, $(i_1, i_2, \ldots, i_r)$ is a permutation ρ on $(1, 2, \ldots, r)$, "$|$" can be read as "subject to," and the expression for $N_c(S)$ is read from left to right. That is, the number of signals cutting c_{ij} is minimized without changing the signals cutting lines $c_{i_{j-1}}, c_{i_{j-2}}, \ldots, c_{i_1}$. Hence $c_{i_1}, c_{i_2}, \ldots, c_{i_r}$ represents an ordered sequence of cut lines, c_{i_1} being the first cut line processed and c_{i_r} the last. By appropriate selection of ρ we can specialize the placement algorithm to fit a specific carrier geometry.

In the remainder of this paper we will discuss and illustrate heuristic procedures for minimizing $N_c(S)$, as well as present several useful orderings for the cut lines.

III. TWO BASIC STRUCTURES FOR MIN-CUT PLACEMENT ALGORITHMS

A. A block and block division

Consider the carrier shown in Figure 4. Assume that within the rectangular subarea A we associate a set of moveable elements $\xi(A)$, and a set of open slot locations $\mathcal{L}(A)$. Each element in $\xi(A)$ is to be assigned to a location in $\mathcal{L}(A)$. Cut line c is seen to divide A into areas A_1 and A_2, and $\mathcal{L}(A_1)$ and $\mathcal{L}(A_2)$ can be determined in a natural way (we assume that a cut line does not go through any location). By an optimal partitioning or assignment of set $\xi(A)$ we mean the determination of sets $\xi(A_1)$ and $\xi(A_2)$ such that $\xi(A_1) \cup \xi(A_2) = \xi(A)$, $\xi(A_1) \cap \xi(A_2) = \varphi$, $|\xi(A_i)| \le |\mathcal{L}(A_i)|$ for $i = 1,2$, and that the total number of signals on the carrier cut by c is minimized. The determination of an optimal partition of $\xi(A)$ is a complex problem and will be discussed in Section V. Note that elements not in $\xi(A)$ must be considered in determining $\xi(A_i)$.

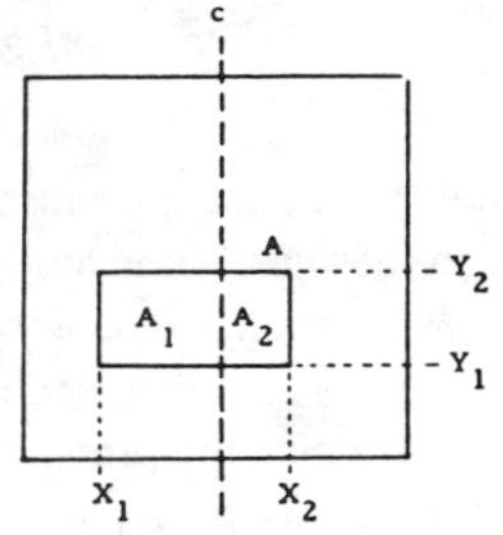

Figure 4. Carrier and a block.

The original area A, along with the slots and elements within A is called a block, and is denoted by B. Note that a cut line divides a block into two new blocks. The process of dividing a block into two new blocks is called block division. A moveable element is not considered placed until the block it is in has only one open slot. At this time it is said to be placed or assigned to that slot. We thus assume that the locations of the cut lines have been selected so that they can divide the carrier down to one slot blocks.

B. Cut and block oriented min-cut placement algorithms

There are numerous variations and parameters concerning a min-cut placement algorithm. In this section we will briefly illustrate a few of these. It is important to note that two main aspects of a min-cut placement algorithm, namely (1) it's main objective is to reduce track usage and (2) it is programmable, i.e. the algorithm can be easily modified to fit the attributes of the carrier geometry being considered. This is done by selecting the appropriate cut sequencing operator (permutation ρ).

Let β be a set of disjoint blocks. Initially let the entire carrier be block B_1, and set $\beta = \{B_1\}$. Then the general form of a min-cut placement algorithm for minimizing $N_c(S)$ is as follows.

Algorithm 1: Cut oriented min-cut placement algorithms for $N_c(S)$.
1. Select a sequence (permutation ρ) for processing the cut lines.
2. Select next cut line c in sequence to process.
3. Assume c cuts across a subset of blocks $\beta' = \{B_{i_1}, B_{i_2}, \ldots, B_{i_t}\}$. Determine an optimal partition of elements within these blocks such that $\nu(c)$ is minimized.
4. In a natural way, form two new blocks from each of the blocks cut by c.
5. If there are no more cut lines to process, or if every block contains at most one moveable element, then exit, else return to step 2. ∎

The sequence in which cut lines are to be processed can be either fixed or adaptive.

Example: Consider the carrier shown in Figure 5a along with the five cut lines. Assume the lines are to be processed in the sequence c_1, c_4, c_5, c_2, c_3. Starting with the initial block β, consisting of the entire carrier and all elements, we process B_1 with respect to c_1, hence minimizing $\nu(c_1)$. Block B_1 is now divided into two new blocks, denoted by B_1 and B_2 (this is a new block B_1). Note that once an element has been assigned to the left (right) of c_1 and B_1 is divided, the element can never be moved to the other side of c_1, no matter where the resulting cut lines occur. We now process B_1 and B_2 (simultaneously) with respect to c_4, producing the four blocks

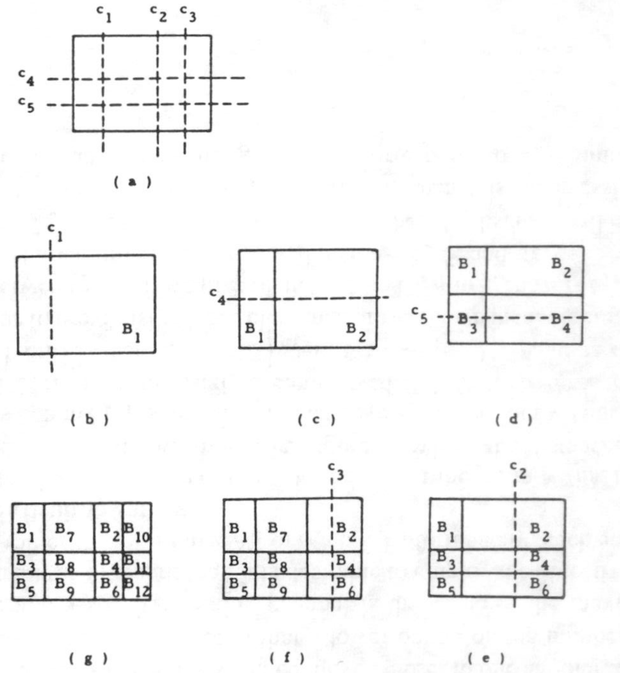

Figure 5. Min-cut algorithm 1 processing B_1 with respect to the sequence c_1, c_4, c_5, c_2, c_3.

shown in Figure 5d. Next we process c_5 which only intersects B_3 and B_4. Note that even though the moveable elements in B_1 and B_2 may not yet have been placed, they are still confined to remain above c_4 and hence can never be moved below c_5. Therefore the actual locations of these elements need not be known when calculating the value of c_5. Continuing the processing of the cuts we obtain the resulting 12 blocks as shown in Figure 5g. ∎

It is clear that this procedure realizes the objective function $N_c(S)$.

In practice, it is not always desirable to execute Algorithm 1 exactly as specified. This occurs for the following two reasons. Consider Figure 5c. In processing c_4 we must process B_1 and B_2 simultaneously. To reduce computation time it is advantageous to process B_1 and B_2 sequentially. That is, first process B_1 and then B_2. When processing B_1 we must ignore the moveable elements in B_2 since they have not yet been assigned to be above or below c_4. Once B_1 has been processed, that is elements assigned to be above and below c_4, we can now use this information in processing B_2. Once the elements in B_2 have been partitioned, we can now reprocess B_1 with respect to c_4 using this information. Hence we can iteratively process B_1 and B_2 until no new partitions occur. At this time the blocks can be divided. When dealing with a large number of blocks, this scheme will significantly decrease computation time. This occurs because the process of optimally partitioning a set of elements is of complexity $O(n!)$ for a block of n elements. Heuristic procedures for carrying out this process are approximately of complexity $O(n^3)$. Hence, for example, rather than process one block of n elements it is usually significantly faster to process, possibly several times, k blocks of size n/k each.

In general, we have adapted the strategy of processing a set β' of blocks sequentially, and iteratively, rather than simultaneously. The disadvantage of this type of processing is that we may achieve a local minimal rather than a global minimal.

When solving the functional assignment problem one must usually deal with many function types (e.g. NAND gates, flip flops, etc.) which are not physically interchangeable in their respective chips. One can reduce CPU time as well as the complexity of the data base and algorithm by dealing with each function type separately, in a sequential fashion, rather than all together. This modification may also lead to nonoptimal results.

Another modification to the procedure just presented deals with the problem of where to cut a block. One useful criteria is to cut a block so that half of the open slots are on either side of the cut line. This is called *bisecting*. Again consider blocks B_1 and B_2 (Figure 5). Assume B_1 has many open elements above c_4 while B_2 does not. Then it may occur that bisecting of B_1 and B_2 cannot be done by the same cut line.

A simple example is shown in Figure 6. Here c_1 bisects B_1 while c_2 bisects B_2. To process these cuts and blocks, we could iteratively process B_1 with respect to c_1 and then B_2 with respect to c_2. When processing B_2 we could use the information about elements above c_1 in B_1 (shaded area), and when processing B_1 we could use the

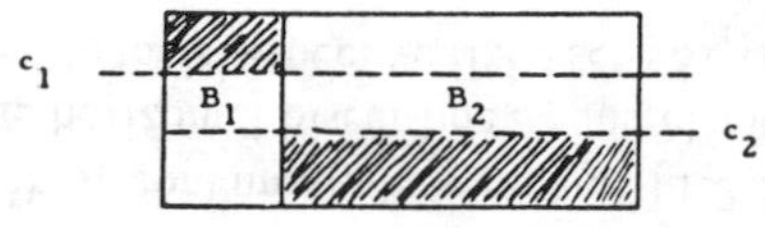

Figure 6. Bisecting blocks B_1 and B_2.

information about elements below c_2 in B_2 (shaded area). Note that we are no longer working with the function

$$\min v(c_2)|\min v(c_1).$$

We see that it is possible to dynamically select cut lines as one proceeds, based upon the location of fixed and open slots in the various blocks encountered. Next we present a revised version of our initial algorithm which allows for this dynamic handling of cut lines. For this algorithm we can no longer write our objective function in a simple concise form.

Algorithm 2: Block oriented min-cut placement algorithm.
1. Select next block B (or set of blocks β') for division.
2. Select cut lines for these blocks.
3. Partition moveable elements in B or β' such that the number of signals cut is minimized. The blocks in β' are sequentially processed, and if desired, iteratively processed.
4. Subdivide blocks, forming a new set of blocks β.
5. If all blocks now contain a single moveable element then exit, else return to step 1. ∎

We will now discuss two aspects of min-cut placement algorithms.

C. Block and cut line selection

Block selection

Associate with each block a level number. Dividing a block at level j produces two blocks at level $j+1$. Initially let the entire carrier be one block at level 1. Blocks can be processed in any order. Two natural orderings are referred to as depth first and breadth first. In breadth first we first divide all blocks at level j before dividing a block at level $j+1$. In depth first, we select as the next block to be divided a block which is (1) divisible, and (2) has a maximum level number.

Cut line selection

Normally the location of each allowable cut line can be specified in advance, e.g. between rows and or columns of slots. The actual problem of selecting a cut line is thus in determining in what sequence to process these lines. The selection criteria is usually a function of the carrier geometry and predicted routing density. For fixed ordering we have found that two types of cut lines are quite effective; they are referred to as a *slice cut* and a *bisection cut*. A *slice cut* c of a block B is a cut line which isolates a fixed number (K) of slots of B to one side of c and the

elements 1 and 2. This figure shows the optimal placement given a 4×4 array of slots on a carrier. For this configuration element i is placed in location L_i, for $i = 1, 2, ..., 16$.

We will now employ the Quadrature placement algorithm to this problem. The results of some of the computations are illustrated in Figure 8. In processing c_1, it is known *a priori* that the connectors a and b are to the left of c_1, and hence are shown in a special set, called the fixed set, and represented by the dotted outline. Elements in fixed sets are used in calculating $v(c)$, but can never be moved from one side of c to the other in an attempt to reduce the value of $v(c)$. The results of processing the cut line c_1 on the block $B_1 = \{L_1, L_2, ..., L_{16}\}$ is shown in Figure 8a. Block division forms blocks B_2 and B_3.

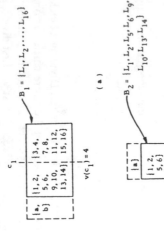
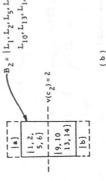
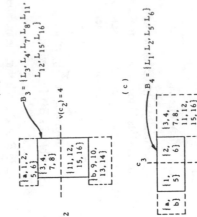

We next bisect block B_2 containing the elements $\{1, 2, 5, 6, 9, 10, 13, 14\}$. Now, element a is above the horizontal bisect line c_2 and element b is below it, hence these elements are put into their respective fixed sets. Since at this time it is not yet

Figure 8. Block evolution in Quadrature Placement Procedure.

---

remaining slots to the other side of c. A *bisection cut* c of a block is a cut line which tends to evenly divide the open slots in B to either side of c. By ''tends to'' we mean that there exists no other cut line c' which more closely divides the open slots.

The order in which cuts are selected can also be determined dynamically by the program itself. One such scheme for doing this is based upon the concept of track utilization. For example, assume the elements have been placed using some initial placement algorithm, and now a min-cut algorithm is to be used to improve this placement. Let U_i be the number of routing tracks on the carrier which cross line c_i. Set $D_i = U_i - v(c_i)$. Then D_i is a lower bound on the number of available tracks crossing c_i. We can then order the c_i such that

$$D_{i_1} \leq D_{i_2} \leq \cdots \leq D_{i_n}.$$

IV. THREE FIXED SEQUENTIAL MIN-CUT PLACEMENT ALGORITHMS

In this section we will illustrate three specific min-cut placement procedures corresponding to Algorithm 2. They differ in the order in which blocks are processed and the type of cut lines employed. We refer to these algorithms as

(1) Quadrature placement
(2) Bisection placement
(3) Slice/Bisection placement.

A. Quadrature Placement Procedure

In this procedure the original block (carrier) B_1 is first bisected by a vertical cut line producing two new blocks B_1 and B_2. These two blocks are then cut by horizontal bisecting cut lines thus forming up to four blocks. These blocks are then cut by vertical and horizontal cut lines. This process is repeated, alternating between vertical and horizontal cut lines, until each element is placed. Note that this procedure processes blocks breadth first. The quadrature placement algorithm is designed to process carriers having a high density of routing in their center. By first processing cut lines in the center of the carrier we attempt to push interconnections away from this region, and hence produce a placement which can be routed with a more uniform density.

We will now illustrate this placement procedure on a simple example.

Example: Consider the problem shown in Figure 7. Here we have 16 moveable elements (nodes 1, 2, ..., 16), two fixed elements (I/O connectors a, b), and 26 signals, represented by line segments. For example, one signal net consists of

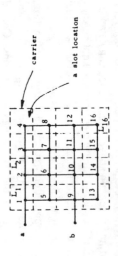

Figure 7. Placement problem showing elements, connections (nets) and optimal placement.

110

have been assigned to a row. The elements are then assigned to columns via vertical bisecting.

Of course, variations of this scheme can be used, such as first creating vertical slices and then using horizontal bisection, or by bisecting before slicing. One can also "grow" slices from both the top (test point connector) and bottom (I/O connector) of a PC board. The technique is best suited to carriers where there is a high interconnect density at the terminals.

V. FORMALIZATION AND IMPLEMENTATION ASPECTS OF MINIMAL CUT PLACEMENT PROCEDURES

A. A block

In this section we present a more detailed version of the three placement procedures described in Section IV. The basic data structure behind a min-cut placement procedure is that of a *block*.

Formally, a *block* B_i consists of a 7-tuple $<Z_i, FE_i, FLOC_i, f_i, ME_i, ALOC_i, F_i>$ defined as follows.

1. $Z_i = (X_1^i, X_2^i, Y_1^i, Y_2^i)$ — the coordinates defining the physical borders associated with B_i (see Figure 5).
2. FE_i — a finite set of fixed elements located within Z_i.
3. $FLOC_i$ — a set of locations each of which is within Z_i, where $|FE_i| = |FLOC_i|$.
4. f_i — a function assigning each element in FE_i to a unique element in $FLOC_i$.
5. ME_i — a set of moveable (unassigned) elements in B_i.
6. $ALOC_i$ — a set of available locations in Z_i for placing the elements in ME_i. We assume that $|ME_i| = |ALOC_i|$. If there are actually more locations than elements, then dummy elements are defined.
7. F_i — a flag which indicates whether or not a block has been processed. This flag is used when a set of blocks must be iteratively processed.

A vertical cut line c of a block B is said to define two pseudo blocks B' and B'', where B' and B'' are defined as follows. In this definition c has x-coordinate X_0, where $X_1 \leq X_0 \leq X_2$.

1. $Z' = (X_1, X_0, Y_1, Y_2)$ and $Z'' = (X_0, X_2, Y_1, Y_2)$.
2. FE' (FE'') is the subset of elements in FE to the left (right) of c.
3. $FLOC'$ ($FLOC''$) is the subset of slots in $FLOC$ to the left (right) of c.
4. f' (f'') is the restriction of f to the domain FE' (FE'').
5. ME' (ME'') is the subset of elements in ME to the left (right) of c.
6. $ALOC'$ ($ALOC''$) is the subset of locations in $ALOC$ to the left (right) of c.
7. $FLAG'$ ($FLAG''$) — flags associated with B' and B'' initially set to 0.

B' and B'' are said to be *pseudo blocks* since their moveable elements can be re-assigned from B' to B'' and vice versa. That is they can be re-assigned from one side of c to the other.

A *vertical assignment* of a block B consists of
a) determining a vertical cut line c for B;
b) constructing pseudo blocks B' and B''; and

known which of the elements $\{3,4,7,8,11,12,15,16\}$ in block B_3 are above or below c_2, these elements are ignored. The result of processing B_2 places elements $\{1,2,5,6\}$ in the block above c_2, and $\{9,10,13,14\}$ in the lower block, and results in $v(c_2)=2$.

We now repeat this procedure on block B_3. Now we know that elements $\{a,1,2,5,6\}$ are above the bisect and $\{b,9,10,13,14\}$ are below the bisect, hence we again construct the two fixed sets as shown. The results are shown in Figure 8c.

We can now go back and reprocess block B_2 using the result that $\{3,4,7,8\}$ are above c_2 and $\{11,12,15,16\}$ are below c_2. Processing this block gives the same result as before, so the repetitive processing of blocks B_2 and B_3 is terminated.

Continuing the division of the remaining blocks we obtain the optimal placement shown in Figure 7. ∎

B. Bisection Placement Procedure

Another placement strategy is to first divide the original block into columns (rows), and then into rows (columns). Here we first find a vertical (horizontal) bisection of the carrier. Then each block is further bisected vertically (horizontally) until finally each moveable element is assigned to a column (row) of slots. Once in a column (row), an element cannot be reassigned to another column (row).

Next each block is bisected horizontally (vertically). Each resulting block is repeatedly divided until finally each element is assigned to a specific row (column). At this time, the elements have been placed. Blocks can be processed either depth first or breadth first.

Again, fixed elements can be used to influence the partitioning.

Note that this procedure is quite similar to the placement algorithm of Alia et. al. [14] who first assigned elements to columns and then permuted elements in columns in order to get a good final placement. The technique can be effectively applied to the polycell channel routing problem discussed earlier, where only the vertical or horizontal cut lines are used. However, this procedure does not guarantee the minimization of the function min (Max $\{v(c)\}$).

C. Slice/Bisection Placement Procedure

In this procedure (see Figure 9), we first divide the initial set of n elements into a set of K and n-K elements, where K>0. Again v(c) is minimized. These K elements represent the bottom row or slice of components on a carrier. This procedure is repeated on the remaining (n-K) elements, again dividing them into a set of K elements, and a set of (n-2K) elements. This process is repeated until all elements

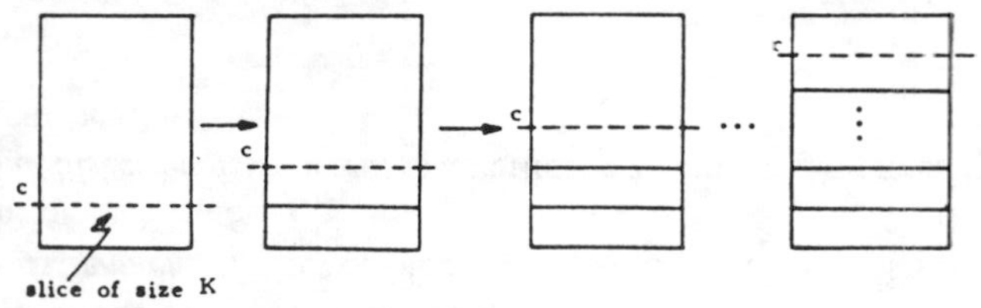

slice of size K

Figure 9. Slice/Bisection Placement Scheme — Growing Horizontal Slices.

c) assigning the elements of ME to the sets ME' and ME'' such that some objective is minimized, such as the number of signals cut by c.

A *vertical assignment* of a set of blocks $\beta=\{B_1, B_2,\dots,B_n\}$ consists of a vertical assignment of each $B_i \in \beta$.

A *vertical bisection assignment* is a vertical assignment generated by a vertical bisection cut line.

B. Partitioning

The key problem in carrying out a vertical assignment is that of assigning the elements in ME to the sets ME' and ME'' such that we minimize the number of signals cut by c. This problem is a generalization of the following partitioning problem. Given a graph G having n nodes, partition the set of nodes of G into two disjoint sets N_1 and N_2 of nodes having n_1 and n_2 elements respectively, where $n_1+n_2=n$, and such that the number of edges between N_1 and N_2 is minimal. There are $\frac{1}{2}\binom{n}{n_1} = \frac{1}{2}(n!/n_1!n_2!)$ assignments of the nodes of N into N_1 and N_2. For large n no computationally efficient procedure for finding an optimal solution to this problem has been found. Kernighan and Lin [15] have described a heuristic procedure which appears to produce very fine results for n large. The complexity of this algorithm is proportional to $n^2\log n$.

This procedure starts with an initial (arbitrary) partition of N into sets N_1 and N_2, and computes sets $A \subset N_1$ and $B \subset N_2$, $|A|=|B|$ such that interchanging A and B reduces the number of edges between the resulting set of nodes to a "near" minimum. The resulting partition of N is thus $N_1'=N_1-A+B$ and $N_2'=N_2-B+A$, where "$-$" and "$+$" refer to set operations. The procedure can then be repeated starting with N_1' and N_2' as the initial partition. This interative procedure is halted according to some user specified termination rule.

To apply the Kernighan-Lin procedure to our problem we must extend it in two ways. First, the graph model they pose corresponds to two element signal nets. However, most signal nets have more than two elements. Hence instead of cutting an edge between two nodes we must deal with cutting signals between sets of nodes. This extension is quite trivial to implement and has been previously discussed by Schweikert and Kernighan [16]. The second extension required is that we must restrict some elements in N_1 and N_2 to be unavailable for interchange. These elements in N_1 and N_2 correspond to our fixed elements. Again this is a simple efension to implement. The extension of the Kernighan-Lin procedure including these two generalizations is referred to as the Generalized Kernighan-Lin procedure. This procedure takes an initial partition $\{F',M'\}$ and $\{F'',M''\}$ of elements, and produce a "min-cut" partition $\{F',M^*\}$ and $\{F'',M^*\}$, where the F's are fixed sets of elements and the M's are moveable sets of elements.

We can now employ this partitioning procedure in the construction of a vertical bisection assignment for a set of blocks $\beta'=\{B_1,B_2,\dots,B_q\} \subseteq \beta$. Next we describe a heuristic procedure for creating this assignment, where the objective is to attempt to minimize the total number of signals cut. This procedure has two modes of operation, namely iterative or noniterative. In the iterative mode the blocks in β' are repeatedly processed until no new reductions in the cut values occurs.

Algorithm 3: Vertical-Bisection Assignment.

Step 1: [Initialization] For $i = 1,2,\dots,q$ set $F_i = 0$ and construct a vertical bisect line c_i for B_i. If c_i does not exist so note. The x-coordinate of c_i is X_0^i. Set $FIRST = 1$. Read in MODE (iterative or non-iterative).

Step 2: [V bisection assignment of B_i] For $i = 1,2,\dots,q$ do the following:

Step 2.1: If c_i exists go to step 2.2, else set $F_i=2$.

Step 2.2: (Construction of F' and F'')

a) For each fixed element e in any block of β to the left (right of X_0^i), pute into F' (F'').

b) For each block B_j such that $X_2^j \leq X_0^i$ ($X_2^j \geq X_0^i$), put all moveable elements into F' (F'').

c) For every block B_j such that $F_j \neq 0$ and X_0 cuts (intersects) B_j, if $X_0^j \leq X_0^i$ set all elements in ME_j' into F', otherwise set all elements in ME_j'' into F''.

Step 2.3: If $FIRST = 1$ go to step 2.4 else go to step 2.5.

Step 2.4: [Construction of initial M' and M''] Partition ME_i into disjoint sets having $|ALOC'_i|$ and $|ALOC''_i|$ elements each, and assign these elements to M' and M'' respectfully. Go to step 2.6.

Step 2.5: [Construction of M' and M''] Set $M'=ME'$ and $M''=ME''$. (Here we use the old value of ME' and ME'' for the initial value of M' and M''.)

Step 2.6: [Optimal partitioning] Construct M^* and M^{**} from $\{F',M'\}$ and $\{F'',M''\}$ using the Generalized Kernighan-Lin procedure.

Step 2.7: Set $ME_i'=M^*$ and $ME_i''=M^{**}$ (i.e. an "optimal" assignment of ME_i has been made such that $v(c_i)$ is minimized). Set $F_i=1$ if $M^*=M'$, else $F_i=2$. ($F_i=2$ implies a new assignment of B_i has been computed, i.e. $M^*\neq M'$, while $F_i=1$ implies no new assignment has occurred.)

Step 3: MODE=iterative?

Yes: Set $FIRST=0$. If $F_i=2$ for some i, set $F_i=0$ for all i and go to step 2, otherwise EXIT.

No: EXIT. ∎

This procedure is finite since whenever a block is processed, either $v(c)$ is reduced in value or else $M'=M^*$. Since $v(c)$ cannot be indefinitely reduced, eventually $M_i'=M^*_i$ for all i and we exit the procedure.

Once β' has been processed by Algorithm 3, each pseudo block is made into a block. For example, the pseudo block B_i' defined by the 7-tuple $< (X_1,X_0,Y_1,Y_2)$, FE_i, $FLOC'_i$, f_i, ME_i, $ALOC_i$, $F_i >$ now defines a new block B_i, defined by the parameters $(X_{j1},X_2,Y_1,Y_2) = (X_1,X_0,Y_1,Y_2)$, $FE_i = FE'_i$, $FLOC_i =FLOC'_i$, etc. We refer to the process of defining pseudo blocks as real blocks as *block division*. Hence, by carrying out block assignment followed by block division, a set $\beta' = \{B_1,B_2,\dots,B_q\}$ is transformed into a new set of blocks $d(\beta') = \{B_1',B_2',\dots,B_m'\}$, where $m \leq 2q$. Note that since some blocks do not have a cut line, m can be less than $2q$.

In a completely analogous manner to the preceding discussions we can define the corresponding block assignment and division procedures for horizontal cut lines.

112

In step 2.4 of algorithm 3 we require that the initial set of elements be partitioned into two disjoint subsets M' and M'', to be used as initial partitions in the Generalized Kernighan-Lin procedure. This initial partition can be formed in numerous ways, one of which is via a random assignment. Kernighan and Lin [15] suggest that one start with several random initial assignments, and select the best result obtained. Because we often have a number of fixed elements, we have found it beneficial to use an initial constructive procedure to create M' and M''. That is, we consider F' and F'' to be ''seed'' elements, and allow elements in ME_i to be assigned to either M' or M'' based upon their connectivity to these elements. Our technique is a simple constructive initial partitioning procedure based upon maximum-conjunction minimum disjunction [17].

Consider the set of blocks $\beta = \{B_1, B_2,...,B_n\}$ shown in Figure 10a. If we vertically slice assign B_n and then divide the resulting block we obtain the set of blocks $\beta = \{B_1, B_2,...,B_{n+1}\}$ shown in Figure 10b. We call this process a vertical slice division of β.

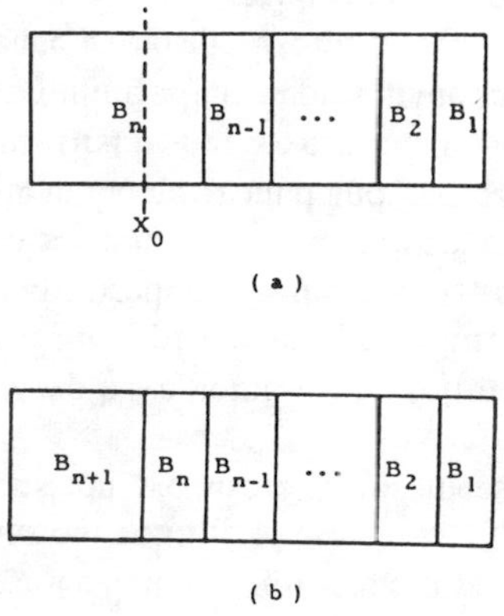

Figure 10. Development of a slice: (a) Before slice, (b) After slice.

Note that the vertical slice assignment of B_n can be obtained by a simplified form of Algorithm 3, namely by setting $\beta' = B_n$, employing a slice cut rather than a bisectional cut, and by setting the MODE to non-iterative.

C. *Min-cut placement algorithms*

Using the previously described procedures for block assignment and division (partitioning), min-cut placement procedures can be easily implemented. Next we give a brief description of our quadrature placement algorithm.

Quadratic Placement Procedure

Step 1: Set $\beta = B_1 = $ initial problem
        Set MODE = iterate

Step 2: If any block in β can be vertically bisected then go to step 3, else go to step 4.

Step 3: Vertically bisect assign β and then divide blocks.

Step 4: If any block in β can be horizontally bisected then go to step 5 else go to step 6.

Step 5: Horizontally bisect assign β and then divide blocks. Go to step 2.

Step 6: If any block in β can be vertically bisected then go to step 3, else exit.

Of course, the order of vertical and horizontal bisecting can be interchanged. In a similar fashion, our Bisection and Slice/Bisection placement algorithms, as well as other min-cut placement algorithms, can be simply described.

VI. EXPERIMENTAL RESULTS

Some of the concepts described in this paper have been implemented, and in this section we will briefly describe a few of the results obtained.

The programs are written in PL/1 and run on an IBM 370/165. They have been used extensively for the layout of PC cards. These carriers are $5'' \times 5''$, have two signal layers, and contain up to 50 IC's layed out in 5 columns of 10 IC's each. Routing is carried out on 25 mil centers.

The physical design of the card is usually carried out in the following order. All steps not modified by the word ''manual'' are done automatically. The process begins with an initial assignment of logic functions (elements) to IC's. This step can be done either manually by the logic designer or automatically by a constructive assignment algorithm. After this step we iteratively use the min-cut placement procedures to first place IC packages and then re-assign elements to packages. The same program can carry out both functions.

At the conclusion of this step a constructive placement procedure is employed to assign each element assigned to an IC to the optimal portion of an IC. For example, a SN 70410 has three 3-input NAND gates, each of which is considered a portion. We do not employ our min-cut algorithms for this assignment since on such a small local level we desire to optimize other routing criteria beside signals cut. This procedure is then followed by a constructive procedure which assigns signal nets to IC pins. Finally the card is automatically routed using a router based upon the Lee algorithm.

In Table 1 we summarize the results for three ''difficult'' cards using our slice/bisection algorithm. For card no. 1 we see that the given manual placement has a half-perimeter ($N_d(P)$) of 838 (inches), and when routed produced 37 failures (signal nets not successfully routed). This same card was then processed using automatic placement. The initial random placement used by the system has a half-perimeter of 888. Three different runs are documented in Table 1. The first employs the min-cut placement algorithm once on the IC's. Run 2 employs this algorithm twice on IC's and once on the elements. Run 3 employs this procedure three times on the IC's and twice on the elements.

Finally, cards 2 and 3 show similar improvement in routing due to this placement program.

To process a card through initial placement of elements to packages, two passes through IC placement and one pass through element re-assignment, gate to IC

Experimental results indicate that min-cut placement procedures perform very well in terms of both high quality and low cost.

There are a number of open problems related to min-cut placement algorithms, a few of which are summarized below.

1. Using block division concepts, devise an algorithm to minimize $N_c(\sigma)$.
2. Devise an algorithm to minimize some function of both the sum of the $v(c)$'s and the variance of the $v(c)$'s.
3. Devise an algorithm to minimize min $(\text{Max } \{v(c)\})$.
4. Extend the concepts so that blocks can be re-combined (block addition) and redivided by a new cut line.
5. Evaluate various cut ordering schemes, and in particular dynamic orderings.
6. Determine a way to predict routability (number of failures) from the placement results, say $N_c(\sigma)$). Using this measure one can determine when to stop the process of placement improvement (i.e., iterating on the element placement process) and enter the routing module.

REFERENCES

[1.] A. Feller, "Automatic Layout of Low-Cost Quick-Turnaround Random-Logic Custom LSI Devices," *Proc. 13th Design Automation Conference*, June 1976, pp. 79-85.

[2.] G. Persky, D. N. Deutsch, and D. G. Schweikert. "LTX-a system for the directed automatic design of LSI circuits." *Proc. Design Automation Conference*. June 1976. pp. 399-407.

[3.] R. L. Mattison, "Design automation of MOS artwork." *Computer*. vol. 7, January 1974. pp. 21-27.

[4.] S. Akers, "Routing," in *Design Automation of Digital Systems: Theory and Techniques*, Prentice-Hall, 1972 (M.A. Breuer, ed.). Chapter 6.

[5.] D. W. Hightower, "The interconnection problem: a tutorial," *Computer*, vol. 7, April 1974, pp. 18-32 and *D. A. Workshop.* 1973.

[6.] J. M. Kurtzberg and M. Hanan, "Placement Techniques," in *Design Automation of Digital Systems: Theory and Techniques*, Prentice-Hall, 1972 (M. A. Breuer, ed.). Chapter 5.

[7.] I. E. Sutherland and D. Oestreicher, "How big should a printed circuit board be?" *IEEE Trans. on Computers*, vol. C-22, May 1973, pp. 536-541.

[8.] L. E. Druffel, D. C. Schmidt, and R. A. Wagner. "A simple, efficient design automation processor," *Proc. 11th Design Automation Workshop*, June 1974, pp. 127-136.

[9.] D. C. Schmidt and L. E. Druffel, "An iterative algorithm for placement and assignment of integrated circuits," *Proc. 12th Design Automation Conference*, June 1975, pp. 361-368.

[10.] A. K. Hope, "Applications of interactive computer techniques and graph-theoretic methods to printed wiring board design," Computer Aided Design Project Application Report 2. University of Edinburgh, July 1973.

[11.] D. C. Schmidt, private communication.

[12.] "PRANCE," (brochure). Automated Systems, Inc. 999 Sepulveda Blvd., El Segundo, California 90245.

[13.] B. W. Kernighan, D. G. Schweikert, and G. Persky, "An optimal channel routing algorithm for polycell layouts of integrated circuits," *Proc. 10th Design Automation Workshop*. 1973. pp. 50-59.

[14.] G. Alia, G. Frosini, and P. Maestrini, "Automated module placement and wire routing according to a structured biplanar scheme in printed boards," *Computer Aided Design*, vol. 5, July 1973, pp. 152-159.

[15.] B. W. Kernighan and S. Lin, "An efficient heuristic procedure for partitioning graphs," *Bell Sys. Tech. J.*, vol. 49, February 1970, pp. 291-308.

[16.] D. G. Schweikert and B. W. Kernighan. "A proper model for the partitioning of electrical circuits," *Proc. 9th Design Automation Workshop.* June 1972, pp. 57-62.

[17.] U. Kodres, "Partitioning and Card Selection," in *Design Automation of Digital Systems: Theory and Techniques*. (M. A. Breuer. ed.). Prentice-Hall, 1972, pp. 173-212.

Card	Manual Placement Automatic Routing		Automatic Placement and Routing	
	½P	# of Failures	½P	# of Failures
1	838	37	Run #1 658 Run #2 646 Run #3 634	19 15 11
2 289 nets	-	28	554	5
3 233 nets	655	35	616	6

Table 1. Placement/Routing Experimental Results.

portion assignment, and IC pin assignment, requires, on the average, 50 CPU seconds at a prime time cost of $25 per minute.

We have found from experience that if ½P is less than 620, which corresponds to a routing density of about 0.35, we obtain almost no failures. Hence we usually iterate the element — IC placement procedure until either we reach a ½P value of less than 620, or else if we see that no significant improvement in the placement is possible. We then go into our routing phase.

In the following discussion we use $P/2$ for measuring the "quality" of a placement.

For our card geometry we have come to the following conclusions.

(1) Starting each of our three placement procedures with a vertical cut rather than a horizontal cut always produces better results, e.g. vertical bisection/slice rather than slice/vertical bisection. Since we have approximately the same number of routing tracks available in the horizontal and vertical directions, the advantage for processing vertical cuts first is that we can line up connected IC's in columns, and thus get the benefit of routing adjacent pins in the same net by short vertical pin to pin segments, and hence avoid the use of a track.

(2) Quadrature usually gives better results than the others two procedures. This is probably due to the fact that we have the highest density of route in the center of the board.

(3) For the second iteration of IC placement, dynamic selection of cut lines usually produces a slightly better results than a fixed ordering. We believe that this is due to the fact that the program can now tailor its strategy on the specific characteristics of a board, and hence work on the high density routing areas first.

Comments

We see that by using the simple structures of blocks and pseudo-blocks, as well as various classes of cut lines such as slices and bisections, numerous min-cut placement procedures can be created. These procedures are potentially quite powerful since (1) they directly deal with an important aspect of routing, namely carrier regions of potentially high routing density, and (2) they are programmable, i.e., they can be easily adapted to a specific board configuration. This is done by selecting the appropriate location and sequence for processing the cut lines and blocks.

Module Placement Based on Resistive Network Optimization

CHUNG-KUAN CHENG AND ERNEST S. KUH, FELLOW, IEEE

Abstract—A new constructive placement and partitioning method based on resistive network optimization is proposed. The objective function used is the sum of the squared wire length. The method has the feature which includes fixed modules in the formulation. The overall algorithm comprises the following subprograms: optimization, scaling, relaxation, partitioning and assignment. The method is efficient because it takes advantage of net-list sparsity and has a complexity of $O[n^{1.4} \log n]$. Another added special feature is that irregular-size modules within cell rows are allowed. Thus the method is particularly useful in standard-cell and gate-array designs. Experimental results on four 4K gate-array placements are illustrated, and they are far superior than manual placements.

I. INTRODUCTION

THE problem of module placement is central to automatic layout design in microelectronics. A good placement is essential to a good layout design. The force-directed method introduced by Quinn and Breuer is a good constructive placement method which leads to initial placement [1]. In their formulation, point modules are assumed, and a force-model is used to determine the state of equilibrium. Hook's Law gives the forces of attraction for modules connected by signal nets, and repulsive forces are used to keep modules apart for those which are not connected. The algorithm amounts to solving a large set of nonlinear equations, which is time consuming. An improvement has been proposed by Antreich, Johnnes, and Kirsch using the same force-directed method but with a more systematic formulation of equations [2].

In this paper we propose a more efficient method based on resistive network analogy of the placement problem. The idea of using network analogy to attack layout problems was first introduced by Charney and Plato [3]. They proposed a method of module clustering according to the sensitivity of a network analogy for the purpose of partitioning. In the present paper, we first solve the optimum placement problem in a systematic way by network analogy. The general formulation of the problem of placing modules on slots involves optimization with nonlinear constraints. However, if only the linear constraints are considered, the problem amounts to solving a linear sparse resistive network. Thus sparse matrix techniques can be used. Because of its computational efficiency, the procedure is repeated in the overall algorithm of partitioning and module assignment. In the formulation, a key feature is that we allow some modules to be fixed in position. Fixed modules could

Manuscript received June 16, 1983; revised January 16, 1984. This work was supported by the National Science Foundation under Grant ECS-8201580 and the Hughes Aircraft Company, Newport Beach, CA.

The authors are with the Department of Electrical Engineering and Computer Sciences and the Electronics Research Laboratory, University of California, Berkeley, CA 94720.

represent I-O pads; but they also play an important role in solving each optimization problem in succession in the overall algorithm.

In Section II we give a detailed formulation of our approach to the problem. Section III is divided into subsections of optimization, scaling, relaxation, and partitioning and assignment. Section IV briefly discuss the problem of multimodule nets, the computation complexity and experimental results. The application to standard-cell and gate-array placements is next presented with computed results on 4K gate-array chips.

II. FORMULATION OF THE APPROACH

Consider the module placement problem in chip layout. With reference to Fig. 1 where movable modules together with fixed modules represent I-O pads are shown. The movable modules are to be placed on slots where horizontal and vertical lines intersect. The net interconnection specification is given by a net list relating nets and modules. We assume first that all nets are 2-module nets and multimodule nets have been preprocessed and replaced with 2-module nets [4]. Futhermore, all modules are assumed to have zero dimension, thus their shape, size, and pin locations are initially ignored.

2.1. Objective Function

Let the two dimensions on the chip be specified by the x and y coordinates. Let there be a total of n modules located at (x_i, y_i), $i = 1, 2, \cdots, n$. Let c_{ij} denote the connectivity between module i and module j, i.e., the number of wires between them. Thus $c_{ii} = 0$. In the literature, the objective function used for placement is usually the sum of wire length. However, because of network analogy, we choose an objective function which is the sum of squared wire length. The ultimate measure of placement is of course ease in routing. In Section V, we will give some comments of our results in terms of the routability of a gate-array chip. Let the objective function be given by

$$\Phi(x,y) = \frac{1}{2} \sum_{i,j=1}^{n} c_{ij} l_{ij}^2$$

$$= \frac{1}{2} \sum_{i,j=1}^{n} c_{ij} [(x_i - x_j)^2 + (y_i - y_j)^2] \qquad (1)$$

where l_{ij} is the Euclidean distance between module i and module j. It is straightforward to show that (1) can be written as follows [5]:

$$\Phi(x,y) = x^T B x + y^T B y \qquad (2)$$

where

$$B = D - C \qquad (3)$$

Reprinted from *IEEE Trans. Computer-Aided Design of Integrated Circuits and Systems*, vol. CAD-3, pp. 218–225, July 1984.

115

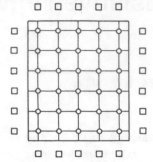

Fig. 1. An example with movable modules to be placed on slots within the square and fixed modules on the boundary represented I-O pads.

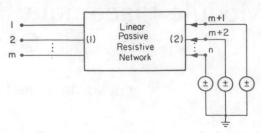

Fig. 2. An n-terminal linear, passive resistive network whose first m nodes are floating and the remaining n-m nodes are connected to voltage sources.

is an $n \times n$ symmetric matrix, $C = [c_{ij}]$ is the connectivity matrix and D is a diagonal matrix whose ith element d_{ii} is equal to $\sum_{j=1}^{n} c_{ij}$.

With the symmetry between x and y in (1), we need to consider only the one-dimensional problem insofar as optimization is concerned. Thus we dispense with the y coordinate until the end of Section III where we discuss partitioning and assignment.

2.2. Network Analogy

For those who are familiar with circuit theory, B in (3) is seen to be of the same form as the indefinite admittance matrix of an n-terminal linear passive resistive network. We will model the coordinate of module i, x_i with a node voltage v_i at node i. The reference coordinate $x = 0$ is thus the datum voltage. The term $-c_{ij}$ in (3) is then the mutual admittance between node i and node j, and $d_{ii} = \sum_{j=1}^{n} c_{ij}$ is the self-admittance at node i.

The power dissipation in the resistive network is

$$P = v^T Y_n v \tag{4}$$

where v is an n-vector representing the node voltage vector and Y_n is the indefinite admittance matrix which is symmetric. Thus the objective function of the placement problem becomes the power dissipation in the linear passive resistive network. It is well known that in a passive resistive network the current distributes itself in such a way that the power is minimum [6]. That is, any other current distributions which are not the solution of the network would have a larger power dissipation. In other words, the problem of solving network equations is equivalent to that of minimizing a well-selected function which represents power.

2.3. Boundary Constraints

Consider the n-terminal resistive network shown in Fig. 2. The first m nodes are floating and their voltages are denoted by an m-vector v_1. The remaining n-m nodes are connected to voltage sources denoted by an $(n-m)$-vector v_2. Thus the coordinates of the n modules are represented by an n-vector

$$v = \begin{bmatrix} v_1 \\ v_2 \end{bmatrix}$$

where the coordinates of the fixed modules are specified by v_2 and the coordinates of the movable modules which are to be determined are represented by v_1.

The network equations are

$$0 = y_{11} v_1 + y_{12} v_2 \tag{5a}$$

$$i_2 = y_{21} v_1 + y_{22} v_2 \tag{5b}$$

where y_{11}, $y_{12} = y_{21}^T$, and y_{22} are the familiar short-circuit admittance submatrices of the indefinite admittance matrix, Y_n. From (5a), we obtain

$$v_1 = -y_{11}^{-1} y_{12} v_2 \tag{6}$$

which gives the solution of the movable modules in terms of the fixed modules and the admittance submatrices.

Remarks

(1) y_{11} is the short-circuit driving-point admittance submatrix of a passive resistive network and is positive definite; thus y_{11}^{-1} always exists.

(2) The solution of (6) must fall inside the region defined by the smallest and largest voltages of the voltage sources. This is because in a passive resistive network, node voltage can not lie outside the range of voltage sources [6].

(3) The dissipated power obtained from the solution in (6) is the minimum among all possible v_1. Any deviation from the solution will result in an increase in power.

2.4. Slot Constraints

Up to now we have not imposed the constraint that the movable modules must be located on slots. This means that the voltage vector v_1 when finally determined must represent a set of prescribed discrete voltages called the *legal values*. Let us designate the prescribed slots in terms of the permutation vector $p = [p_1, p_2, \cdots, p_m]^T$ where p_i is the i-th legal value and m is the total number of the movable modules. Thus the permutation of the m legal values must be assigned to the m modules of v_2. To express this in terms of our optimization problem, let $v_1 = [x_1, x_2, \cdots, x_m]^T$, i.e., x_i denotes the coordinate of module i or the voltage at node i. We claim that the following set of equations represents the constraints on the modules which are required to be on slots:

$$\sum_{i=1}^{m} x_i = \sum_{i=1}^{m} p_i$$

$$\sum_{i=1}^{m} x_i^2 = \sum_{i=1}^{m} p_i^2$$

$$\vdots$$

$$\sum_{i=1}^{m} x_i^m = \sum_{i=1}^{m} p_i^m. \tag{7}$$

The first equation can be written as

$$1^T v_1 = 1^T p \equiv d \tag{8}$$

where 1 is a unit vector and d is a constant which is equal to the sum of the m legal values.

116

Proof:

⇒ Let $[x_1, x_2, \cdots, x_m]$ equal to any permutation of $[p_1, p_2, \cdots, p_m]$, (7) is automatically satisfied.

⇐ Let us define

$$f(x) = \prod_{i=1}^{m} (x + x_i).$$

Then the coefficients of the variable x are multivariable polynomials of $[x_1, x_2, \cdots, x_m]$. Through simple algebraic operations [7] and by using (7), we can show that

$$f(x) = \prod_{i=1}^{m} (x + p_i)$$

which implies that all modules are on slots.   Q.E.D.

III. PROPOSED METHOD

The proposed method can be divided into subproblems of optimization, scaling, relaxation, and partitioning and assignment. The main idea is to solve a simple optimization problem using linear resistive network analogy repeatedly and, in the process, the movable modules are assigned to slots. We shall use node voltages and module coordinates interchangeably in the ensuing discussion, for sometimes it is more intuitive to make statements in terms of voltages; while in dealing with the actual placement problem it is more convenient to use the coordinates.

3.1. Optimization

From (4) and (5), we wish to minimize the power dissipation

$$P = v^T Y_n v = \begin{bmatrix} v_1^T, v_2^T \end{bmatrix} \begin{bmatrix} y_{11} & y_{12} \\ y_{21} & y_{22} \end{bmatrix} \begin{bmatrix} v_1 \\ v_2 \end{bmatrix} = v_1^T y_{11} v_1$$

$$+ 2 v_1^T y_{12} v_2 + v_2^T y_{22} v_2 \tag{9}$$

subject to the complete set of constraint equations in (7). This is clearly not feasible. Therefore, we propose to use only the first equation in (7), which is a linear constraint expressed by (8).

The solution to the optimization problem of minimizing P in (9) subject to the linear constraint in (8) is given by the well-known Kuhn–Tucker conditions:

$$v_1 = y_{11}^{-1} [-y_{12} v_2 + i_1] \tag{10a}$$

where

$$i_1 \equiv \frac{d + 1^T y_{11}^{-1} y_{12} v_2}{1^T y_{11}^{-1} 1} \, 1. \tag{10b}$$

It is seen that the first term in (10a) is precisely that given by (6) for which there is no constraint on slots. The second term of (10a) can be viewed as a correction term which attempts to put the solution on slots. In terms of electric network, we may use current sources to interpret the effect as shown in Fig. 3. Thus we have a linear resistive network with both voltage and current sources. In addition, we know that the network is sparse because of the inherent nature of the placement problem. Using well-known sparse matrix algorithms, we can greatly

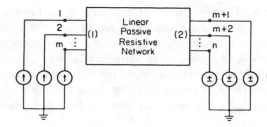

Fig. 3. Network interpretation of the optimization problem with linear constraints.

reduce the computation time in comparison with those that use attraction and repulsion forces [1], [2].

As mentioned previously, because only the linear constraint equation is used, the solution will not put modules on slots. As a matter of fact the result will lead to modules more or less confined to the center of the region. Therefore we must introduce ways to spread the modules so obtained apart and then to bring them to the legal positions. Thus the next step in our overall method is scaling which will distribute the solution more evenly over the entire region. However, let us first analyze the effect of module movements to changes in power dissipation. Let us assume that we deviate away from the solution v_1 of (10) by δv_1 under the constraint of (8), i.e.,

$$1^T \delta v_1 = 0. \tag{11}$$

We claim that the power dissipation is increased by

$$\delta v_1^T y_{11} \delta v_1.$$

Proof:

From (9), we have

$$\Delta P = P(v_1 + \delta v_1) - P(v_1)$$

$$= 2\delta v_1^T y_{11} v_1 + \delta v_1^T y_{11} \delta v_1 + 2\delta v_1^T y_{12} v_2.$$

From (10),

$$y_{12} v_2 = -y_{11} v_1 + i_1$$

and using (11), we obtain

$$\Delta P = \delta v_1^T y_{11} \delta v_1$$

                                                                  Q.E.D.

Furthermore, it is possible to derive an upper bound on the increase in power dissipation in terms of y_M, the largest diagonal element in y_{11}. From the Theorem of Gerschgorin [8], we know that the eigenvalues of y_{11} are not larger than $2y_M$, then

$$\Delta P = \delta v_1^T y_{11} \delta v_1 \leqslant \|y_{11}\| \|\delta v_1\|^2 \leqslant 2y_M \sum_{i=1}^{m} \delta v_i^2 \tag{12}$$

Therefore, the increase in power dissipation has an upper bound which is proportional to the norm of the deviation δv_1.

3.2. Scaling

The result of the optimization with linear constraint leads to solutions which have movable modules concentrated at the center of gravity of all modules. The linear constraint dictates the mean position of the modules. The only forces which attempt to scatter the modules are the fixed modules at the boundary. Therefore, in order to be able to partition the modules we will introduce scaling to redistribute the modules at the expense of increasing the power dissipation. The method used here is to minimize the increase of power ΔP under the constraints which include both the first- and second-order equations in (7). Fortunately, by using the norm of δv_1 in (12), we again can resort to the well-known Kuhn-Tucker conditions.

Let us assume that in the region there are k modules the legal values are given by the permutation vector $[p_1, p_2, \cdots, p_k]$. Let $[x_{01}, x_{02}, \cdots, x_{0k}]$ denote the solution obtained from optimization and let $[x_{n1}, x_{n2}, \cdots, x_{nk}]$ denote the new solution after scaling. Thus our problem is to minimize

$$\sum_{i=1}^{k} (x_{ni} - x_{0i})^2 \qquad (13)$$

under the constraints

$$\sum_{i=1}^{k} x_{ni} = \sum_{i=1}^{k} p_i \qquad (14)$$

and

$$\sum_{i=1}^{k} x_{ni}^2 = \sum_{i=1}^{k} p_i^2. \qquad (15)$$

The solution is given by the Kuhn-Tucker conditions, namely: For $i = 1, 2, \cdots, k$

$$x_{ni} = \frac{x_{0i} - c_0}{a_0} a_n + c_n \qquad (16)$$

where

$$c_n = \frac{1}{k} \sum_{i=1}^{k} p_i \qquad (17)$$

$$a_n = \left[\frac{1}{k} \sum_{i=1}^{k} (p_i - c_n)^2 \right]^{1/2} \qquad (18)$$

$$c_0 = \frac{1}{k} \sum_{i=1}^{k} x_{0i} \qquad (19)$$

and

$$a_0 = \left[\frac{1}{k} \sum_{i=1}^{k} (x_{i0} - c_0)^2 \right]^{1/2}. \qquad (20)$$

Thus c_0 is the mean position of the computed module positions and a_0 is the root mean square amplitude measured from c_0. If a_0 turns out to be very small which approaches zero, so does $x_{0i} - c_0$ in (16); then (16) must be replaced by

$$x_{ni} = c_n. \qquad (21)$$

After scaling the norm of deviation becomes

$$\sum_{i=1}^{k} (x_{ni} - x_{0i})^2 = k [(a_n - a_0)^2 + (c_n - c_0)^2]. \qquad (22)$$

The result of scaling represents an improvement from the

result of optimization as far as module location is concerned, but it gives an increase in power dissipation.

3.3. Relaxation

Before undertaking partitioning and assigning of modules to slots, we need to perform relaxation to be described below. This will greatly improve the preliminary results from optimization and scaling. The method calls for repeated use of scaling and optimization over subregions to be specified by designers. This tends to spread the modules out more evenly over the entire region. It is important to note that when a pertinent subregion is considered, modules outside are always kept fixed.

We propose to choose subregions in the following way: First we start from one end of the region, then the other end and, finally, the middle. After the initial optimization over the entire region, three such steps of scaling and optimization over subregions are carried out. The result tends to settle down and is ready for partitioning. Thus we have as

Input:
A one-dimensional region with coordinates of movable modules x_i, $i = 1, 2, \cdots, m$ obtained from initial optimization in the entire region with specified fixed modules x_i, $i = m + 1, m + 2, \cdots, n$ on the boundary. A parameter β is to be chosen by the designer with $0 < \beta < 50$ percent.

Relaxation:
(1) Order the modules from left to right according to coordinates with the smallest one first.
(2) Choose $[\beta m]$ modules from the left, let other modules be fixed and do scaling in the left β region.
(3) Fix the modules so determined in the left β region and release the modules in the right $(1 - \beta)$ region. Do optimization.
(4) Choose $[\beta m]$ modules from the right, let other modules be fixed and do scaling in the right β region.
(5) Fix the modules in the right β region and release the modules in the left $(1 - \beta)$ region. Do optimization.
(6) Choose $[\beta m]$ modules from the left, let other modules be fixed and do scaling in the left β region.[1]
(7) Set modules in both the left β region and the right β region fixed and release the modules in the center subregion. Do optimization.

Output:
A one-dimensional region with m modules and new coordinates x_i, $i = 1, 2, \cdots, m$.

3.4. Partitioning and Assignment

We next partition the region into two. The ratio of the left subregion to the right subregion is $[m/2]/[m/2]$ where $[k]$ denotes the largest integer which is smaller than k. We do scaling once more for the left subregion and then for the right subregion. As before, in scaling for a subregion we always keep those modules outside fixed. The result of this gives two partitioned subregions together with their associated modules. We next repeat the process for each subregion, i.e., perform independently for each subregion, optimization, relaxation and partitioning.

---

[1] $\lceil k \rceil$ denotes the smallest integer which is larger than k.

In the following we will reinstate the y coordinate to summarize the two-dimensional partitioning and assignment problem.

Input:

A two-dimensional region to be partitioned into rectangles each containing a module, a set of m movable modules together with their coordinates and a set of n-m fixed modules.

Assignment:

(1) Do optimization on both the x coordinate and the y coordinate of the movable modules.

(2) While each region contains more than one module

Do

    Choose the direction of the cut-line.
    Cut the longer side of region.
    List all current regions.
    For each region do partitioning.

(3) For each region, assign the module to the legal value.

IV. DISCUSSION

4.1. Multi-Module Nets

As mentioned in the introduction section, we assume that all nets are 2-module nets in our treatment. Since multi-module nets are always present, we use the following to deal with the them:

(1) At the beginning we use a clique to simulate a multi-module net. If there are r modules in a net, the weight of each edge on the clique is $2/r$.

(2) After the relative module position is determined, we use a chain to connect the modules. Consider the x direction, we order the modules according to their coordinates; we then link the modules by a chain in this order. In the experiments to be discussed in Section IV-4.3, the lengths of wires are measured according to the chain model in x and y directions, respectively.

4.2. Computation Complexity

The optimization algorithm amounts to a linear resistive network computation. Using sparse matrix technique, we have the computation complexity $O(m^{1.4})$ where m is the number of movable modules. The scaling operation is linear with k where k is the number of modules in a subregion.

As to partitioning and assignment, in each iteration, all current regions are divided into two subregions. It takes $\log_2 n$ iterations to make all the necessary divisions. Thus the total complexity is $O(n^{1.4} \log_2 n)$.

4.3. Experimental Results

A 20-module example is designed to illustrate the procedure of our algorithm. As shown in Fig. 7, four modules are fixed on the four corners of the chip, and in the optimal placement every module is connected to the neighboring modules only. Figs. 4–7 illustrate how module locations evolve from the initial placement onto slots. In the figures, module positions are indicated by points with module numbers. The connectivity among modules is represented by linking lines.

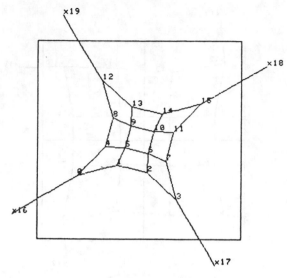

squared   length = 17.629627
manhattan length = 25.4815

Fig. 4. Result of assignment step (1) of the 20 module example. The module positions are optimal under the constraint that the center of gravity of the modules is at the center of the chip.

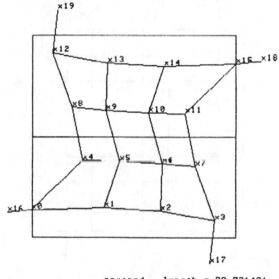

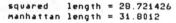

squared   length = 28.721426
manhattan length = 31.8012

Fig. 5. Result of first level partitioning and scaling after relaxation is carried out in the vertical direction.

Fig. 4 is the result of initial optimization. The module positions are optimal under the constraint that the center of gravity of the modules is at the center of the chip. Relaxation is next carried out and the modules spread out over the entire region in the vertical direction. Next partitioning and scaling are used to relocate the modules into two subregions as shown in Fig. 5. Fig. 6 is the result of second level of optimization, relaxation and partitioning using a vertical cut-line; hence, module spread out in the horizontal direction. Fig. 7 is the solution of the final assignment. It is seen that all modules are located on slots.

To evaluate the effectiveness of our method, we use the example given by Steinberg [5], [9]. However, because we always assume that there exist fixed modules in our formula-

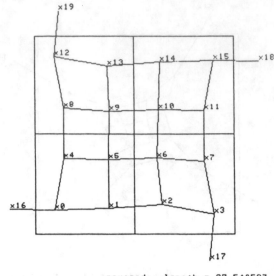

squared   length = 27.540583
manhattan length = 29.6487

Fig. 6. Result of second level partitioning and scaling.

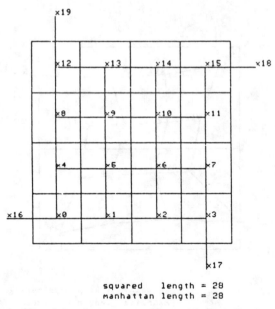

squared   length = 28
manhattan length = 28

Fig. 7. A 20-module placement problem with four fixed modules specified.

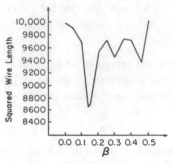

Fig. 8. Result on Steinberg's example, with different values of β in relaxation.

	9	2	
16	8	3	17
10	4	18	5
1	7	13	6
15	20	11	12
28	19	14	27
32	29	23	21
33	30	22	25
34	31	24	**26**

Fig. 9. Result on Steinberg's example, with β equal to 0.125.

Measure	Steinberg	Hall	Cheng-Kuh
Squared length	11875	9699	8596
Manhattan length	N.A.	5139	5316
Euclidean length	4894.54	4419.13	4358.36

Fig. 10. Example 1: Steinberg's example.

Measure	Stevens	Quinn and Breuer	Cheng-Kuh
Squared length	N.A.	8794	7521
Manhattan length	2733	2558	2495

Fig. 11. Example 2: Placement of ILLIAC IV Board IC136.

tion, we modified Steinberg's example by fixing the position of the two modules (34 and 26 shown in Fig. 9) in the bottom row. In relaxation, we tried different values of β to compare the results. These are shown in Fig. 8 where we plot the sum of the squared length for different values of β. It is clear that $\beta = 0$ implies no relaxation. The placement for $\beta = 0.125$ which leads to the smallest squared wire length is shown in Fig. 9. This 34 modules, 172 nets example took 13.1 s CPU time and 169K memory on VAX 11/780 machine. For comparison with Steinberg and Hall, we also calculated the sum of the Manhattan length and the sum of Euclidean length. The table in Fig. 10 summarizes the comparison.

As a second example we use the ILLIAC IV PC Board problem given by Stevens [10]. Again we fix the I–O Pads according

to the placement result of Quinn and Breuer [1]. The result with $\beta = 0.25$ is given in the table of Fig. 11 together with those of Stevens and, Quinn and Breuer. This 136 modules, 432 nets example took 104.2 s CPU time and 480K memory on VAX 11/780 machine.

In both examples it is seen that in terms of our chosen objective function, i.e., the sum of squared length, our method yielded the best results by far.

V. GATE-ARRAY AND STANDARD-CELL PLACEMENT

In gate-array and standard-cell designs, it is often necessary to consider modules of different sizes. Also, pin locations need to be considered in dealing with some large modules. The method introduced above based on resistive network optimization can be easily modified to allow such a fexibility. Consider a large module shown in Fig. 12(a) where pins 1, 2, and 3 shown are to be connected to modules 4, 5, and 6, respectively. We can model the module with voltage sources connected to center of the module shown in Fig. 12(b). Using standard circuit-analysis techniques, we can easily include the fixed voltage sources in the formulation of circuit equations. Weights are then introduced to represent the different sizes of

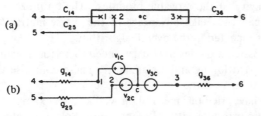

Fig. 12. Network model of a large module with given orientation and pin position. (a) The module with three pins connected to other modules. c is the center point of the module. (b) The network model of the module. Constraints on pin locations is modeled by branch voltage sources.

Chip #	manual placement	automatic placement
1	5.98×10^9	1.31×10^9
2	3.70×10^9	3.37×10^9
3	2.91×10^9	1.83×10^9
4	2.16×10^9	0.94×10^9

Fig. 13. Placement results of four 4K gate-array chips in terms of sum of the squared wire length.

Chip 1: 317 module instances, 676 nets and 2284 pins.
Chip 2: 255 module instances, 916 nets and 2049 pins.
Chip 3: 442 module instances, 983 nets and 3012 pins.
Chip 4: 484 module instances, 1030 nets and 1969 pins.

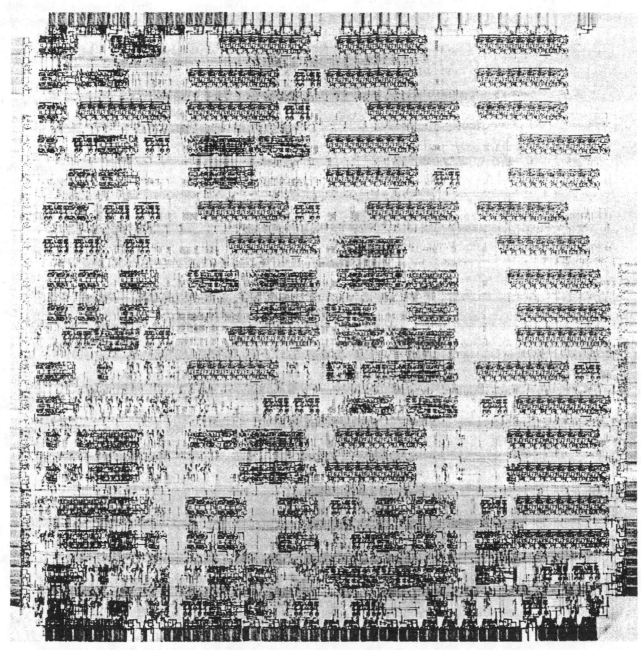

Fig. 14. Picture of the chip placement.

modules in optimization and partitioning. After the last level of partitioning, the modules are scanned row by row. A decompaction scheme is then used to take care of the overlapping modules.

The above method has been implemented and tested with 4K gate-array designs used at the Hughes Aircraft Company. The results of placement of four chips in comparison with that of the manual designs are shown in Fig. 13. Fig. 14 gives

Percentage of Track Demand vs. Supply

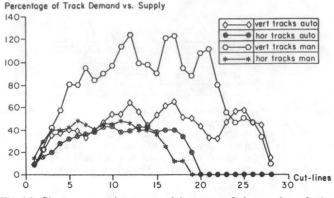

Legend:
- ◇——◇ vert tracks auto
- ●——● hor tracks auto
- ○——○ vert tracks man
- ∗——∗ hor tracks man

Fig. 15. Placement results measured in terms of the number of wires crossing the cut-lines on chip 1.

the solution of automatic placement of chip 1. Fig. 15 exhibits results in terms of wirability of chip 1 for both the automatic and manual designs. The abscissa represents cut-lines in the horizontal and vertical directions. The ordinate represents percentage of routing track demand over supply. It is seen that in the horizontal direction both the manual and automatic placements yield results which are easily routable. However, in the vertical direction, the manual placement requires over a hundred percentage of track demand over supply, which is clearly unroutable; while the automatic placement requires a peak percentage of less than 70. In general, automatic placement tends to distribute the wires more evenly and thus achieves better routability. The cpu time of the above placement is about one minute on an Amdahl V8.

VI. Conclusion

The module placement problem has been formulated in terms of linear resistive network optimization. The objective function used is the sum of squared wire length which corresponds to power dissipation in the resistive network. Fixed modules become nodes with constant voltages. Movable modules then correspond to nodes whose voltage are to be determined. Since modules must be put on slots, a set of constraint equations are imposed on the modules. We consider only the first-order linear constraint which, in essence, fixes the center of gravity of the movable modules. The optimization calculation can thus take advantage of the sparse matrix technique, and is repeated in the overall algorithm.

To assign modules to slots, we need to perform scaling, relaxation, partitioning and assignment. These comprise the overall algorithm.

We have tried our method on well-known examples and compared our results with other methods. So far, we always obtain the least squared wire length as we expected.

An important extension from point modules to modules of different sizes on cell rows has been implemented. The results applied to Hughes Aircraft Company 4K gate-array chips lead to far superior results than that of manual placements. We believe that our method can be used effectively in both the gate-array and the standard-cell designs. Extension to the placement of general cell or building block design also appears possible.

Acknowledgment

The authors wish to acknowledge Dr. B. S. Ting and Dr. B. N. Tien for their assistance and advices.

References

[1] N. R. Quinn, Jr. and M. A. Breuer, "A force directed component placement procedure for printed circuit boards," *IEEE Trans. Circuits Syst.*, vol. CAS-26, pp. 377–388, June 1979.

[2] K. J. Antreich, F. M. Johnnes, and F. H. Kirsch, "A new approach for solving the placement problem using force models," in *Proc. IEEE Int. Symp. Circuits Systems*, pp. 481–486, 1982.

[3] H. R. Charney and D. L. Plato, "Efficient partitioning of components," in *Proc. 5th Ann. Des. Automation Workshop*, pp. 16-1-16-21, 1968.

[4] D. G. Schweikert and B. W. Kernighan, "A proper model for the partitioning of electrical circuits," *Proc. 9th Annual Design Automation Workshop*, pp. 56–62, June 1972.

[5] K. M. Hall, "An *r*-dimensional-quadratic placement algorithm," *Management Sci.*, vol. 17, no. 3, pp. 219–229, Nov. 1970.

[6] C. A. Desoer and E. S. Kuh, *Basic Circuit Theory*. New York: McGraw-Hill, 1969.

[7] N. Jacobson, *Basic Algebra*. San Francisco, CA: Freeman, pp. 133–135, 1974.

[8] R. S. Varga, *Matrix Iterative Analysis*. Englewood Cliffs, NJ: Prentice-Hall, 1962.

[9] L. Steinberg, "The backboard wiring problem: A placement algorithm," *SIAM Rev.*, vol. 3, no. 1, pp. 37–50, Jan. 1961.

[10] J. E. Stevens, Fast Heuristic Techniques for Placing and Wiring Printed Circuit Boards, Ph.D. dissertation, Comp. Sci. Dep., Univ. of Illinois, 1972.

An Efficient Algorithm for the Two-Dimensional Placement Problem in Electrical Circuit Layout

SATOSHI GOTO, MEMBER, IEEE

Abstract—This paper deals with the optimum placement of modules on a two-dimensional board, which minimizes the total routing length of signal sets. A new heuristic procedure, based on iterative improvement, is proposed. The procedure repeats random generation of an initial solution and its improvement by a sequence of local transformations. The best among the local optimum solutions is taken as a final solution. The iterative improvement method proposed here is different from the previous one, in the sense that it considers interchanging more than two modules at the same time and examines only a small portion of feasible solutions which has high probability of being better. Experimental results show this procedure gives better solutions than the best one up to now. The computation time for each local optimum solution grows almost linearly with regard to the number of modules.

I. INTRODUCTION

IN THE design of a large scale electronic system, one of the most important problems is the layout problem, which involves the placement and interconnection of a large number of components and subsystems to satisfy a given specification. The problem is of particular significance in the present day design of VLSI chips, as a huge number of components is mounted on one chip and interconnected more complicatedly. Because of its complexity, the layout problem is usually divided into two stages, namely: "placement" and "routing." In the placement stage, the locations of individual modules are decided on a board to facilitate routing. In the routing stage, the interconnections on external leads, called pins, are made in such a manner as to satisfy various physical constraints.

This paper will deal with only the placement problem. The final layout design goal is to achieve 100-percent routing. However, it is far too difficult to consider the final goal itself in the placement stage. The usual goal will be adopted of minimizing the total routing length defined in an appropriate way. From the computational complexity point of view, this problem is considered to be a hard combinatorial problem in the sense that the computation time required to obtain the real optimum solution increases in exponential order when the problem size, i.e., number of modules, increases. Hence, algorithms based on heuristic rationales have to be employed.

Until now, several algorithms have been proposed. All of these algorithms are either constructive or iterative [1]–[7]. Hanan and Kurtzberg survey known algorithms in [3] and showed some experimental results in [7]. They concluded that the force-directed pairwise relaxation algorithm, called FDPR, operating on the associated quadratic assignment problem, yielded the best result in near minimum time.

This paper proposes an algorithm which is more efficient than the best one up to now. The algorithm, as usual, consists of two phases, the initial placement phase, called SORG, and the iterative improvement phase, called GFDR. The Sub-Optimum-Random-Generation ($SORG$) method sequentially selects unplaced modules according to their connectivity to other modules and places them so as to minimize the total routing length. The SORG method differs from those in pertinent literature [3], in that it produces a good initial solution randomly by local random selections. The Generalized-Force-Directed-Relaxation ($GFDR$) method performs iterative improvement to reduce the total routing length. It is different from the FDPR method in [7], which tries to interchange only a pair of modules, whereas the GFDR method interchanges more than two modules at the same time. The GFDR method is considered to be a generalization of the FDPR, combined with the technique of realizing a λ-optimum solution proposed in [8].

II. PRELIMINARIES

Consider a two-dimensional board on which modules are to be placed. The *board* is characterized in terms of a finite array of slots. A matrix location, or *slot* may be represented by a point in an *x–y* coordinate system. *Modules* are the entities which are to be assigned to slots on the board. It is required that one module can occupy one and only one slot, and on each slot not more than one module is allowed.

Modules contain pins for connection by physical wires to form *signal sets*. In this study, the pins on the modules are ignored and the distance is measured from the center of the module. Hence, a signal set becomes a subset of modules, and a signal set specification defines the connection of all modules on a specific board.

Let the board have *m* rows and *n* columns. It may be assumed that the number of modules is equal to $m \times n$

Manuscript received November 30, 1979; revised May 29, 1980.
The author is with Central Research Laboratories, Nippon Electric Company, Ltd., Kawasaki, 213 Japan.

Reprinted from *IEEE Trans. Circuits Syst.*, vol. CAS-28, pp. 12–18, Jan. 1981.

without loss of generality, because dummy modules, which are not connected, can always be introduced. The distance between two adjacent slots is defined as, either vertical or horizontal, one unit length.

The routing length of a signal set is defined as half-perimeter of the smallest rectangle, which encloses the modules in the signal set [5], [6]. The placement problem is now defined in the following.

> Given a set of modules with signal sets defined on subsets of these modules and a set of slots, place all the modules on the slots so that the total routing length over all signal sets is minimum.

III. MEDIAN OF A MODULE

Preceding a placement algorithm, it is necessary to introduce a concept, called median of a module, and present an algorithm to find it, since the present placement algorithm depends on it. Let us consider a board on which every module is placed. Pick one module, denote it by M. Move only module M on the board, while the other modules remain fixed. The routing length of a signal set does not change, as long as the signal set is not connected to module M. Therefore, consider the signal sets connected to module M only and the sum of the routing length of these signal sets. This value is referred to as *the routing length associated with module M*.

Now, define the median of module M. Module M may be placed on $m \times n$ different positions. *The module M median* is defined as a position where the routing length associated with module M is minimum. Next, sort all the routing lengths associated with module M with respect to the module M position in ascending order. In this order, choose ϵ elements from the minimum one. The set of these ϵ positions is defined as the *ϵ-neighborhood for module M median*.

Now consider how to find a median of a module. Let i ($i = 1, 2, \cdots, r$) designate a signal set which is connected to module M. For each signal set i, consider the smallest rectangle which encloses the module in the signal set. Here, module M is excluded from the signal set when forming the rectangle. Let us denote the rectangle by l_i and its figure by parameters (x_i^a, y_i^a) and (x_i^b, y_i^b), where x_i^a and x_i^b are the minimum and maximum values in the x-direction on the rectangle, respectively. The same definitions are pertinent for y_i^a and y_i^b in the y-direction.

The routing length associated with module M, which is required to place module M in position (x, y), is given by

$$F(x, y) = \sum_{i=1}^{r} (f_i(x) + f_i(y))$$ (1)

where

$$f_i(x) = \begin{cases} x_i^a - x, & x < x_i^a \\ 0, & x_i^a \le x \le x_i^b \\ x - x_i^b, & x > x_i^b \end{cases}$$ (2)

$$f_i(y) = \begin{cases} y_i^a - y, & y < y_i^a \\ 0, & y_i^a \le y \le y_i^b \\ y - y_i^b, & y > y_i^b. \end{cases}$$ (3)

The problem is to find a pertinent position (x, y) on the board, such that $F(x, y)$ is minimized. Since the function $F(x, y)$ has a separable form with respect to variables x and y, $F(x, y)$ can be calculated independently from each other for x and y. The y-component can be found in the same way as the x-component. Thus only the x-component will be discussed in the following.

Equation (2) is transformed as

$$f_i(x) = \frac{1}{2} \left\{ |x - x_i^a| + |x - x_i^b| - (x_i^b - x_i^a) \right\}.$$ (4)

The problem is reduced to finding a position where

$$\sum_{i=1}^{r} (|x - x_i^a| + |x - x_i^b|)$$

is minimum, since $x_i^b - x_i^a$ is a constant value. The value $|x - x_i^a| + |x - x_i^b|$ indicates the sum of the distances from x to x_i^a and x to x_i^b, thus the problem is to find a point x such that the total sum of the distances from x to each point x_i^a, x_i^b ($i = 1, 2, \cdots, r$) is minimum.

In general, it is necessary to solve the following problem. There are α_i points on position i ($i = 1, 2, \cdots, n$) along the line. Find a position x on the line such that

$$\sum_{i=1}^{n} \alpha_i |x - i|$$

is minimum. This problem is a particular case of finding an absolute median of a graph presented in [9]. The present problem, treating only a linear tree instead of a general graph, can be easily solved by using the following theorem.

> *Theorem 1:* Point q is the median if
>
> $$\sum_{i=1}^{q-1} \alpha_i \le \frac{N}{2} \le \sum_{i=1}^{q} \alpha_i$$ (5)
>
> holds, where $N = \sum_{i=1}^{n} \alpha_i$.

Fact 1: The total distance decreases monotonically from point 1 to a median. From a median to point n, it also increases monotonically. This fact is quite useful when more than one median are interested.

Fact 2: The x-component and y-component of a median can be calculated independently from each other. Each component has the characteristics mentioned in Fact 1. Therefore, the median on the two-dimensional board can be found easily. When interest is in finding the kth minimum, instead of the first minimum, it is possible to use the efficient algorithm reported in [10]. Thus the ϵ-neighborhood of a module can be easily obtained.

124

IV. A Placement Algorithm

The problem treated here is considered to be NP-complete[1] in the sense of Cook and Karp [11]. Hence, some heuristic procedure should be applied in order to obtain a nearly optimum solution in reasonable computation time.

There are, in general, two types of heuristic methods for this kind of problem. One is a *constructive method*, which obtains a solution using heuristic rules, often in sequential, deterministic manner. The other is an *iterative improvement method*, which improves a solution by means of local transformations.

The algorithm proposed here is composed of these two methods. The constructive one is called *SORG*, and the other one is called *GFDR*. The algorithm proceeds in the following way.

A Placement Algorithm

Step 1: Randomly generate a good initial solution by SORG.

Step 2: Improve the initial solution to obtain a locally optimum solution by GFDR.

Step 3: If available computation time has not been exhausted, go to *Step 1*; otherwise choose the best one among the locally optimum solutions obtained in *Step 2*.

This kind of hard combinatorial problem usually has many locally optimum solutions. One locally optimum solution might not be good enough. Therefore, repeating random generation of an initial solution and its improvement, we obtain a set of locally optimum solutions. The best one among them is chosen as a final solution.

A. SORG Method

SORG method selects modules, one at a time, based on an evaluation function which measures signal set connectivity to modules already or not yet selected, and then decides which slot the selected module will be placed on. Once a module is fixed into a position, it is not moved.

The conventional evaluation function, called IOC in [5], is adopted here. The module with the first or the second highest IOC is the logical candidate to be selected. Each of them is selected at *random* as the module to be put in place. The selected module is placed on the slot which yields the minimum total routing length among available slots. All available slots need not be examined here. Only a small part of them is examined, by calculating the ϵ-neighborhood of the median.

B. GFDR Method

Let S be the set of all feasible solutions and let x be a feasible solution, $x \in S$. Consider the neighborhood of x, denoted by $X(x)$, which is a subset of S. In the first step,

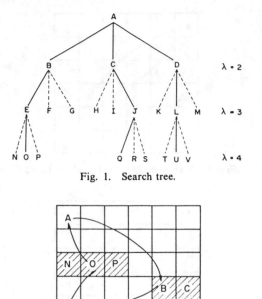

Fig. 1. Search tree.

Fig. 2. Trial interchange of modules ($\lambda = 4$).

x is set to a feasible solution and a search is made in $X(x)$ for a better solution x' to replace x. This process, which is referred to hereafter as a *local transformation*, is repeated until no such x' can be found. A solution is said to be *locally* optimum if x is better than any other elements of $X(x)$.

A lot of definitions may be considered for the neighborhood of a solution. In [14], the set of solutions transformable from x by exchanging not more than λ elements is regarded as the neighborhood of x. A solution x is said to be λ-*optimum*, if x is better than any other solutions in the neighborhood in this sense.

Although the λ-optimum solution gets better as λ increases, the computation time easily goes beyond the acceptable limit, when an exhaustive search is performed for large λ. The following method does not examine all the elements in the neighborhood, nor does it guarantee a λ-optimum solution. However, it is very efficient in the sense that it can be applied for a large value of λ with limited searches in the neighborhood.

The present search procedure is illustrated along with the search tree shown in Fig. 1, where each node represents a module and each edge represents a trial transformation. The root node of the tree A is a module chosen to initiate the trial interchange, it is referred to as the *primary module*. A path connecting node A and one of the other nodes defines a possible interchange. For example, the path $A \rightarrow B \rightarrow E \rightarrow K$ refers to the trial interchange of four modules, as shown in Fig. 2. Here, module A is placed on the slot of B, B is placed on E, E on K, and K on A, in a round robin sequence. Although this transformation is a quadruple interchange, it includes a pairwise interchange as a special case, i.e., paths $A \rightarrow B$, $A \rightarrow C$, and $A \rightarrow D$, as shown in Fig. 3. Value λ indicates the number of modules to be interchanged.

---

[1]This is not yet proved. However, the quadratic assignment problem, which is a particular case of the present problem, is proved to be NP-complete [12].

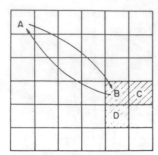

Fig. 3.   Trial interchanges of modules ($\lambda = 2$).

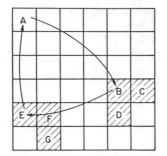

Fig. 4.   Trial interchange of modules ($\lambda = 3$).

The search tree is examined as follows. In this example, ϵ is fixed as 3. First, module A is interchanged with either one of the modules on trial in the ϵ-neighborhood of A median ($\lambda = 2$). The ϵ-neighborhood modules are B, C, and D, thus pairwise interchanges between A and B, A and C, and A and D are performed. (See Fig. 3.) The trial interchange is accepted if it results in the reduction of the total routing length. If more than one reduction occurs in these transformations, the interchange with the greatest reduction is selected for acceptance. If no interchange contributes to reducing the total routing length, the next step ($\lambda = 3$) is initiated.

Module A is placed on the slot of B. Then the median of B and its ϵ-neighborhood are calculated. In this case, the ϵ-neighborhood module are E, F, and G. Thus interchanges $A \rightarrow B \rightarrow E$, $A \rightarrow B \rightarrow F$, and $A \rightarrow B \rightarrow G$ are tried. (See Fig. 4.)

These trial interchanges are accepted if one of them results in the reduction of the total routing length. Otherwise, consider the three interchanges of paths $A \rightarrow B \rightarrow E$, $A \rightarrow B \rightarrow F$, and $A \rightarrow B \rightarrow G$, and choose the best one (least total routing length) for the later tree search. Here, $A \rightarrow B \rightarrow E$ is chosen, and $A \rightarrow B \rightarrow F$ and $A \rightarrow B \rightarrow G$ are omitted.

The solid lines in the tree search shown in Fig. 1 indicate which searches are to be continued. Broken lines show the searches which are to be terminated. Therefore, no more search efforts are made along paths $A \rightarrow B \rightarrow F$ and $A \rightarrow B \rightarrow G$. There is only one solid line under any node, except for root node A. Triple interchanges are performed for the other ϵ-neighborhood modules, C and D, of root node A. Tree search will be continued following J or L, whereas no search will be accomplished through H, I, K, and M. The tree search is continued, i.e., a path from node A is extended as long as λ is no greater than λ^*, which is given as a parameter.

TABLE I

	Example 1	Example 2	Example 3	Example 4	Example 5
#Modules	67	108	116	136	151
#Internal Modules	52	93	101	121	136
#External Modules	15	15	15	15	15
#Signal Sets	132	277	329	432	419
Av. #Signal Sets per Module	6.32	6.41	7.30	7.86	5.98
Av. #Modules per Signal Set	3.43	2.78	2.66	2.73	2.35
Range of Signal Set Size	2 - 9	2 - 9	2 - 9	2 - 9	2 - 9
Placement Grid Size	5 x 15	8 x 15	8 x 15	10 x 15	11 x 15

Each selection of a primary module is identified with an interchange *cycle*. Cycles are iterated until there is no reduction in the total routing length.

The GFDR method is different from the FDPR method in [7] which tries to interchange only a pair of modules. The GFDR, on the other hand, tries interchanges of more than two modules at the same time. If the interchanges are performed randomly in the GFDR, like in the PI method of [7], the gain will not compensate for the comparative great increase in computation time. The GFDR method examines and performs trial interchanges for subsets of modules which have a large possibility for improvement. This limited trial interchanges enable us to find a good locally optimum solution quickly.

Values ϵ and λ^* greatly affect the computation time and the total routing length. For a greater value of ϵ and λ^*, a better solution is obtained at the expense of computation time. In order to find a better solution within a given amount of computation time, the key problem is to determine how to set values ϵ and λ^*. In the following section, several experimental results are shown to point to the best values of ϵ and λ^*.

V. EXPERIMENTAL RESULTS

In order to check and compare the results of the present algorithm with others, the algorithm was programmed and tested for five examples in a real problem. The program was written in FORTRAN and run on NEAC ACOS-77/700.

Example logic graphs were obtained from [13]. They represent five boards of the ILLIAC IV computer. The actual dimensions of ILLIAC IV boards are used as the grid size, i.e., there are 10×15 internal card locations and 15 I/O connector locations in one row on the bottom of the board. Grid size was reduced to the value shown in Table I. Statistics for the example logic graphs are also shown in Table I.

Experiment 1

The placement problem treated here has many locally optimum solutions. Thus different initial solutions lead to

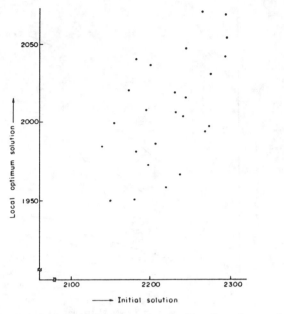

Fig. 5.   Relation between initial solutions and locally optimum solutions.

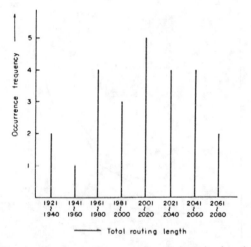

Fig. 6.   Occurrence frequency for total routing length.

different solutions followed by an iterative improvement. It has been debated in the literatures whether it is better to use a random start or to use a constructive-initial solution, followed by an iterative-improvement. Experimental results in [6] and [7] showed that the constructive-initial start approach is superior to the random start in both solution value and computation time. Taking this fact into account, this constructive method generates *good solutions randomly* as initial solutions. In order to know the relation between the initial solution and its locally optimum solution, Example 5 was provided as a test with $\epsilon = 4$ and $\lambda^* = 4$, and 25 different initial solutions were generated by SORG and improved by GFDR.

Fig. 5 shows the relation between initial solutions and their locally optimum solutions. A better initial solution does not always result in a better locally optimum solution. This result does not recommend generating many good random starts and following *only the best one* with iterative improvement. But it recommends repeating ran-

dom generation of an initial solution and its improvement to obtain *a set of locally optimum solutions*. The best one among them is chosen as a final solution.

Fig. 6 shows the occurrence frequency for the total routing lengths, which may be approximated by a normal distribution. The average value is 2007 and the standard deviation is 39.

Experiment 2

A simple transformation, such as pairwise interchange with smaller ϵ in general, does not yield better locally optimum solutions than more complicated ones with larger ϵ or λ^*. However, from the computational complexity point of view, the latter one takes much more time than the former one. In order to explore the influence of value ϵ or λ^* on computation time and the solution, the program was run by changing the value of ϵ or λ^*.

In Fig. 7, total routing length versus computation time curves are plotted for five values of ϵ operating on Example 5. Here, λ^* is fixed as 4. Each mark on the curves represents one cycle of GFDR method. The curve with smaller ϵ results in steeper descent, whereas it does not converge to a better solution. On the other hand, the curve with larger ϵ converges to a better solutions at the expense of computation time. A fairly small value of ϵ, i.e., $\epsilon = 4$ or 5 is sufficient to lead to good solutions.

In Fig. 8, total routing length versus computation time curves are plotted for five values of λ^* operating also on Example 5, where $\epsilon = 4$. The same discussion can be made for the value of λ^*, as for ϵ. The value $\lambda^* = 3$ or 4 seems to be enough to have good solutions.

The aim of the overall scheme is to generate as many locally optimum solutions as possible within some time interval. Then, the best one among them is chosen. Let P be the probability that the cost of the locally optimum solution is less than V and let T_0 be the running time to produce the local optimum solution. Then, the best locally optimum solution produced within time interval T has a probability

$$\hat{P} = 1 - (1 - P)^{T/T_0} \qquad (6)$$

of being better than V [14], [15].

Random start were repeated as many times as possible within $T = 30$ min for a value of ϵ. Here, Example 5 was provided as a test and λ^* was fixed as 4. Let $\hat{P}'_\epsilon$ be the probability of (6) associated with ϵ, and calculated $\hat{P}'_\epsilon$ of being less than 2000. The experimental results show the $\hat{P}'_2 = 0.68$, $\hat{P}'_3 = 0.71$, $\hat{P}'_4 = 0.95$, $\hat{P}'_5 = 0.92$, $\hat{P}'_8 = 0.70$. This procedure performs best at $\epsilon = 4$.

The same analysis was done for finding the value of λ^*. Let $\hat{P}^*_\lambda$ be the probability of (6) associated with λ^* and calculated $\hat{P}^*_\lambda$ of being less than 2000. Here, ϵ was set to 4. We have the results that $\hat{P}^*_2 = 0.74$, $\hat{P}^*_3 = 0.93$, $\hat{P}^*_4 = 0.98$, $\hat{P}^*_5 = 0.89$, and $\hat{P}^*_8 = 0.75$. The best value was $\lambda^* = 4$. The same tendency was observed for the other four examples.

Note 1: The algorithm with $\lambda^* = 2$ did not result in good solutions in the meaning of both the appropriateness for solution value and computation time. This algorithm is called as the FDPR method in [7] and considered to be best among existing algorithms.

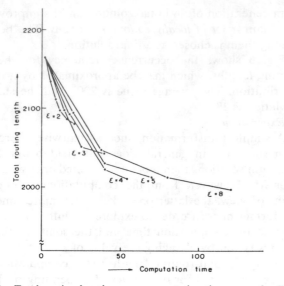

Fig. 7. Total routing length versus computation time curve for Example 5, varying the value of ϵ.

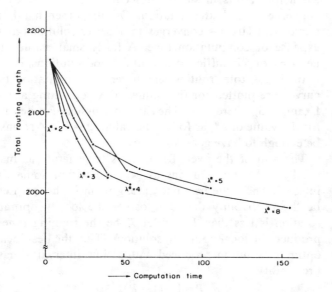

Fig. 8. Total routing length versus computation time curve for Example 5, varying the value of λ^*.

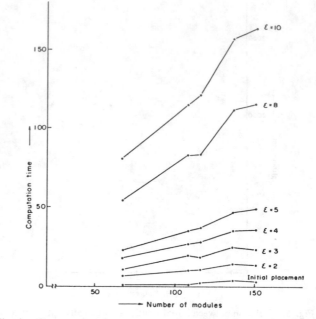

Fig. 9. Computation time versus number of modules, varying the value of ϵ.

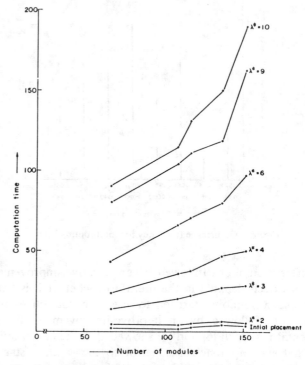

Fig. 10. Computation time versus number of modules, varying the value of λ^*.

Note 2: The algorithm with larger ϵ did not result in good solutions either. The method with the testing of all possible interchanges is exactly equal to the proposed algorithm with the largest value of ϵ. In this sense, the testing of all possible interchanges is said to be inferior to limiting the number of possible exchanges proposed here.

Experiment 3

In order to explore the influence of the number of modules on the computation process, 5 examples were examined.

The graph in Fig. 9 shows computation time versus number of module for various values of ϵ. The computation time increases quite rapidly for $\epsilon \geqslant 6$, when the number of modules increases. Therefore, it seems infeasible to set $\epsilon \geqslant 6$ for large scale problems.

The graph in Fig. 10 also shows computation time versus number of modules, while varying the value of λ^*.

From a practical point of view it may be reasonable to set $\lambda^* \geqslant 5$ for large scale problems.

The computation time to obtain a locally optimum solution with $\epsilon = 4$ and $\lambda^* = 4$ was linearly proportional to the number of modules.

VI. CONCLUDING REMARKS

In this paper, a new algorithm based on heuristic approach was proposed for the two-dimensional placement problem. The algorithm consists of two phases, namely:

IEEE TRANSACTIONS ON CIRCUITS AND SYSTEMS, VOL. CAS-28, NO. 1, JANUARY 1981

initial constructive placement and iterative improvement. After repeating the random generation of an initial solution and its improvement by a sequence of local transformation, the best one among them is chosen as the final solution. The algorithm is greatly affected by the values of two parameters, ϵ and λ^*. From the present computational results, the algorithm performs best at $\epsilon = 4$ and $\lambda^* = 4$ in the meaning of both the appropriateness for solution value and computation time.

It should be noted that the algorithm with $\lambda^* = 2$ did not result in good solutions, which is called as FDPR method in [7] and considered to be best among existing algorithms.

The most important point in the two-dimensional placement problem is to investigate various objective functions for optimum placement. Most of the present placement algorithms use routing length as an objective function, as was proposed in this paper. The "utilization of the most crowded channel", or "maximum density" is more meaningful for certain applications. One meaningful and perhaps more important criterion is to consider routability in placement.

ACKNOWLEDGMENT

The author is greatly indebted to Dr. E. S. Kuh, Dr. T. Ohtsuki, Dr. K. Kani, and H. Kawanishi for their helpful discussions and encouragements. Finally, he gratefully acknowledges the constructive comments of the reviewers who contributed to the improvement of an earlier version of this paper.

REFERENCES

[1] L. Steinberg, "The background wiring problem: A placement algorithm," *SIAM Rev.*, vol. 3, no. 1, pp. 37–50, Jan. 1961.

[2] K. M. Hall, "An *r*-dimension quadratic placement algorithm," *Management Sci.*, vol. 17, no. 3, pp. 219–229, Nov. 1970.

[3] M. Hanan and J. M. Kurtzberg, "Placement techniques," in *Design Automation of Digital Systems; Theory and Techniques.* Vol. 1, (M. A. Breuer, ed.), Englewood Cliffs, NJ: Prentice-Hall, ch. 5, pp. 213–282, 1972.

[4] D. C. Schmidt and L. E. Druffel, "An iterative algorithm for placement and assignment of integrated circuits," in *Proc. 12th Design Automation Conf.*, pp. 361–368, 1975.

[5] D. G. Schweikert, "A two-dimensional placement algorithm for the layout of electrical circuits," in *Proc. 13th Design Automation Conf.*, (San Francisco, CA), pp. 409–414, 1976.

[6] S. Goto and E. S. Kuh, "An approach to the two-dimensional placement problem in circuit layout," *IEEE Trans. Circuits Syst.*, vol. CAS-25, pp. 208–214, Apr. 1978.

[7] M. Hanan, P. K. Wolff, and B. J. Anguli, "Some experimental result on placement techniques," in *Proc. 13th Design Automation Conf.*, (San Francisco, CA), pp. 214–224, 1976.

[8] S. Lin and B. Kernighan, "An effective algorithm for travelling-salesman problem," *Oper. Res.*, vol. 11, pp. 498–516, 1973.

[9] S. L. Hakimi, "Optimum locations of switching centers and the absolute centers and medians of a graph," *Oper. Res.*, vol. 12, pp. 450–459, 1964.

[10] D. B. Johnson and T. Mizoguchi, "Selecting the kth element in $X + Y$ and $X_1 + X_2 + \cdots + X_m$," *SIAM J. Computing*, vol. 7, no. 2, pp. 141–143, May 1978.

[11] R. M. Karp, "Reducibility among combinatorial problems," in *Complexity of Computer Computations*, (R. E. Miller and J. W. Thatcher, eds.), New York: Plenum, 1972.

[12] S. Sahni and T. Gonzales, "*P*-complete approximation problem," *J. ACM*, vol. 23, no. 3, pp. 555–565, July 1976.

[13] J. E. Stevens, "Fast heuristic techniques for placing and wiring printed circuit boards," Ph.D. dissertation Comp. Sci., Univ. of Illinois, 1972.

[14] S. Lin, "Heuristic programming as aid to network design," *Networks*, vol. 5, pp. 33–43, 1975.

[15] T. Yoshimura, "An algorithm for designing multi-drop teleprocessing networks," in *Proc. ICCC-78*, Oct. 1978.

Part IV
Routing

ON STEINER'S PROBLEM WITH RECTILINEAR DISTANCE*

M. HANAN†

1. Introduction. This paper is concerned with the following type of problem. Given n cities, construct a network of roads of *minimum total length* so that a traveler can get from one city to any other. Roads may cross each other outside of the city limits and these points are called junction points. (Roads which cross within the cities, however, will not be referred to as junctions.) It is assumed that junction points add no extra cost to the construction of the network so that there may be as many as necessary to minimize the total length. Usually, the roads are straight-line connections and the distance between two points is the Euclidean distance. In this paper, however, the rectilinear distance is used. The rectilinear distance $d(p_1, p_2)$ between two points p_1 and p_2 is defined as

$$d(p_1, p_2) = |x_1 - x_2| + |y_1 - y_2|,$$

where (x_i, y_i) are the coordinates of p_i.

Rectilinear distance has application in printed circuit technology where n electrically common points must be connected with the shortest possible length of wire and the wires must run in the horizontal and vertical directions. The junction points of the wires are analogous to the above-mentioned road junctions.

Actually this is a well-known problem due to Steiner (cf. [2] and [4]) and is now formally stated.

STEINER'S PROBLEM. *Given n points in the plane find the shortest tree(s) whose vertices contain these n points.*

A tree with m vertices is a connected graph with $m - 1$ edges. (For graph-theoretic terminology see Berge [1].) Several necessary conditions about the solution of this problem are known when the distance between two points is taken to be the Euclidean distance. In this paper several necessary conditions are given for any n, using rectilinear distance. Some of these conditions are analogous to the problem with Euclidean distance, some hold only for rectilinear distance, and some are invariant with respect to the metric. Exact solutions are constructed for $n \leq 5$.

Since rectilinear distance is not invariant with respect to rotations in the plane, the statement of Steiner's problem must be properly interpreted. Hence it is assumed throughout this paper that when n points in the plane

* Received by the editors March 15, 1965.

† Thomas J. Watson Research Center, International Business Machines Corporation, Yorktown Heights, New York.

are given, a Cartesian coordinate system is also given and the rectilinear distance is defined with respect to this coordinate system.

We now state two problems which are related to Steiner's problem and whose solutions we will have occasion to use in this paper. To distinguish these we refer to Steiner's problem as S_n and we now define P_n and T_n.

P_n : *Given n points $(p_1, \cdots, p_n)$ in the plane, find a point q such that the sum of the distances from q to p_i, $i = 1, \cdots, n$, is a minimum.*

T_n : *Given n points in the plane, find the shortest tree whose vertices are these n points.*

(We have departed slightly from the notation used by Melzak [6].) The P_n problem has been solved for both Euclidean distance (cf. [7]) and rectilinear distance (cf. [3]). The T_n problem has also been solved (cf. [5] and [8]) and the method of solution is independent of the metric used.

2. Steiner's problem with three points.

2a. Euclidean distance. Given three points in the plane, let T be the triangle whose vertices are these three points. If every angle of T is less than 120°, then the point q of P_3 lies inside T and the lines from p_i to q, $i = 1, 2, 3$, meet at 120° at q. If an angle of T is greater than or equal to 120°, then q coincides with that vertex. (See [4] for a proof and a construction of the solution.)[1] It is not difficult to see that P_3 yields the same solution as S_3. Also, if an angle of T is greater or equal to 120°, then the solution of S_3 is identical to the solution of T_3. See Fig. 1.

2b. Rectilinear distance. Using rectilinear distance, the solution to S_3 (or equivalently P_3) is simpler to construct than the corresponding problem using Euclidean distance. In place of the triangle T, we consider the enclosing rectangle R which we now define, in general, for n points.

DEFINITION 1. *Given n points in the plane enclosing rectangle R is the smallest rectangle whose sides are parallel to the x and y axes and which includes the n points either within or on its boundary.*

We refer to the solution of the three-point problem throughout this paper and therefore state the result as a separate theorem.

THEOREM 1. *Let (x_i, y_i) be the coordinates of the given points p_i; $i = 1, 2, 3$. The q-point of P_3 is located at (x_m, y_m) where x_m and y_m are the medians of $\{x_i\}$ and $\{y_i\}$, respectively.*

As stated earlier, this special case of P_n is solved in [3]. The following theorem relates P_3, S_3 and T_3. Let d_{s_n}, d_{P_n}, and d_{T_n} be the total (rectilinear) distance in the solutions of S_n, P_n, and T_n, respectively.

[1] We would like to thank the referee for suggesting two other references to this problem: E. GOURSAT, *A Course in Mathematical Analysis*, vol. 1, Dover, New York, 1959, p. 130; H. S. M. COXETER, *Introduction to Geometry*, John Wiley, New York, 1961, p. 21.

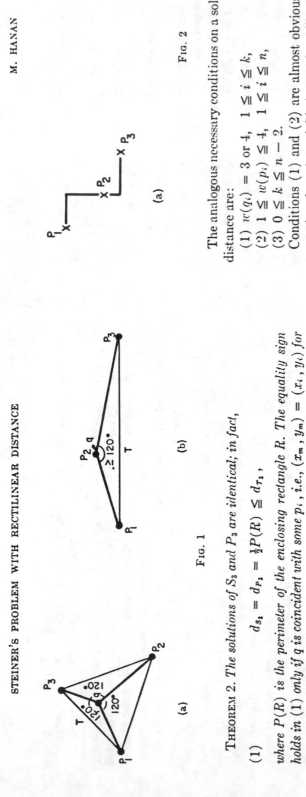

(a)   (b)

FIG. 1

THEOREM 2. *The solutions of S_3 and P_3 are identical; in fact,*

(1)
$$d_{s_3} = d_{P_3} = \tfrac{1}{2}P(R) \leq d_{r_3},$$

where $P(R)$ is the perimeter of the enclosing rectangle R. The equality sign holds in (1) only if q is coincident with some p_i, i.e., $(x_m, y_m) = (x_i, y_i)$ for some $i = 1, 2, 3$.

The proof of Theorem 2 is straightforward. It follows from Theorem 1 and the fact (which we prove later for S_n, in general) that the minimum tree solution to S_3 can have either zero or one additional vertex. See Fig. 2.

3. Necessary conditions on a solution to S_n. We use the following notation: p_i are the given n points and q_i, $i = 1, \cdots, k$, are the additional k vertices in the solution G of S_n. When we are referring to the vertices of G, we speak of p-vertices or q-vertices. When we are referring to the location of these vertices in the coordinate system we speak of p-points or q-points. We use the notation p_i (or q_i) interchangeably for a vertex of G or the location of that vertex. Its meaning should be clear from the context. We let P be the set of p-points or p-vertices and Q be the set of q-points or q-vertices. When we speak of a vertex a_i, $i = 1, \cdots, n + k$, we mean either p_i or q_i. We let $w(a_i)$ be the local degree of the vertex a_i, that is, the number of vertices adjacent to a_i and $C(a_i)$ be this set of vertices. (Two vertices are adjacent if they have an edge in common.) The following essentially sums up the present knowledge about the solution to S_n using Euclidean distance (cf. [2] and [6]).

(1) $w(q_i) = 3$, $1 \leq i \leq k$,
(2) $w(p_i) \leq 3$, $1 \leq i \leq n$,
(3) $0 \leq k \leq n - 2$,
(4) each q_i, $1 \leq i \leq k$, is the q-point of $C(q_i)$.

These conditions are easy to prove. In fact (4) can be replaced by the stronger statement that every connected subtree of a solution (G of S_n is a minimum tree of those $m \leq n$ points.

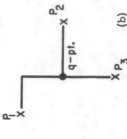

(a)

(b)

FIG. 2

The analogous necessary conditions on a solution G of S_n using rectilinear distance are:

(1) $w(q_i) = 3$ or 4, $1 \leq i \leq k$,
(2) $1 \leq w(p_i) \leq 4$, $1 \leq i \leq n$,
(3) $0 \leq k \leq n - 2$.

Conditions (1) and (2) are almost obvious. In fact if $w(a_i) = 4$, then two pairs of vertices of $C(a_i)$ must be collinear and a_i is at the intersection of the straight lines connecting those pairs. See Fig. 3.

To prove the inequality on the right side of (3), assume that there are k q-vertices in G and find the least number of p-vertices possible. Assume the worst case, that is, $w(q_i) = 3$ and the q-vertices form a subtree with $k - 1$ edges. Since each edge counts twice in the total degree of the q-vertices,

$$n \geq 3k - 2(k - 1) = k + 2,$$

or

$$k \leq n - 2.$$

To show that zero is a true lower bound, we can easily construct an example where $k = 0$. Since $\tfrac{1}{2}P(R)$ is a lower bound for d_{s_n}, and, in the example shown in Fig. 4, $d_{s_n} = \tfrac{1}{2}P(R)$, we have found a minimum tree with $k = 0$.

We state condition (4) as a separate lemma for future reference.

LEMMA 1. *Given n points in the plane, let G be a solution of S_n. If G' is a connected subgraph of G with m vertices, then G' is a solution to S_m.*

The following theorem has an analog in Euclidean geometry where the triangle T replaces the rectangle R. However, we have not seen it stated in the literature.

THEOREM 3. *If q is any q-vertex of G with degree three, then q can be the only vertex of G inside the enclosing rectangle R of C(q).*

We note first that there may be vertices (including q itself) on the boundary of R. To prove the theorem, assume the contrary, that is, assume that there is another vertex a_m inside R. There must exist some path from

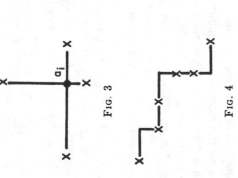

FIG. 3

FIG. 4

a_m to one of the vertices $a_i \in C(q)$, say a_1. Now consider the problem S_3 consisting of the three points $\{a_m, a_2, a_3\}$ and let R_1 be the enclosing rectangle of these three points. Since a_m lies inside R, $\frac{1}{2}P(R_1) < \frac{1}{2}P(R)$, so that we can replace the subtree on the vertices (a_1, a_2, a_3, q) with the new subtree on (a_m, a_2, a_3, q_1), where q_1 is the new q-point. This subtree is connected to the rest of the graph by the path from a_m to a_1. Hence we have found a new graph G_1 with a smaller total distance than G, contradicting the hypothesis that G is a solution of S_n.

In general, a solution G to S_n is not unique, that is, there is more than one set Q which yields a minimum tree. Let $N(n, k)$ be the number of sets Q. The main result of Melzak [6] is that $N(n, k)$ is finite for all n and k in Euclidean geometry and there exists a finite sequence of Euclidean constructions yielding all minimizing trees of the problem S_n. In rectilinear geometry this is not true. (We will give an example in §4 where $N(4, 2)$ is infinite.) Hence we cannot guarantee that we can find (by construction) all solutions G to S_n. However, we now prove a theorem which does guarantee finding a finite subset of solutions. The theorem proves, in effect, that there always exists a solution G such that all the vertices in the set Q are located at a predetermined finite set of possible locations.

THEOREM 4. *Let $\{x_p\}$ and $\{y_p\}$ be the sets of x and y coordinates of the given n p-points. If (x_{q_j}, y_{q_j}) are the coordinates of any vertex $q_j \in Q$, then there exists a solution G to the problem S_n such that $x_{q_j} \in \{x_p\}$ and $y_{q_j} \in \{y_p\}$ for all $j = 1, 2, \cdots, k \leq n - 2$.*

If straight lines are drawn parallel to the x and y axes through all the given points, a grid is imposed on the plane. Theorem 4 states that there exists a solution G such that all the q-vertices are on the intersections of

these grid lines. Let I be the set of intersection points or, when referring to the vertices, the set of vertices of G which are located at these intersections. By definition, $P \subset I$. For any $q \in Q$, if $C(q)$ contains only p-vertices then $q \in I$. This last statement is an immediate consequence of Theorem 1 and Lemma 1 when $w(q) = 3$ and is obvious when $w(q) = 4$.

To facilitate the proof of Theorem 4, we first prove two lemmas.

LEMMA 2. *Let G be any solution of S_n and let q_j be any vertex in Q such that $C(q_j)$ contains two vertices in the set I, say i_1 and i_2. If $q_j \notin I$, then a solution G' can be obtained from G such that there is no vertex of G' located at the point q_j and if a vertex q_j' of G' is connected to both i_1 and i_2 then $q_j \in I'$.*

Lemma 2 states, in effect, that given a tree G, certain of the q-vertices can always be "moved" to new locations which are at the intersections of the grid. We first note that if $w(q_j) = 4$ then clearly $q_j \in I$, so we assume that $w(q_j) = 3$. Let i_1, i_2, and a be the three vertices of $C(q_j)$. If $a \in I$ then $q_j \in I$ (Theorem 2 and Lemma 1) so that $a \in Q$. Hence let us call this vertex q_1. The locations of these vertices with respect to this vertex q_j must be essentially as shown in Fig. 5. (We have drawn the connection from q_j to i_2 in the way shown for future use.)

By Lemma 1 and Theorem 1, at least one of the vertices i_1 or i_2 must be on the horizontal line through the point q_j. (Clearly the figure can be rotated through an angle of $m\pi/2$, $m = 1, 2, 3$. There is, of course, no loss of generality in assuming this configuration.) Assuming that i_1 is the vertex on this horizontal line, then i_2 can be anywhere in the quadrant $x > x_{q_j}$ and $y \geq y_{q_j}$. Again, by Lemma 1 and Theorem 1, at least one vertex of $C(q_1)$, say a_1, must lie on the line $y = y_{q_1}$. There are now two possibilities which must be considered: (i) a_1 is to the right of i_2, and (ii) a_1 is in the interval between q_j and i_2.

Considering (i) first, the line joining q_j to q_1 can be moved parallel to itself to the line $x = x_{i_2}$ as indicated in Fig. 6. By making this move, the new graph G' is also a connected tree and its length is the same as G. Hence G' is also a solution to S_n. Clearly $q_j \in I$ and there is no vertex of

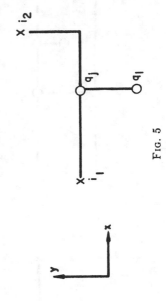

FIG. 5

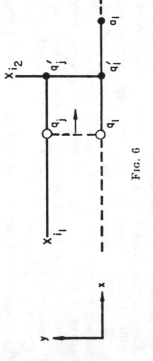

FIG. 6

G' located at the point q_j. The graph G' may have more, fewer, or the same number of vertices as the original graph G. For example, if i_2 were on the line $y = y_{q_j} = y_{i_1}$, then no q-vertex of G' would be generated at the point designated q_j since this point would be occupied by i_2. Also if $w(q_i) = 3$ then G' has no vertex at q_1 and if $w(q_1) = 4$ then G' has vertices both at q_1 and q_1'. This completes the proof of Lemma 2 for the case (i).

We now examine case (ii). First, if $a_1 \in I$, we move the line joining q_j to q_1 parallel to itself to the line $x = x_{a_1}$ so that $q_j' \in I$ and there is no vertex of G' at the point q_j. Now assume $a_1 \in Q$ and $a_1 \notin I$. It is not difficult to see that no vertex in $C(a_1)$ can be in the region $y > y_{a_1}$. For, referring to Fig. 7, this implies that the subtree connecting this vertex to $\{q_j; q_1; a_1\}$ is not minimum, contradicting Lemma 1. Hence one vertex of $C(a_1)$, say a_2, must be on the line $y = y_{a_1}$ and to the right of a_1. We can now use the same arguments as above, with a_2 replacing a_1, to find another vertex a_3 on the line $y = y_{a_1} = y_{a_2}$ and to the right of a_2. Continuing this argument, the process must eventually end. Either $a_i \in P$ or a_i is to the right of i_2. In either case we can move the line joining q_j to q_1 such that a tree G' is generated with a vertex $q_j' \in I$ and no vertex at the point q_j. (Actually we can prove a stronger result, that is, $l \le 2$, but this is not essential to the proof of the Lemma.) This completes the proof of Lemma 2.

LEMMA 3. *If Q_1 is the set of vertices, which are not in I, of a minimum tree G then either Q_1 is empty or it contains at least one vertex adjacent to two vertices in I.*

Assume Q_1 is nonempty and let k_1 be the number of vertices in Q_1. To prove the lemma, assume the contrary, that is, assume all vertices in Q_1 are adjacent to at most one vertex in I. Let $E(Q_1)$ be the number of edges in the subgraph with the Q_1 vertices. Since $w(q_i) \ge 3$ for all $q_i \in Q_1$,

$$E(Q_1) \ge \sum_{i=1}^{k_1} \frac{w(q_1) - 1}{2} \ge k_1.$$

This implies that there exists a cycle in the subgraph with the Q_1 vertices, which is absurd.

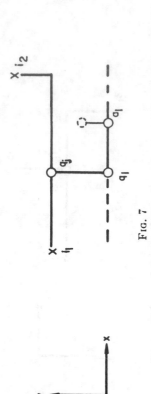

FIG. 7

The proof of Theorem 4 now follows immediately by successively applying Lemmas 2 and 3. Given a solution G_1, partition the vertices into two disjoint sets Q_1 and I_1. If Q_1 is not empty, then, by Lemma 3, at least one vertex in Q_1 has two adjacent vertices in I_1. By Lemma 2 this vertex can be moved to a new position which is in I_1. Partition the vertices of this new solution tree G_2, generated by this move, into two disjoint sets Q_2 and I_2. If Q_2 is empty, the theorem is proved. If Q_2 is nonempty, apply Lemma 3. By continuing this process, Q_1 must be empty for some finite l since the set of points I is certainly finite.

4. The cases $n = 4$ and $n = 5$. In this section we concentrate on the solutions implied by Theorem 4, that is, those solutions which have all their vertices in I, although many of the statements made here are applicable to all solutions of S_4 and S_5.

We begin the study of S_4 by first solving the special case where the given four points are located on the corners of the enclosing rectangle R. By Theorem 4, there exists a solution G with no q-vertices since the four intersection points are occupied by p-vertices. Hence, in this case, there exists a solution to Steiner's problem which is the same as the solution to the minimum spanning tree problem. We state this as a separate lemma for future reference.

LEMMA 4. *If the four points of S_4 are located at the corners of the enclosing rectangle R, then*

$$d_{s_4} = d_{T_4} = l + 2w,$$

where l and w are the length and width[2] of R.

This is the simplest example where the number of sets Q is infinite. For, referring to Fig. 8, the line joining q_1 to q_2 can be moved parallel to itself anywhere in the interval $y_{p_1} \le y \le y_{p_2}$ which implies an infinite number of possible locations for q_1 and q_2.

_____
[2] We assume throughout this paper that $w \le l$, i.e., we say the width, by definition, is the smaller of the two numbers.

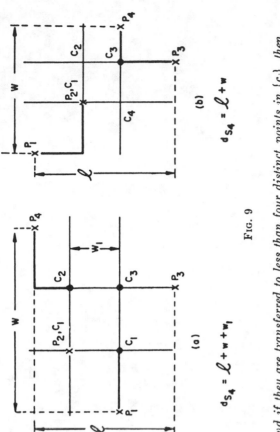

(a)

$$d_{S_4} = \ell + w + w_1$$

(b)

$$d_{S_4} = \ell + w$$

Fig. 9

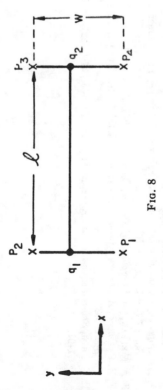

Fig. 8

We now show that when the four points of S_4 are located anywhere in the plane, the problem can always be reduced either to the above case or to a Steiner problem with less than four points. We first note that there cannot be any q-vertices on any side of R unless there are two p-points on this same side. (This can be deduced easily by Theorem 1 and Lemma 1.)

Therefore, if there is a side s_1 of R with only one vertex on it, that vertex can be "moved" perpendicularly to s_1 to the closest intersection point.

It is not difficult to see that, in doing this, we have reduced the general four-point Steiner problem to one where Lemma 4 is applicable (see Fig. 9a) or to a Steiner problem with less than four points (see Fig. 9b). In order to state these results more succinctly, some new terminology is introduced.

First order both (separately) the x- and y-coordinates of the given p-points in increasing order. (In doing this (x_i, y_i) no longer corresponds to the point p_i.) Then by drawing lines parallel to the y-axis through x_2 and x_3 and lines parallel to the x-axis through y_2 and y_3, this defines, in general, four points in I which we call $c_1, \cdots, c_4$. The rectangle which has these four points at its corners is called the *inner rectangle* R_1. Consider the four quadrants U_{c_i}, exterior to R_1, formed by the extended lines of R_1 and each of the c_i. If there is a point p_j in a quadrant U_{c_i}, then we say that p_j is *transferred to the point* c_i. By construction, the p-vertex may, of course, be at the point c_i. The inner rectangle R_1 may degenerate to a straight line. There may or may not be vertices of G located at the points c_i.

A solution to S_4 can be found by applying Lemma 4 and the following theorem.

THEOREM 5. *Given four points in the plane, let l and w be the length and width of the enclosing rectangle R and let w_1 be the width of the inner rectangle R_1. If the p-vertices are transferred to four distinct points in $\{c_i\}$, then*

$$d_{S_4} = l + w + w_1,$$

and if they are transferred to less than four distinct points in $\{c_i\}$, then

$$d_{S_4} = l + w.$$

It is not difficult to see that, in general, d_{S_n} has a lower bound of $\frac{1}{2}$ the perimeter of the enclosing rectangle R, that is,

$$d_{S_n} \geq \tfrac{1}{2}P(R) = l + w.$$

In fact, the following can easily be deduced from Theorem 5 and Lemma 4.

COROLLARY 1. *If $d_{S_4} = l + w$ then the solution to S_4 is unique; and if $d_{S_4} > l + w$ then there exists an infinity of solutions.*

The five-point Steiner problem can be treated in essentially the same way as the four-point problem. In this case, lines are drawn parallel to the y-axis through x_2, x_3, and x_4 and parallel to the x-axis through y_2, y_3, and y_4, so that there are in general nine c-points. The inner rectangle R_1 is the largest rectangle defined by these nine lines. The p-vertices are transferred to the c-points in a manner similar to the above, except that the concept of quadrants must be generalized to include the points c_i which are not at the corners of R_1. Hence, by transferring the p-points, we can always reduce the problem S_5 to the case where at least two p-vertices are on each side of the enclosing rectangle. The following theorem (the proof of which we omit) can then be used to find a solution to S_5 when the points are located anywhere in the plane.

THEOREM 6. *Given five points in the plane with at least two points on each side of the enclosing rectangle R. If four of the five points are at the corners of the enclosing rectangle R, then*

137

R, then

$$d_{s_6} = d_{s_4} = l + 2w \leqq d_{\tau_5},$$

where S_4 is the Steiner problem with these four corner points. If all five points are on the boundary of R, then

$$d_{s_5} = d_{\tau_5} \leqq l + 2w,$$

where l and w are the length and width of R.

5. General comments. An algorithm which incorporates several of the necessary conditions stated in §3 has been developed. It yields "good" approximate solutions to the n-point problem and exact solutions for $n \leqq 4$. The algorithm is easy and fast to do both by hand and on a computer.

The algorithm is rather elementary in concept and it is anticipated that a more sophisticated algorithm can be devised which incorporates almost all the results presented in this paper. For example, Theorem 4 states that there always exists a solution to Steiner's problem where the q-vertices are located at a predetermined finite set of points. Therefore, if the number of points n is not too large, we can find an exact solution by an exhaustive search procedure. These ideas will be investigated in the future.

Acknowledgment. The author wishes to thank P. H. Oden for many interesting discussions relating to this problem.

REFERENCES

[1] CLAUDE BERGE, *The Theory of Graphs*, John Wiley, New York, 1961.
[2] R. COURANT AND H. ROBBINS, *What is Mathematics?*, Oxford University Press, New York, 1941.
[3] RICHARD L. FRANCIS, *A note on the optimum location of new machines in existing plant layouts*, J. Indust. Engrg., 14 (1963), pp. 57–59.
[4] HUA LO-KENG ET AL., *Application of mathematical methods to wheat harvesting*, Chinese Math., 2 (1962), pp. 77–91.
[5] J. B. KRUSKAL, *On the shortest spanning subtree of a graph*, Proc. Amer. Math. Soc., 7 (1956), pp. 48–50.
[6] Z. A. MELZAK, *On the problem of Steiner*, Canad. Math. Bull., 4 (1961), pp. 143–148.
[7] F. P. PALERMO, *A network minimization problem*, IBM J. Res. Develop., 5 (1961), pp. 335–337.
[8] R. C. PRIM, *Shortest connecting networks*, Bell System Tech. J., 31 (1952), pp. 1398–1401.

The α-β Routing

T. C. HU AND M. T. SHING

Abstract—An algorithm is proposed to find a minimum cost path connecting two nodes in a grid-graph. The cost of traveling an arc is α and there is an additional cost of β if the path turns 90°. The cost of a path is the sum of costs of arcs and corners in the path. The α-β algorithm has a worst-case running time of $O(n \log n)$. Possible modifications of the α-β algorithms are also discussed.

I. INTRODUCTION

THERE are basically three type of routing methods:

1) Maze-runner (the Lee–Moore method or its modifications, see for example, [1], [3], [7], [9], [10]).
2) Line-expansion methods (see [5], [8]).
3) Special-methods for channel or river routing.

The first two types of methods can be used in global or final routing, while the third type is used mainly in a channel or "river" and will not be discussed here. The Lee–Moore method (or its modifications) is essentially a breadth-first-search method (BFS) [6] applied to a grid-graph. It is currently the only method that guarantees to find a path with minimum wire-length. The major drawback of this method is that it will always pick a shorter path which may have many vias instead of a longer path with fewer vias, even though the latter path is more preferable in many practical situations.

The line-expansion type algorithm is designed to be a fast algorithm to find a path with very few vias. It is essentially a depth first search (DFS) method. Unfortunately, it does not guarantee to find a path even if such a path exists and its running time is hard to analyze. Recently, Heyns *et al* [4] suggested an algorithm to generate all possible escape lines from a given line until a solution is found. However, the solution obtained may not be optimum with respect to some predefined criteria such as shortest wire-length, minimum vias, or a combination of both.

In this paper, we propose an algorithm to find a minimum-cost path between a pair of nodes in a grid-graph G where some of the arcs in G are missing. The cost of traveling along an arc is α and there is an additional cost of β for each 90° turn along the path. The cost of a path is given by the sum of the costs of the arcs in the path plus the additional costs of turning around corners.

Assume, for the time being, that $\beta = 0$ and we want to find a shortest path from the source node v_s to the sink node v_t. In most shortest-path algorithms, we basically construct a rooted tree with the root at v_s. There is a unique path from the root to every node on the tree, and the unique path is also a shortest

Manuscript received January 7, 1985.

T. C. Hu is with the Dept. of Elect. Eng. and Computer Science, University of California, San Diego, La Jolla, CA 92093.

M. T. Shing is with the Department of Computer Science, University of California, Santa Barbara, CA 93106.

path from the root to the node. The rooted tree consists of only the source node v_s initially and is expanded until the sink node v_t becomes a node on the tree.

Different shortest-path algorithms have different ways of expanding and modifying the tree. In BFS-type algorithms, we always expand the tree by selecting a node which is closest to v_s and repeat the process until the node v_t is included into the tree. The BFS algorithm always finds a minimum cost path if one exists.

In the DFS-type algorithm like the line-expansion method, we keep expanding the rooted tree along a horizontal (or a vertical) path, and then among the vertical (horizontal) paths which intersect the horizontal (vertical) path, pick the one which has the best *chance* of reaching v_t in the cheapest way and expand the tree along that path. The exact order was not specified [5], and when v_t is reached, the path may *not* be of minimum cost. The line-expansion method has the advantage of finding a path quickly. However, unless systematic backtracking is used, the line-expansion method may not find a path from v_s to v_t even if such a path exists.

The choice between BFS and DFS depends on the graph and the objectives of the user. When $\beta \neq 0$, these shortest-path algorithms no longer work. (We need to construct a directed acyclic graph (DAG) instead of a rooted tree.)

II. THE α-β ROUTER

The regularity of a grid-graph G warrants special algorithms. Before we introduce one such algorithm, we first introduce some terminology. For convenience, we assume $\alpha = 1$ from now on.

Two nodes joined by an arc in the grid-graph G are *neighbors* to each other. For any node (x_i, y_i) in the grid-graph, we call the lower bound on the cost of traveling from (x_i, y_i) to the sink (x_t, y_t), namely

$$|x_t - x_i| + |y_t - y_i|,$$

as the *future cost* of the node (x_i, y_i). When we actually find a minimum-cost path from the source (x_s, y_s) to a node (x_i, y_i), we call the minimum cost of reaching (x_i, y_i) the *current cost* of the node (x_i, y_i). Furthermore, we define the *potential cost* of a node as the sum of its current cost and its future cost.

Assume that the sink is at (x_t, y_t). For any given node (x_i, y_i), we can move away from the node in 4 directions. We call the directions which lead to neighbors with nonincreasing future costs as *forward* directions, and the directions which lead to neighbors with increasing future costs as *backward* directions. For example, if $x_i < x_t$ and $y_i < y_t$, the north and the east will be the forward directions at (x_i, y_i) while the south and the west are the backward directions at (x_i, y_i). On the other hand, if $x_i > x_t$ and $y_i > y_t$, the south and west will be the forward directions while the north and the east will be the backward directions. If the sink is at (10, 10) and the given node is at (10, 0), then north is the only forward direction, and all other three directions are backward directions.

An arc is called an *h*-arc if its two adjacent nodes have the same *y*-coordinates. An arc is called a *v*-arc if its two adjacent nodes have the same *x*-coordinates. Two nodes are *h*-connected (*v*-connected) if they can be connected by a straight line consisting entirely of *h*-arcs (*v*-arcs). Two nodes are called *line-neighbors* if they are either *h*-connected or *v*-connected.

All nodes which are line-neighbors to each other and have the *same* potential costs along a path from v_s to the nodes are called *brothers*. And the one with less future cost is *younger* than the one with more future cost. If v_i and v_j are line-neighbors and v_i has lower potential cost along a path from v_s to v_j, then v_i is a *father* of v_j. If v_j has a brother v_k, then v_i is also a father of v_k.

The proposed algorithm, called the α-β router, is a labeling algorithm. Let v_i be a labeled node and v_j be its unlabeled neighbor. To label v_j from v_i means to draw a directed arc from v_i to v_j and assign a potential cost to v_j. (In the original grid-graph, all arcs are undirected.)

For example, let the source be at (0, 0) and the sink be at (10, 10). Initially, the source is the only labeled node with its current cost = 0 and future cost = $|10 - 0| + |10 - 0| =$ 20. Hence the potential cost of the source = $0 + 20 = 20$. If we label the node (1, 0) from (0, 0), then the current cost of (1, 0) = 1 and the future cost of (1, 0) = $|10 - 1| + |10 - 0| =$ 19. So the potential cost of (1, 0) is still $1 + 19 = 20$. If we label (1, 1) from (1, 0), the current cost of (1, 1) will be $1 + 1 + \beta = 2 + \beta$ while it future cost will be $|10 - 1| + |10 - 1| = 18$. So the potential cost of (1, 1) = $2 + \beta + 18 = 20 + \beta$, an increase of β in the potential cost.

Assume that we have labeled v_i first, then we label v_j from v_i and there are three unlabeled neighbors to v_j. The potential costs of the three unlabeled neighbors depend on the location of v_j and how v_j was labeled. In Fig. 1, the solid arrows indicate the various positions of the directed arc from v_i to v_j, and the numbers beside the dotted arrows indicate the additional potential cost to be assigned to the unlabeled neighbors of v_j. The black node in the center of Fig. 1 represents the relative position of the sink node.

We shall first assume that $\beta = 0$ and describe the Dijkstra's algorithm [6] for finding a minimum-cost path in a grid-graph, then we shall describe the modifications needed to handle the case where $\beta \neq 0$. (Note that the Dijkstra's algorithm does not work if $\beta \neq 0$.) The Dijkstra's algorithm is a labeling algorithm. It constructs a rooted tree by drawing directed arcs between the nodes and assigning labels (or costs) to the nodes. There are two kinds of labels, *temporary* labels and *permanent* labels. (Only temporary labels of a node can be changed.)

The Dijkstra's Algorithm:

Step 0: The source node v_s is the only node with a permanent label zero and is the only node in the

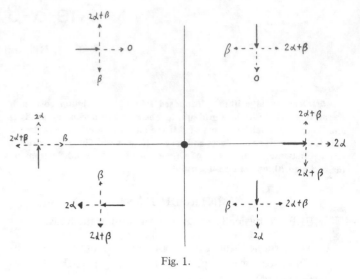

Fig. 1.

rooted tree initially. Label all neighbors of v_s with temporary current cost α and all other nodes with temporary current cost ∞.

Step 1: Among all nodes not in the rooted tree and with temporary costs $< \infty$, pick the one with minimum temporary cost. Mark the label permanent and include it into the rooted tree. Stop if there is no node with temporary cost $< \infty$ or if the sink has received a permanent label.

Step 2: Let v_j be the node that just received a permanent label in Step 1. Assign to each neighbor v_k of v_j which is not in the rooted tree

min (old temporary cost of v_k, permanent cost of $v_j + \alpha$)

as the temporary label of v_k. Return to Step 1.

Now, we shall discuss necessary modifications to find a minimum-cost path where $\beta \neq 0$.

In the Dijkstra's algorithm, let v_j be a node in the rooted tree with its current cost c, and it has three unlabeled neighbors. All three neighbors will have the same temporary current cost $c + \alpha$. In the modified algorithm we do not want to label a node by its current cost but by its potential cost. If v_j is a node with a permanent label and its has three unlabeled neighbors, the three neighbors will have different temporary potential costs depending on the location of v_j and how v_j was labeled. The increases in potential costs among the neighbors would be $0, 2\alpha, \beta, 2\alpha + \beta$, as shown in Fig. 1.

We must also replace the notion of a rooted tree from the source by a directed acyclic graph (DAG) in the modified algorithm. There could be two paths with different costs to the same node v_j and we have to keep track of both paths. For example in Fig. 2, the path $v_s - v_1 - v_j$ may be cheaper than the path $v_s - v_2 - v_j$ and yet the path $v_s - v_1 - v_j - v_k$ can be more expensive than $v_s - v_2 - v_j - v_k$ because there is an additional cost of β in turning 90° at v_j for the first path. Thus, we may need to record two potential costs for each node v_j and

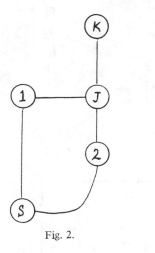

Fig. 2.

keep track of how v_j was labeled. We now describe the α-β router.

Each node v_j in the grid-graph has two potential costs (or simply called labels). We shall use $p_j(q_j)$ to denote the potential cost of a path from v_s to v_j where v_j is h-connected (v-connected) to the last intermediate node in the path. Every label is always in one of the following two states:

1) *temporary*
2) *permanent*.

Step 0: The DAG is empty and all labels are temporary initially.

0a/ Set p_j and q_j to ∞ for all nodes v_j in the grid-graph.

0b/ Assume that $(|x_s - x_t| + |y_s - y_t|) = k$ for some $k \geq 0$.
Label the source node v_s with

$$p_s = q_s = k.$$

Mark the labels permanent and include v_s into the DAG.

0c/ Set p_j of all v_s's h-connected brothers v_j to k. Starting from the oldest to the youngest, mark the labels p_j permanent and include the nodes v_j into the DAG.

0d/ Set q_j of all v_s's v-connected brothers v_j to k. Starting from the oldest to the youngest, mark the labels q_j permanent and include the nodes v_j into the DAG.

Step 1: Scan all the nodes v_j which have newly marked permanent labels as follows.
Let v_k be a neighbor of v_j.

1a/ Set $p_k = \min(p_k, p_j, q_j + \beta)$ if v_k is h-connected from v_j in the forward direction.

1b/ Set $p_k = \min(p_k, p_j + 2\alpha, q_j + \beta + 2\alpha)$ if v_k is h-connected from v_j in the backward direction.

1c/ Set $q_k = \min(q_k, q_j, p_j + \beta)$ if v_k is v-connected from v_j in the forward direction.

1d/ Set $q_k = \min(q_k, q_j + 2\alpha, p_j + \beta + 2\alpha)$ if v_k is v-connected from v_j in the backward direction.

Step 2:

2a/ Stop if there is no temporary label $< \infty$ or if the sink becomes a node in the DAG; else pick the smallest temporary label. In case of tie, pick the one with minimum future cost. Mark the label permanent and include the node, say v_j, into the DAG if it is not already in the DAG.

2b/ If p_j is chosen in Step 2a, set p_k of all v_j's h-connected brothers v_k to p_j. Starting from the oldest to the youngest, mark the labels p_k permanent and include the nodes into the DAG if they are not already in the DAG.

2c/ If q_j is chosen in Step 2a, set q_k of all its v-connected brothers v_k to q_j. Starting from the oldest to the youngest, mark the labels q_k permanent and include the nodes into the DAG if they are not already in the DAG.

2d/ Return to Step 1.

In Fig. 3, we give a numerical example where the source is at $(0, 0)$ and the sink at $(5, 5)$. The numbers beside the grid points are the potential costs of these nodes. The minimum cost path is

$$(0, 0), (0, 8), (8, 5), (5, 5).$$

Note that the $(5, 6)$ has the temporary costs $(p_j, q_j) = (21, 22)$ but q_j is used in the final minimum cost path.

III. PROOF OF CORRECTNESS AND ANALYSIS OF ALGORITHM

Before proving that the α-β router in Section II does indeed give a minimum cost path, we shall first state several simple lemmas.

Lemma 1: If the path from v_s to v_k is of minimum potential cost, then the path is also of minimum current cost, i.e., it is also a cheapest path from v_s to v_k.

Proof: The lemma follows from the fact that the future cost of any node is unique for that node. $\square$

Lemma 2: The α-β router computes the potential cost of the cheapest path from v_s to every node in the DAG.

Proof: We shall prove by induction on the number of nodes in the DAG that for each node v_j in the DAG, $\min(p_j, q_j)$ equals the potential cost of a cheapest path from v_s to v_j. Furthermore, for all v_j not in the DAG, $\min(p_j, q_j)$ is the potential cost of the cheapest path from v_s to v_j with the father of v_j lying within the DAG.

Basis: Min (p_s, q_s) of the source node and min (p_j, q_j) of v_s's brothers all equal to the lower bound $(|x_s - x_t| + |y_s - y_t|)\alpha$ and hence must be minimum. Furthermore, Step 1 correctly initializes the sons of v_s and the sons of v_s's brothers.

Inductive Step: (Case 1) Suppose v_j is included into the DAG in Step 2a but min (p_j, q_j) is not the potential cost of the cheapest path from v_s to v_j, then there must exist a cheaper path P. The path P must contain some vertex v_i other than v_j and v_i is also not in the DAG. Furthermore, v_i's father is

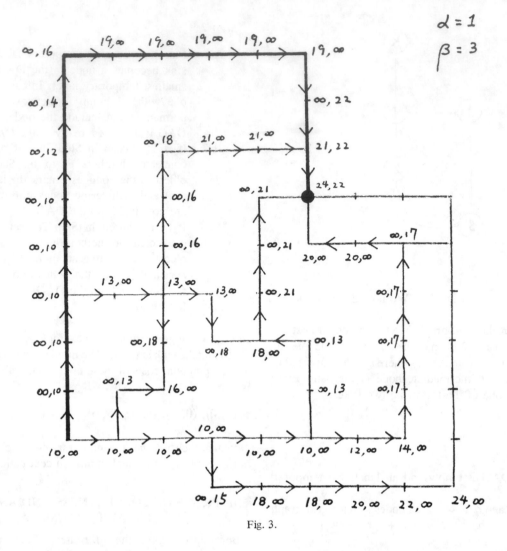

$\alpha = 1$
$\beta = 3$

Fig. 3.

included into the DAG before v_j. Thus by the inductive hypothesis, min $(p_i, q_i) <$ min (p_j, q_j) when v_j is selected, a contradiction.

(*Case* 2) Suppose v_j is included into the DAG in Step 2b or 2c but min (p_j, q_j) is not the potential cost of the cheapest path from v_s to v_j. We can reduce this case to Case 1 by considering the situation when the oldest brother of v_j is added to the DAG.

The second part of the inductive hypothesis, that min (p_j, q_j) remains correct, follows from Step 1 of the algorithm. □

Theorem 1: The α-β router correctly computes the cost of the cheapest path from v_s to v_t.

Proof: This follows from Lemmas 1 and 2. □

Theorem 2: The α-β router has a worst-case running time of $O(n \log n)$, where n is the number of nodes in the grid-graph.

Proof: Since at least one label becomes permanent each time when Step 2 is executed, Steps 1 and 2 can each be executed at most $O(n)$ times.

In Step 2, we have to compare at most $O(n)$ temporary labels and pick the minimum one. This can be done in $O(\log$ $n)$ time using a priority queue. Hence, the total time spent on Step 2 is bounded by $O(n \log n)$.

In Step 1, we have to update the temporary labels of the neighbors of the nodes with new permanent labels. Since we have to update at most 3 neighbors per permanent label, the total time spent on Step 1 is bounded by $O(n)$.

Hence, the algorithm has a worst-case running time of $O(n \log n)$. □

IV. CONCLUSION

The use of potential costs in searching for a path was suggested in [2]. Here, we focus our attention on a special kind of graphs, namely the grid-graphs, and express the potential costs as explicit numerical functions. Due to the fact that β may not be zero, we have to keep two potential costs at each node.

With the help of potential costs and future costs, the algorithm biases the search towards the sink node and may reduce the total number of nodes being labeled. The labeling process is further speeded up by including all brothers of a given node into the DAG once the node is included into the DAG.

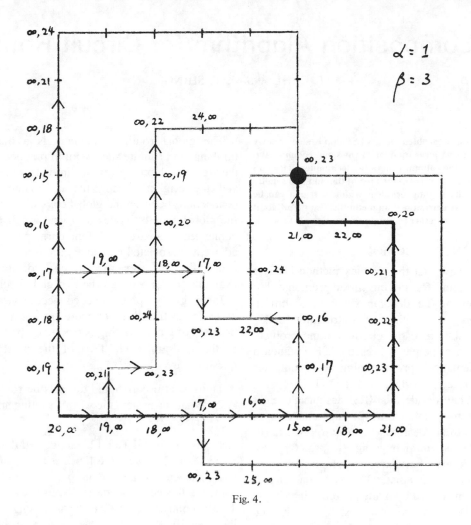

$\alpha = 1$
$\beta = 3$

Fig. 4.

The α-β router can be generalized easily to handle the case where each vertical arc costs α_1 and each horizontal arc costs α_2 for some $\alpha_1 \neq \alpha_2$. We can also obtain other heuristic cost functions by defining the potential cost of a path differently. For example, if we define the potential cost of a node as

current cost + (2 × future cost),

the potential costs of the line-neighbors of a node will be decreasing in the forward direction. If we always expand the DAG by including the node with minimum potential cost, we will still find a path from v_s to v_t. However, the path obtained may not be of minimum current cost. In Fig. 4, we have an example showing that (0, 0), (7, 0), (7, 4), (5, 4), (5, 5) is the path with minimum potential cost while the path (0, 0), (0, 8), (8, 5), (5, 5) is the path with minimum current cost.

Using different weighing factors to the future costs, we can have algorithms with different trade-offs similar to those between the BFS and DFS algorithms.

REFERENCES

[1] S. Hanan, "On Steiner's problem with rectilinear distance," *SIAM J. Appl. Math.*, vol. 14, no. 2, pp. 255–265, Mar. 1966.

[2] P. E. Hart, N. J. Nilsson, and B. Raphael, "A formal basis for the heuristic determination of minimum cost path," *IEEE Trans. Syst. Sci. Cybern.*, vol. SSC-4, no. 2, pp. 100–107, July 1968.

[3] S. Heiss, "A path connection algorithm for multi-layer boards," in *Proc. 5th Design Automation Conf.*, pp. 6–14, 1968.

[4] W. Heyns, W. Sansen, and H. Beke, "A line-expansion algorithm for the general routing problem with a guaranteed solution," in *Proc. 17th Design Automation Conf.*, pp. 243–249, 1980.

[5] D. W. Hightower, "The interconnection problem—A tutorial," *Computer*, vol. 7, no. 4, pp. 18–32, Apr. 1974.

[6] T. C. Hu, *Combinatorial Algorithms.* Reading, MA: Addison-Wesley, 1982.

[7] C. Y. Lee, "An algorithm for path connections and its applications," *IRE Trans. Electorn. Comput.*, pp. 346–365, Sept. 1961.

[8] K. Mikami and K. Tabushi, "A computer program for optimal routing of printed circuit connections," *IFIPS Proc.*, pp. 1475–1478, 1968.

[9] E. F. Moore, "The shortest path through a maze," in *Proc. Int. Symp. on Theory of Switching, Part II*, pp. 285–292, 1959.

[10] J. Soukup, "Circuit layout," *Proc. IEEE*, vol. 69, no. 1, pp. 1281–1304, Oct. 1981.

A Decomposition Algorithm for Circuit Routing

T. C. HU AND M. T. SHING

Abstract—The circuit routing problem on a VLSI chip is an extremely large linear program with a very large number of rows and columns, too large to be solved even with the column-generating techniques. Based on the distribution of nets, we recursively cut the area of the chip into smaller and smaller regions until the routing problem within a region can be handled by the Dantzig–Wolfe decomposition method. Then the adjacent regions are successively pasted together to obtain the routing of the whole chip.

I. INTRODUCTION

MOST scientists know that the simplex method [4] is a very efficient method for solving linear programs, but few realize the power of the decomposition algorithm in solving large linear programs with special structures [6] [7] [8]. In this paper, a solution to the circuit routing problem using the decomposition principle is presented. For standard terminology in mathematical programming and computer science, the reader is referred to [5], [14], and [18].

For simplicity, we shall divide the VLSI design into eight parts, namely: (1) system specification, (2) functional design, (3) logic design, (4) circuit design, (5) *circuit layout*, (6) design verification, (7) test and debugging, (8) prototype test and manufacture. While most parts are too complicated to be represented by a simple mathematical model, the *circuit layout* problem has been studied by many people using simple mathematical models. We shall describe a simple model and illustrate the decomposition algorithm in the design of a gate-array chip. Such model can also be used in the design of standard-cell chips and custom-designed chips.

In the simplest terms, the circuit layout problem is to place many modules on a board (or a chip) in a nonoverlapping manner, and then connect the pins on various modules by mutually noninterfering wires according to a given wiring list. Thus, we can divide the circuit layout problem into two subproblems:

1) *Placement problem:* How to place the modules on the board?
2) *Routing problem:* After the modules are placed, how to route the wires to connect all the nets on the wiring list?

The chip can be thought of as a grip-graph with horizontal and vertical arcs, where the nodes are the potential positions of the pins and holes and the arcs are the places for wires. Since there are thousands of nodes in the grid-graph, and equally many nets, the routing problem is further divided into global and detailed routing. In this paper, we consider global routing only.

Manuscript received January 11, 1985.

T. C. Hu is with the Dept. of Elec. Eng. and Computer Science, University of California, San Diego, La Jolla, CA 92093.

M. T. Shing is with the Department of Computer Science, University of California, Santa Barbara, CA 93106.

A shorter version of this paper will be published in *Mathematical Programming Studies,* Elsevier Science Publishers B. V.

The global routing technique is a hierarchical way of thinking, and is quite successful in practice. In global routing, the area of a chip is partitioned into global cells (see [23], [24]) and the original wiring list is condensed into a list of connections between the global cells. Again, we can represent the global cells by nodes of a grid-graph G. Two nodes are connected by an arc in G if the corresponding global cells are adjacent. Associated with every arc is an nonnegative number, called the arc capacity, which indicates the number of available tracks between the two global cells.

In this grid-graph G, a net connecting two pins is a path and a net connecting three or more pins is a spanning tree. The problem of global routing is then to embed the paths and trees in the grid-graph G. Two of the most commonly asked questions are

1) In what order should the nets be routed?
2) What is a systematic way of routing and re-routing of the nets?

Recently, Ting and Tien [25] suggested that all nets should be routed as if they were the first net to be routed, i.e., each net has equal chance to find its best pattern. This would make some of the cell boundaries overcrowded (or overflowed). These boundaries as well as the nets using these boundaries are then identified and represented as vertices of a bipartite graph. The nets which use many overflowed boundaries are re-routed. If the re-routing causes new overflowed boundaries, the same procedure will be repeated until the level of overflow can be tolerated.

In this paper, we would like to formalize some of the intuitive ideas and formulate the problem in more vigorous terms.

The circuit routing problem is very much like a traffic problem. The pins are the origins and destinations of the traffic. The wires connecting the pins are the traffic, and the channels are the streets. If there are more wires than the number of tracks in a given channel, some of the wires have to be re-routed just like the re-routing of traffic.

For the real traffic problem, every driver wants to go to his destination in the quickest way, and he may try a different route every day. Eventually, every driver selects his best possible route and the traffic pattern become stabilized. Intuitively, we can do the same for the circuit routing problem. For example, let us associate a cost of routing a wire to each arc. The cost of routing a path is then the sum of the costs of all the arcs along the path. Now, every net is to be connected in the cheapest way and we may try different ways of connecting each net until every net is connected in its cheapest possible way.

We can immediately raise two questions for this approach. First, what kind of cost function should we assign to each arc?

Second, there are so many ways of connecting a net, and there are so many nets, how can we select the best way to connect each net and make all the connections compatible? Intuitively, we can let the cost function be proportional to the number of wires in the arc, the ratio of wires to the tracks available, or inversely proportional to the difference between the number of wires and the number of tracks available in the arc, etc. For the second question, the usual approach is to try various configurations based on the heuristic cost functions until a satisfactory configuration is found.

Many heuristic cost functions have been suggested, and many ingenious heuristic algorithms have been proposed and implemented quite successfully. However, the fundamental question is: Is there an efficient algorithm which can be analyzed mathematically? (That is, to explain mathematically why the algorithm works.) Here, we discount the algorithms, such as backtrack or branch-and-bound, which can be classified as implicit enumerations.

Amazingly, the answer is "yes." The algorithm was invented by G. B. Dantzig and was extended by many other people in the early sixties. The way to select various connections for different nets is called the "column-generating" technique in the mathematical programming community. The *correct* cost function for an arc is called the "shadow price." The reason why most of the heuristic cost functions are wrong is because they are based on parameters associated with the arc itself. The correct cost depends not only on the arc itself by also on its adjacent arcs, in fact, on the arcs adjacent to its adjacent arcs.

We do not claim that the circuiting problem is solved by the linear programming formulation, since there are many particular features of the circuit routing problem that have to be addressed. However, the concepts of the linear programming formulation should be studied very carefully. We shall introduce the basic concepts very briefly in the next section. Serious readers should refer to books on linear programming (e.g., [5], [14]) for more details.

II. GLOBAL ROUTING AS A LINEAR INTEGER PROGRAM

We shall first formulate the global routing problem as a very large linear programming problem and then discuss how to solve the large linear programming problem.

Here, to embed trees (or to pack trees) in the grid-graph G is different from the usual embedding in graph theory, because only the positions of the pins are fixed here and all trees connecting the pins of a net are equivalent. There are many equivalent trees for a given net, and there are many nets. Thus, the linear programming formulation is a very natural one.

In a gate-array chip, the size of the chip is fixed and the main problem of the circuit layout is the feasibility of routing. Neglecting the size of the problem, we can formulate the global routing problem as a very large linear program, very much like a multicommodity network flow problem [8], [14].

There are many ways to connect a net, each way is a spanning tree connecting the given pins in the net. We associate a variable y_j to each tree which connects a net. The variable y_j is equal to 1 if that particular tree is used and y_j is equal to 0 if that particular tree is not used. For example, if there are three ways to connect the first net and five ways to connect the second net, we will set

$$y_1 + y_2 + y_3 \qquad = 1$$
$$y_4 + y_5 + y_6 + y_7 + y_8 = 1. \qquad (1)$$

Note that there is one equation for each net since only one tree is needed to connect a given net.

In principle, we can represent all possible ways by a $(0, 1)$ matrix $[a_{ij}]$. For a grid-graph with m arcs, there will be m rows in $[a_{ij}]$, with the ith row corresponding to the ith arc and each column corresponding to different ways of connecting a net. If there are a thousand nets and there are ten different ways to connect each net, then there will be ten thousand columns in the matrix and we cannot enumerate all possible ways of connecting all the nets in practice. Fortunately, we do not need to know all elements of the matrix to solve the linear integer problem.

For any column j, the ith entry, a_{ij}, equals 1 if that particular connection uses the ith arc and equals 0 otherwise. The fact that the number of wires in any arc must be bounded above by its arc capacity can be expressed as

$$\sum a_{ij} y_j \le c_i \qquad i = 1, \cdots, m; \ j = 1, \cdots, \qquad (2)$$

where c_i is the arc capacity of the ith arc. (Some c_i may be zero if the corresponding regions are forbidden. Note also that we leave the range of the index j open because we do not know how many ways to connect the nets.)

In Fig. 1, we show a grid-graph with four nodes and four arcs, where $c_i = 2$ for $i = 1, \cdots, 4$. There are four nets to be connected and all nets are two-terminal nets. Net 1 connects the node in the upper left corner to the node in the lower left corner, Net 3 connects the node in the upper right corner to the node in the lower left corner, etc. Each net can be connected in two different ways and the two ways of connecting net j will be denoted by y_j and y_j'. The eight different ways of connecting the four nets are shown in the following $(0, 1)$ matrix:

	y_1	y_1'	y_2	y_2'	y_3	y_3'	y_4	y_4'
c_1	1	0	0	1	0	1	0	1
c_2	0	1	1	0	0	1	0	1
c_3	0	1	1	0	1	0	1	0
c_4	0	1	0	1	1	0	0	1

For the moment, let us assume that there are p nets and a total of n possible ways of connecting all the nets. We shall denote the set of y_j's which correspond to various ways of connecting the kth net by N_k.

If we write down all the constraints (1) and (2), we have the typical structure of a linear integer program suitable for

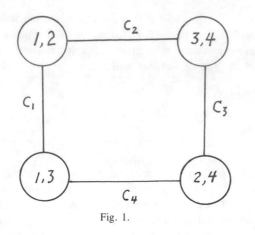

Fig. 1.

decomposition algorithms. Thus the linear integer program is

$$\max \sum b_j y_j \qquad j = 1, 2, \cdots, n \tag{3}$$

$$\text{subject to} \sum_{y_j \in N_k} y_j = 1 \qquad k = 1, 2, \cdots, p$$

$$\sum a_{ij} y_j \le c_i \qquad i = 1, 2, \cdots, m$$

$$0 \le y_j \le 1, \qquad \text{integers.}$$

Note that the number of equality constraints in (3) is equal to the number of nets, and the number of inequality constraints of the type $a_{ij} y_j \le c_i$ is equal to the number of arcs in the grid-graph G.

The constant b_j is the benefit of connecting the kth net by the jth tree. The constants b_j are the same for all $y_j \in N_k$, but the constants are different for different nets. For a net k with many pins, we may want to assign a large positive constant, and for a two-pin net with pins near to each other, we may want to assign a small constant. On the other hand, for a given net k we do not care which tree (y_j) is used to connect the net k. We can also assign all b_k to be the same, then the objective is simply to connect as many nets as possible.

Let π_i be the shadow price of the ith row under the current basis of the linear program (3), and let $\bar{b}_j$ be

$$\bar{b}_j = b_j - \pi_i a_{ij}. \tag{4}$$

To solve the linear program (3), we will select a column j to enter the basis if $\bar{b}_j$ is positive.

Since the b_j's are the same for all spanning trees of any given net, the best way to connect the net corresponds to the column which maximizes $\bar{b}_j$; and among all possible ways of connecting the net, maximum $\bar{b}_j$ corresponds to minimum $\pi_i a_{ij}$. In other words, we want to find a minimum Steiner tree in a grid-graph where the arc lengths are defined by the shadow prices.

Once a new column enters the basis, and a pivot operation is performed, a new set of shadow prices is used for selecting the best way to connect the next net. This process is iterated until no more nets can be connected, or until the objective function is maximized. (Note that y_j are bounded, see [6].)

Note that the grid-graph model presented here is a simplified model which roughly corresponds to double metal layer gate-array chips. A real VLSI chip has multilayers and the wires can be polysilicon and metal, which have different physical characteristics. Thus, all computer programs in industry are based on heuristic algorithms [19]. Intuitively, we will avoid the use of a region if that region is overcrowded, and divert wires to a region which is currently empty or sparse. This amounts to assigning high prices to arcs in crowded regions and low prices to arcs in sparse regions. *This is exactly the role of shadow price.* If some regions are forbidden, we can simply let $c_i = 0$ in those places. The linear programming formulation has several defects even on this simple grid-graph model. However, despite the defects, which we shall discuss later, we want to emphasize that the linear programming formulation is currently the only formulation that can be guaranteed to solve the routing under the grid-graph model (if we discount implicit enumeration methods such as backtrack or branch and bound). Let us list all the defects in the above formulation.

1) Too many columns in the matrix a_{ij}, too many to be written down explicitly.
2) Too many rows in the matrix a_{ij}.
3) Integer constraints on the y_j.
4) How to find a minimum Steiner tree with arcs of different lengths.

We can cure (1) by column-generating techniques [7], [8], [10], and cure (2) by *cut-and-paste* methods discussed later. (Here, we trade computer time and memory space with the possibility of obtaining a suboptimal solutions.) The problem (4) arises as a subproblem of generating the best column. The problem of constructing minimum Steiner trees is common to all methods. However, unlike the linear programming formulation presented here, the existing heuristic algorithms cannot define the arc length exactly.

For (3), we can relax the integer constraints on the variable y_j's and solve the relaxed linear program. (Note that the y_j's are bounded, see [6].) The relaxed linear programming solution may not be integers. However, since the number of nets is much larger than the number of arcs in G', most y_j's will be at their integral upper bounds, which is exactly what we want. The few exceptions can be handled separately after we solve the relaxed linear programming.

The number of rows in the relaxed linear programming is usually still too large to prevent us from using the column-generating techniques. Hence, we propose to get around the above problem using a *cut-and-paste* approach.

II. THE CUT AND PASTE APPROACH

We shall introduce a new way of partitioning the area of a chip into global cells such that we can use the decomposition algorithm to solve the routing problem. If the area of the chip is already partitioned into a large number of global cells, we can use the same technique to solve the global routing problem by decomposition. We call the technique *cut and paste* because we *cut* the chip into regions small enough to be handled by the decomposition algorithm and then *paste* the adjacent regions successively, also by the decomposition algorithm. A similar technique is used by Karp [17] to solve the traveling salesman problem.

A. Cutting

Consider a vertical line in the chip which partitions the area of the chip into left and right regions. There are three kinds of nets: 1) nets with pins all in the left region, 2) nets with pins all in the right region, 3) nets with pins in both regions. We say that the nets in 3) are nets separated by the vertical line. Now we define the ratio R with respect to the vertical line as follows:

$$R = \frac{\text{no. of nets separated by the vertical line}}{\text{no. of available tracks across the vertical line}}.$$

If the ratio is greater than 1, then nets cannot possibly be routed. If the ratio is slightly less than 1, we certainly do not want to use more than one wire across the vertical line to connect a separated net. Thus, in the cutting stage, we look for a vertical line (or a horizontal line) which has the maximum ratio and allow only one wire per net to cross this vertical line. This line will cut the chip into two regions, say the left and the right regions. For the left region, we find the horizontal line with the largest ratios in the left region. This will cut the left region into an upper left and a lower left region. Similarly, we can cut the right region into an upper right and a lower right region. The result after these 3 cuts is shown in Fig. 2. Each region will be cut into smaller regions by vertical and horizontal lines successively until the number of arcs within a region is small enough or the numbers of pins of most nets are less than four in the region. (It is not necessary to cut the area by vertical and horizontal lines alternately.)

We can represent the successive partitioning as a binary tree, where the root of the binary tree corresponds to the whole chip. The two sons of the root are the left and the right regions, and the two sons of the left region are the upper left and the lower left regions, and so on. The small subregions at the end of the partition are called *atomic* regions and they correspond to the leaves of the binary tree. For example, the chip shown in Fig. 3 is partitioned into ten regions and its corresponding binary tree is shown in Fig. 4. We shall call this binary tree the *partition tree* of the chip. (If we do not cut the area by vertical and horizontal lines alternately, then some of the sons of nodes are empty.)

Assume that the computer can handle a matrix $[a_{ij}]$ with four hundred rows, then we partition the regions into smaller and smaller regions until the number of arcs in a region is less than one hundred, approximately one quarter of the capacity. Within a single region, the problem of routing is again a linear program like equation (3).

Note that a net may not have all its pins in one region, e.g., it may have three pins in one region and the fourth pin is in the adjacent region. In formulating the routing problem as a linear program in one region, we use a column of the matrix $[a_{ij}]$ to represent a way of connecting pins inside the region. Thus, we have three problems:

1) how to construct a minimum Steiner tree connecting pins in one region [16];
2) how to connect pins not in the same region; and
3) what to do if there exists a net which cannot be connected.

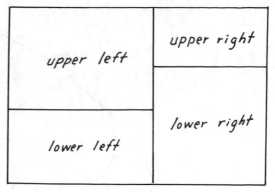

Fig. 2.

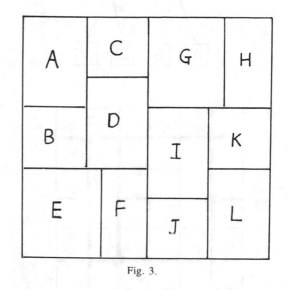

Fig. 3.

B. Routing Within a Region

We can relax the integer constraints and solve the relaxed linear programming using column-generating techniques. Based on the shadow prices, we generate the next column by finding a minimum Steiner tree with arc lengths corresponding to the shadow prices. Unfortunately, the problem of constructing a Steiner tree with minimum total arc length on a grid-graph is NP-complete [9]. However, in circuit routing, the average number of pins per net is between 2.5 and 3.5. Thus, we can assume that a net normally has less than five pins. (Of course, a power line, a ground line, or a bus has many pins, but those are usually fixed ahead of all other nets.) When a net has less than 5 pins, then its minimum Steiner tree on a grid-graph can be easily constructed. For an arbitrary number of pins, Hanan [12] proved that there always exists a minimum rectilinear Steiner tree which uses arcs having the same x (or y) coordinates as one of the pins. For example, there are three pins in Fig. 5 and there always exists a minimum Steiner tree which uses only the heavy-line-segments. We shall call the class of trees which uses only the heavy-line-segments the *main-street class M*. Hanan's theorem says that there exists a rectilinear Steiner tree of minimum total length in the class M where the length of the tree is the number of arcs in the grid-graph. When the arcs have different lengths in the grid-graph (i.e., different shadow prices), Hanan's theorem no longer holds. However, the class M has a very interesting property,

147

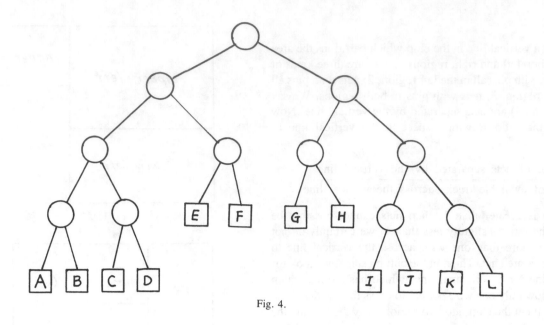

Fig. 4.

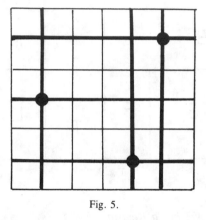

Fig. 5.

namely, the tree has the minimum number of holes or corners among all the trees which span the same set of pins. In VLSI technology, the holes are really metal contacts called the *vias*. For two paths of the same length, the one with fewer number of vias is preferred. The balance between the number of vias and the length of wires can be handled by α-β routing (see [15]). Thus, we shall restrict our search to a minimum Steiner tree within the class M.

If we cannot connect all the pins of a net in an atomic region by the decomposition algorithm, we can either consider it a failure or we can extend our search beyond the main-street class M.

C. Pasting

Now, suppose the pins in each region have been connected by a Steiner tree. We shall paste all the subtrees of the same net in the adjacent regions together, starting at the bottom of the partition tree and working towards the root.

For a net having pins (i.e., subtrees) in two or more atomic regions adjacent to a cut, as shown in Fig. 6, we project all the pins (or the horizontal mainstreets) onto the common boundary. (The projected points on the boundary are shown as *x-marks* in the figure.) Since the common boundary is the line with maximum ratio, we would like to connect these pins in

the two regions with a Steiner tree which crosses the common boundary only once. In other words, we first find a minimum-length connection from each x-mark on the common boundary to the two spanning trees in the two regions and then pick the minimum one among all these minimum-length connections. Here, the connection of the pins in a net is again restricted to its main-streets in both regions.

If there are more than one net in the atomic regions adjacent to a cut, we can formulate the pasting problem as a linear integer program similar to (3). We again associate a variable y_j with each way of connecting the two subtrees of a net via an x-mark on the common boundary of the atomic regions. Each way of connecting the subtrees, which is restricted to the main-streets of the corresponding net in the regions, is represented by a column of a $(0, 1)$ matrix $[a_{ij}]$. We again denote the set of y_j's which correspond to various ways of connecting the subtrees of the kth net by N_k. Assume that there are p nets and a total of n ways connecting all the nets via the x-marks on the cut, we have

$$\max \sum b_j y_j \qquad j = 1, 2, \cdots, n \qquad (5)$$

$$\text{subject to} \sum_{y_j \in N_k} y_j = 1 \qquad k = 1, 2, \cdots, p$$

$$\sum a_{ij} y_j \le c_i \qquad i = 1, 2, \cdots, m$$

$$0 \le y_j \le 1, \qquad \text{integers.}$$

Like the linear program (3), we do not need to write down every column in (5). Let π_i be the shadow price of the ith row under the current basis of the linear program (5), and let $\bar{b}_j$ be $\bar{b}_j - \pi_i a_{ij}$, the maximum $\bar{b}_j$ corresponds to the minimum-length path of connecting two subtrees of a net via a x-mark on the cut. In other words, we want to find a shortest path connecting the two subtrees of a net in the grid-graph where the arc lengths are defined by the shadow prices.

Assume that a net has pins in the atomic regions not adjacent to the cut, say in regions A, E, H, L as shown in Fig. 7. We shall first collapse these regions by projecting the pins in these

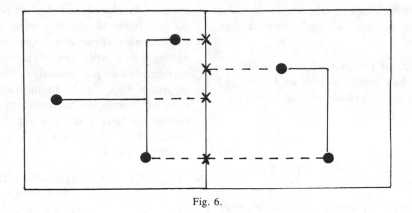

Fig. 6.

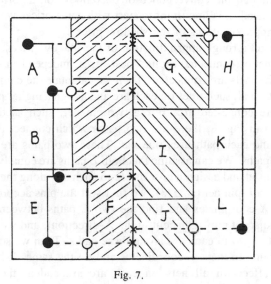

Fig. 7.

regions on to the boundaries of their *sibling* regions which are adjacent to the cut, i.e., the shaded regions in Fig. 7. (Two atomic regions are called *siblings* with respect to a cut if they are at the same side of the cut.) The images of these pins are presented by circles in Fig. 7. Then, we can treat the circles along the boundaries as if they are pins in the shaded regions and solve the pasting problem using linear integer program [6].

Hence, a Steiner tree spanning any number of pins on the chip is always reduced to a Steiner tree in the shaded region, as shown in Fig. 7. If a net has k pins, then there are k x-marks on the cut and the worst case requires $2k$ shortest path problems within the class M [20], [22].

If we cannot connect two subtrees together, we can either consider it a failure, or we can extend the pasting algorithm to handle the situation where the pins on one side of the cut cannot be connected by a spanning tree, but by two disjoint subtrees. In this case, we try to find a connection from the x-marks on the common boundary to the two subtrees on one side of the cut and to the spanning tree on the other side of the cut.

IV. OUTLINE OF THE CUT-AND-PASTE ALGORITHM

Let us summarize the cut-and-paste approach to the routing problem.

⟨1⟩ Cutting

Let the whole chip be the root of the partition tree. We first cut the chip into smaller regions using alternating vertical and horizontal cuts as follows:

While {there exists a region with more than 100 arcs in the corresponding grid-graph} and {there exists a net with more than 4 pins in the region} do

begin

1a. Find the vertical cut (or a horizontal cut) which has the maximum ratio R.

This is equivalent to allowing one wire per net to cross the vertical boundary. Since the vertical cut is the most critical boundary, this is well justified. If we use the cut-and-paste approach for global routing, where the global cells are already determined, we could select the cuts along the boundaries of the global cells. In case of tie, a cut close to the middle part of the region should be selected.

1b. Cut the region into two parts, and record the two subregions as the children of the original region in the partition tree.

end.

⟨2⟩ Routing within a region

For {each atomic regions obtained in Step 1} do

begin

Formulate the routing problem within a region as a linear program (3) and solve it by the Dantzig–Wolfe decomposition method.

This is a maximum packing problem, where all nets (or subnets) within the region compete for the available space. Since there are so many spanning trees for a single net, the number of columns in the linear programming formulation is almost infinite and cannot be written down explicitly. Here the problem of generating the best column to enter the basis is equivalent to constructing a minimum Steiner tree on a grid-graph with arcs having different lengths (shadow prices).

The subproblem of generating a minimum Steiner tree is not hard as most nets have less than five pins. Since the VLSI technology prefers minimum vias in a tree, we search for the minimum Steiner's tree within the class M. This reduces the number of vias as well as the computational effort.

If we cannot connect all the pins of a net within an atomic region, we shall mark that net as failure and ignore these pins for the rest of the pasting process.

end.

⟨3⟩ Pasting

After all nets are connected within each atomic region, we try to connect nets having pins in different regions.

We shall paste the subnets in various regions together following the post-order traversal sequence of the partition tree, i.e., we shall consider the two leaves which have a common father first, and then successively work towards the root of the partition tree.

The pasting is done as follows:

begin

3a. For {each net with pins on both sides of the cut} do

begin

i. For {each pin in an atomic region not adjacent to the cut} do

begin

Project a circle on the boundary of its sibling region which is adjacent to the cut.

end.

ii.

For {each pin and circle in the atomic regions adjacent to the cut} do

begin

Project a x-mark on to the cut.

end.

end.

3b. Formulate the pasting problem as a linear program (5) and solve it by the Dantzig–Wolfe decomposition method.

The problem of generating the best column to enter the basis is equivalent to constructing a shortest path connecting the two subtrees on a grid-graph with arcs having different lengths (shadow prices). Here, we again restrict our search to paths intersecting the cut at the x-mark only. Naturally we could allow more x-marks on the boundary but this would increase the amount of computation.

end.

⟨4⟩ Handling the failure

Assume that the total number of arcs in G' is small enough and we can formulate the whole chip as a linear program-

ming. If the algorithm fails then no feasible solution exists. So we have to do the placement phase again. The suboptimality of our algorithm comes from the cutting and pasting. However, since each cut is a line with the maximum ratio, it is unlikely a feasible solution exists if the algorithm fails. In an atomic region, if a subset of pins cannot be connected, we may leave them disconnected as k forests and later allow k wires to cross the boundaries to connect them. (Here, if $k \geq 3$, it is unlikely that a feasible solution exists.)

V. THE STARTING SHADOW PRICES OF THE LINEAR PROGRAM

Although the convergence of decomposition algorithms is guaranteed, it is important to start with some prices which are first approximations of the shadow prices.

To get a rough approximation of the shadow prices, we let each net be connected by its "best" tree, independent of how all other nets are connected. Then we calculate the cumulative effects of all such trees on each arc and assign a higher price to a more congested arc. For a two-pin net, such as the one hown in Fig. 8, there are two paths which connect the two pins and each path uses one via. These two paths are called "L" paths. We can assume that both L paths are equally likely to be used and assign a cost of 1/2 to each arc along the paths.

For a k-pin net ($k \geq 3$), we first sort the pins according to their X-coordinates and find the two L paths between every two adjacent pins (adjacent in the X-direction), and we assign a cost of 1/4 to each arc along these paths. Then we sort pins according to their Y-coordinates and do the same. When the total effects of all nets on each arc are added, the value associated with each arc indicates its price relative to the prices of the other arcs.

We can refine the approximation by considering two-vias paths between two pins. These are called Z paths (also called "dog legs" in channel routing). There are many Z paths between two pins. For two pins which are p arcs apart in the X-direction and q arcs apart in the Y-direction, there are $p + q$ paths which use two-vias or less. Assume that all these paths are equally likely to be used. It is very easy to calculate the relative price of each arc. A cost of $r/(p + q)$ will be added to an arc if r out of the $(p + q)$ Z paths go through that arc. In Fig. 9, we show the total effects of the seven paths between two pins, with the relative prices written beside the corresponding arcs.

VI. CONCLUSION

We have discussed the problem of routing, i.e., connecting pins with fixed positions on the chip. In practice, before the pins are assigned to fixed positions on the chip, the designer first designs a module (a rectangle with pins on the perimeter of the rectangle) for each unit of the integrated circuits. After all modules are completed, the designer places the modules on the chip. This is called module placement, or placement. Then the designer has to lay down the wires to connect the pins between various modules. Hence, the circuit layout problem consists of placement followed by routing.

The placement problem was formulated as a quadratic

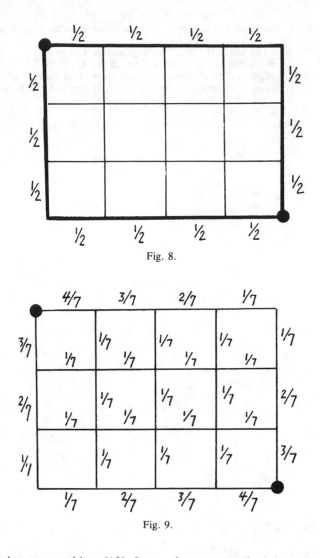

Fig. 8.

Fig. 9.

although the condensation of nodes in a network was used in 1961 [11] and is used here to explain why the global-cell approach works.

We have implemented the algorithm using the programming language "C", and are testing the programs with several large circuit layout files from industry. Based on this cut-and-paste approach, we will test the various heuristics in calculating the starting relative prices of the arcs, and the various heuristics in generating the best column in the decomposition algorithm.

Many problems in layout are maximum packing problems, packing of rectangular blocks or rectilinear trees into limited space. All such problems can be formulated as mathematical programming problems. Naturally, we can propose many sophisticated algorithms and a theory but

The final test of a theory is its capacity to solve the problems which originate it.

—G. B. DANTZIG (*Linear Programming and Its Extensions*, 1963)

REFERENCES

[1] M. Breuer, "Min-cut placement," *J. Design Automation and Fault Tolerant Comput.*, vol. 1, pp. 343–362, 1976.

[2] K. A. Chen, M. Feuer, K. H. Khokhani, N. Nan, and S. Schmidt, "The chip layout problem: An automatic wiring procedure," in *Proc. 14th Design Automation Conf.*, pp. 298–302, 1977.

[3] H. Crowder, E. L. Johnson, and M. W. Padberg, "Solving large-scale zero-one linear programming problems," *J. Opt. Soc. Amer.*, vol. 31, no. 5, pp. 803–834, 1983.

[4] G. B. Dantzig, "Maximization of a linear function of variables subject to linear inequalities," in *Activity Analysis of Production and Allocation*, T. C. Koopman, Ed. New York: Wiley, 1951, pp. 339–347.

[5] G. B. Dantzig, *Linear Programming and Its Extensions*. Princeton, NJ: Princeton University Press, 1963.

[6] G. B. Dantzig and R. M. VanSlyke, "Generalized upper bounded techniques for linear programming," *J. Comput. Syst. Sci.*, vol. 1, pp. 213–226, 1967.

[7] G. B. Dantzig and P. Wolfe, "The decomposition algorithm for linear programming," *Econometrica*, vol. 29, no. 4, pp. 767–778, 1961.

[8] L. R. Ford and D. R. Fulkerson, "Suggested computations for maximal multi-commodity network flows," *Management Sci.*, vol. 5, no. 1, pp. 97–101, 1958.

[9] M. R. Garey and D. S. Johnson, "The rectilinear Steiner tree problem is NP-complete," *SIAM J. Appl. Math.*, vol. 32, pp. 855–859, 1977.

[10] R. E. Gomory, "Large and non-convex problems in linear programming," in *Proc. Symp. Appl. Math.*, 1963.

[11] R. E. Gomory and T. C. Hu, "Multi-terminal network flows," *J. SIAM*, vol. 9, pp. 551–570, 1961.

[12] M. Hanan, "On Steiner's problem with rectilinear distance," *SIAM J. Appl. Math.*, vol. 14, pp. 255–265, 1966.

[13] M. Hanan and J. M. Kurtzberg, "A review of the placement and quadratic assignment problems," *SIAM Rev.*, vol. 14, pp. 324–342, 1972.

[14] T. C. Hu, *Integer Programming and Network Flows*. Reading, MA: Addison-Wesley, 1969.

[15] T. C. Hu and M. T. Shing, "The α-β routing," Tech. Rep. CS-077, Dept. Elec. Eng. and Comput. Sci., Univ. California, San Diego, La Jolla, CA, 1983. (This book; page 139.)

[16] F. Hwang, "On Steiner minimal trees with rectilinear distance," *SIAM J. Appl. Math.*, vol. 30, no. 1, pp. 104–114, 1976.

[17] R. M. Karp, "Probabilistic analysis of partitioning algorithms for the traveling salesman problem in the plane," *Math. Oper. Res.*, vol. 2, no. 3, pp. 209–244, 1977.

[18] D. E. Knuth, *The Art of Computer Programming*, vols. 1–3. Reading, MA: Addison-Wesley, 1968, 1969, 1973.

[19] E. S. Kuh, Ed., "Special issue on routing in microelectronics," *IEEE Trans. Computer-Aided Design*, vol. CAD-2, no. 4, 1983.

[20] C. Y. Lee, "An algorithm for path connections and its applications," *IRE Trans. Electron. Comput.*, vol. EC-10, pp. 346–365, 1961.

assignment problem [13]. Later, the concept of minimum cut was used as a criteria for placement [1]. Currently, people realize the placement problem is an integer programming problem and branch-and-bound algorithms can be used to solve the problem very successfully [26]. Eventually, the success in large zero-one linear programming problems [3] will help to solve the placement problem.

The earlier papers on routing are basically shortest path algorithms (see [20], [22]) which find the connections of nets sequentially. When it fails to connect all the nets, some of the connected nets may have to be re-routed to leave room for other nets. Interested readers should refer to the special issue on routing in microelectronics [19]. This special issue gives a lot of updated information about what is being used in industries today, and there are many diverse and clever heuristic algorithms for the complicated and real design problems.

In this article, we formulate the routing problem as a large linear program and solve it using the decomposition principle. The reason why many people have not used this approach before may be due to the lack of communications among researchers in different fields. Also, the size of the problem forbids the direct application of the decomposition principle published in 1961 [7]. The division of routing into global routing and final routing is a recent event ([2], [23], [24]),

[21] C. Mead and L. Conway, *Introduction to VLSI Systems*. Reading, MA: Addison-Wesley, 1980.

[22] E. F. Moore, "The shortest path through a maze," Bell Telephone Lab. Rep., Monograph 3523, Murray Hill, NJ 1959.

[23] R. Nair, S. J. Hong, S. Liles, and R. Villani, "Global wiring on a wire routing machine," in *Proc. 19th Design Automation Conf.*, 1982.

[24] J. Soukup and J. C. Royle, "On hierarchical routing," *J. Digital Syst.*, vol. 5, no. 3, 1981.

[25] B. S. Ting and B. N. Tien, "Routing techniques for gate array," *IEEE Trans. Computer-Aided Design*, vol. 2, no. 4, pp. 301–312, 1983.

[26] W. Widjaja, "An effective structured approach to finding optimum partitions of networks," *Computing*, vol. 29, pp. 241–262, 1982.

Routing Techniques for Gate Array

BENJAMIN S. TING, MEMBER, IEEE, AND BOU NIN TIEN, MEMBER, IEEE

Abstract—This paper describes the routing techniques used for a Hughes internally developed high-density silicon-gate bulk CMOS gate array family. This layout software can be easily adapted to different array sizes and/or technologies (e.g., bipolar) through a change of parameters. A routing model and hierarchical decomposition schemes are presented to address the routability issue. More specifically, this paper focuses on the formulation and analysis of global routing and vertical assignment problems and gives a systematic breakdown of the routing task into well-defined subtasks. Instead of performing sequential routing, techniques and formulations are introduced to achieve a high degree of order independency in all subtasks. In routing subtasks where iterations are required, independent selection and interconnection are performed to avoid order dependency in typical routing problems. Implementation results are provided to indicate the efficiency of the system.

I. INTRODUCTION

THE USE OF gate arrays as a standard part chip for fast turnaround design has been successfully demonstrated by the introduction of IBM's E and H series mainframe computers. One key element for the gate array product success is its regular structure amenable to layout automation. Gate arrays look more like a standard part than other design methods; they are easier to standardize for packaging and provide a fast design-turnaround cycle. Layout automation systems dealing with gate arrays are well published [1], [2], but not well understood by the industry as can be seen by the lack of efficient software in the industry (except IBM) and sometimes even the wrong implementation philosophy in their programs (e.g., using a force-directed method in placement instead of considering local congestion during placement).

There are several readily identifiable problems in gate array layout automation. In the wiring stage, one of the most troublesome decision-making processes is the ordering of nets to be wired [3], [4]. Sequential routing usually results in later routing difficulties for remaining signal nets due to lack of look-ahead ability. No intelligent scheme has yet been devised to determine a good ordering in routing signal nets and avoid future wiring bottlenecks. Conventional methods such as Moore [5], Lee [6], or line-search [7] types of algorithms generally have limited intelligence to handle the aforementioned problems. Channel routing [8]–[10] at least offers a partial solution in that it packs net segments INDEPENDENT of ordering in a simplified environment. However, many decomposition stages are necessary before channel routing can be applied, and those earlier wiring stages are usually order dependent in both the selection and routing processes.

Manuscript received March 10, 1983; revised April 24, 1983.
The authors are with the Hughes Aircraft Company, Newport Beach, CA.

The global wiring approach [1], [11] was introduced to partially emulate the independent net routing process. All signal nets have equal opportunities to compete for wiring spaces. Congested wiring areas are then analyzed. Signal nets can be selected and rerouted away from the congested areas if there are alternative paths for such rerouting. The order-dependency problem usually comes back during the net selection and rerouting stage.

In this paper, the global-wiring problem is further examined and a formal treatment of the problem is introduced in a graph theoretic approach. Not surprisingly, as one would expect, the problem is in the NP-complete class. Generalized formulations and heuristic algorithms are introduced both as future research topics and/or practical implementation algorithms for solving the gate array layout problem. Specifically, the routing order-dependency problem is addressed, and a procedure is developed to avoid the order-dependent net-selection process and to reroute the nets in an independent fashion. Extensions of ideas presented here can be applied to hierarchical layout system [4], [12] in the routing process to reduce wiring area and even out wiring density distributions on the chip. Instead of using routability models discussed in the literature [3], [14], a more empirical technique is used for routability prediction during design. The concepts of global routing, vertical track assignment, channel routing, and clean-up mazerunner are discussed. The key ingredient in this routing scheme is the proper mix of routing techniques which culminates in an efficient system. Some theoretical formulations, models, and routing results are presented.

II. GATE ARRAY ROUTING ENVIRONMENT

Fig. 1(a) shows a symbolic representation of a portion of a gate array without any interconnection. First-level metal runs horizontally (in the routing channels), parallel to the active area, and second-level metal runs vertically, crossing both the active area and routing channels. The basic gate-array cells are arranged back-to-back in dual-cell rows as shown in both Fig. 1(a) and abstracted in Fig. 1(b). The routing channel area is substantially smaller than the active area in this 3-μm bulk CMOS technology, and the ratio is approximately 3 : 5. More specifically, there are a total of 230 first-level metal routing tracks and 390 second-level metal routing tracks for one of the gate array family of chips containing over 23 000 transistors in a die size slightly less than 300 mils on the side. If all the routing tracks are strung linearly on a line, there are approximately 4 m of routing metal available in this particular gate array. The goal for automatic layout is to achieve, routinely, near 100-percent routing completion for designs using 75 to

Reprinted from *IEEE Trans. Computer-Aided Design of Integrated Circuits and Systems*, vol. CAD-2, pp. 301–312, Oct. 1983.

153

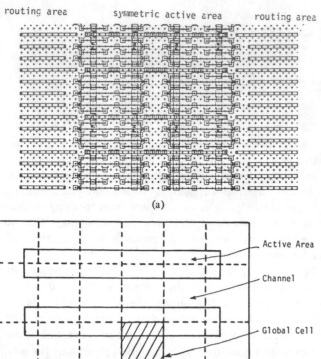

Fig. 1. (a) A symbolic representation of a portion of gate array. (b) Global routing environment.

80 percent of the active area. A macro library design approach (a macro can be a simple inverter or an ALU using large sections of the chip) is assumed. The routing system takes, as control parameters, a set of design constraints (e.g., first-level metal routing tracks, second-level metal routing tracks, design rules, global cell boundaries, generic routing map information for blocked and unblocked areas of the chip, etc.) and uses them to build up the routing environment. This approach allows the routing system to be independent of technology. By changing control parameters the system can adapt to more complex and larger arrays, with different processing technologies, provided the new arrays can be modeled as a two-level interconnection problem.

III. PLACEMENT AND ROUTING CONSIDERATIONS

1. Placement

The first phase of gate array layout problem is the logical gate to physical location mapping or the placement problem. Placement consists of using the elements from an established macro library by placing those elements in a pattern that allows efficient use of the routing area. Clustering, constructive placement, and iterative exchange are the common procedures used in approaching the problem. Because of the complexity

involved in this phase [2], [15], there are two prime considerations: one is to break the problem into manageable size (clustering, limited scope of iterative exchange), and the other is to have an accurate estimate on congestion with a given placement. The approaches discussed in [15] are generally used for minimization of chip area: the key factor there is to minimize wire length or chip area; local wire congestion is not a problem because more routing space can always be provided by expanding the chip to finish 100-percent routing. However, in gate arrays, such a luxury is not available. Redistribution or avoidance of bottlenecks is generally the criterion in gate array layout.

2. Global Routing

The second phase of the gate array layout problem is global routing. The array chip is first superimposed by coarse grids as shown in Fig. 1(b). Each area enclosed by the grid lines is called a global cell. In this step, global routes for all signal nets are generated on coarse grids. The advantage of using a coarse grid is the reduction of problem complexity. Using a detailed routing map for a 23 000-transistor gate array, there are well over 600 000 two-level grids, including the active area. With global grids, the number is reduced to 580 (20 X 29). This gives us a chance to perform computationally intensive operations during this global routing stage. The exact location of each wire during this stage is deferred to the detailed routing stage. The locations of the global cell boundaries are very crucial to the performance of the global routing stage. Different gate array architectures call for different implementation of the global coarse grids (e.g., nonuniform grids), and have to be considered very carefully. Our example shows the basic concept.

3. Detailed Routing

Once the placement and global routing phases are finished, the program will then perform detailed routing assigning each signal a specific physical track. The process of translating global routing information into detailed signal paths involves finalization of both the y- direction and x-direction routes. For each horizontal global cell boundary with signal-nets crossing count (*demand*) less than or equal to the available routing tracks (*supply*), a detailed routing stage will make assignment of boundary crossing locations for those signal nets. This process is generally called Vertical Assignment. This stage fixes the precise vertical-crossing locations for all global routes in each global cell. This decomposes each global route pattern into signal subnets where each subnet is confined within a channel routing area. After the assignment stage, the channel routing stage packs all horizontal wire segments and passes the failed subnets to a clean-up mazeruner for final completion. The mazerunner uses a modified Lee-type algorithm. This maze routing stage finishes up any remaining unconnected subnets in a localized area.

IV. GLOBAL ROUTING AND ROUTABILITY ISSUE

As mentioned earlier, the basic idea in global routing is to first superimpose a coarse grid map onto the chip image. Global routing patterns for signal nets are then assigned over

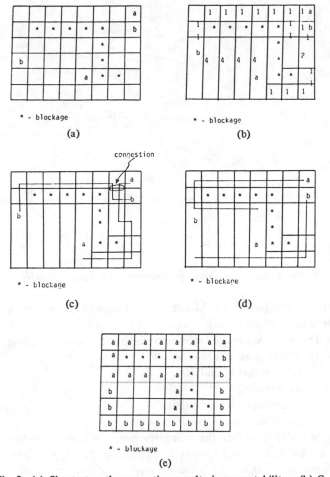

* - blockage

(a)

* - blockage

(b)

congestion

* - blockage

(c)

* - blockage

(d)

* - blockage

(e)

Fig. 2. (a) Shortest path connection results in unroutability. (b) Capacity with a nonuniform global grid pattern. (c) Using shortest path to generate global path, congestion detected. (d) Rerouted global path. (e) A solution.

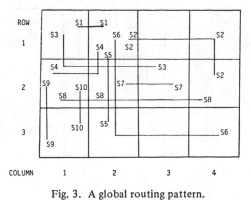

Fig. 3. A global routing pattern.

the global cell map as follows. For each signal net, a corresponding tree is constructed over the global cells. This signal tree can be constructed by either using Steiner tree construction [16], [17] or simply using a Lee-type algorithm to generate a minimum spanning tree. The important point is that each tree over the global cells is constructed INDEPENDENT of others. This gives equal chances for all nets to find their natural interconnect patterns. With the assigned global routes for all signal nets, the demand at each global cell boundary can be precisely calculated. The local connections within each global cell are ignored in this model. Therefore, the routing demand over the global cell boundary represents a lower bound for physical realization with this particular global routing pattern.

A congestion spot is identified when the demand exceeds the supply at any boundary. Rerouting of nets over global cells to reduce the congestion follows. Prior to the actual rerouting stage, two steps are involved: first, congested global cell boundaries are identified, and second, the signal nets over those boundaries are identified for rerouting. Fig. 2 illustrates the concept of global routing with two signal nets. Fig. 2(a) gives a routing problem on a detail grid map. Fig. 2(b) abstracts the problem into variable-size global grids with supply indicated on each boundary, and Fig. 2(c) gives an initial rout-

ing pattern using the shortest path algorithm with local congestion identified. Fig. 2(d) shows a rerouting pattern avoiding the congestion and Fig. 2(e) shows a detailed implementation.

Essentially, if the demand exceeds the available resource at any location, the chip layout cannot be realized due to congestion. Congestion can be alleviated by selecting signal nets and rerouting them away from those locations. If there are congested locations remaining after the global routing stage, then a potential unroutable situation exists. This information can help the designer to either change placement of circuits or change the partition of the original logic into different chips.

Fig. 3 shows an example of a global routing pattern over a global cell map. A boundary-crossing supply matrix over all cells can be calculated for a given placement. It is apparent from the figure that making demand less than or equal to supply is a necessary condition for feasible realization of the global routes.

We shall now discuss in more detail the two phases of global routing, i.e., 1) initial routing phase, and 2) iterative rerouting phase.

1. Initial Routing

The global route for each signal net is a tree in a graph. Initial construction of all signal trees is with a modified Lee-type algorithm. Each tree is constructed INDEPENDENT of others (regardless of routing congestion). Thus after the construction, each signal net tends to be connected with shortest wire length and is in its most "natural" state. This allows each net to compete for routing resources on an equal basis. Hence, no net-ordering problem is encountered at this phase. When all the nets are routed, a boundary-demand matrix is constructed; the overflow matrix is generated from this. Overflow is defined to be the difference between the demand and the supply. Positive overflows indicate routing congestion and demand needs to be redistributed to satisfy the supply constraint. The cell boundary-crossing demand by the signal nets in Fig. 3 can be estimated given the global routes. For instance, at the right-hand side boundary of cell coordinate (column 2, row 2), nets $S3$, $S7$, $S8$ are crossing the boundary: the demand is 3. At the upper boundary, nets $S4$, $S5$, and $S6$ cross: the demand is 3.

Suppose the boundary-crossing supply for each cell is two, except for the outer boundaries where supply is zero. There are three overflow boundaries: the right-side boundary of cell

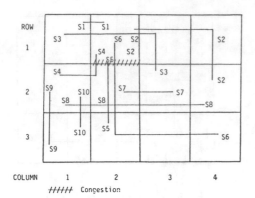

Fig. 4. Congestion cannot be alleviated with supply = 2.

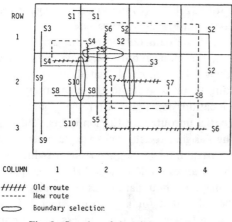

///// Old route
------ New route
⬭ Boundary selection

Fig. 5. Supply = 2 for all boundaries.

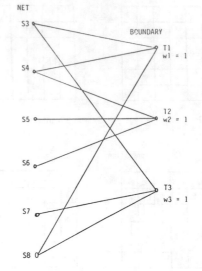

Fig. 6. Bipartite graph representation for net and boundary.

(2,2), the top-side boundary of cell (2,2), and the left-side boundary of cell (2,2). In this case there are three specific global cell boundaries to be dealt with to reduce congestion or to eliminate overflow. In order to facilitate wirability, rerouting of some nets must be performed to eliminate the overflow of those boundaries.

Fig. 4 shows a case where net $S3$ is singled out to reroute. With the particular rerouted pattern, one congested cell boundary remains where nets $S4$, $S5$, and $S6$ are crossing the boundary. Attempts to reroute any of the preceeding three nets will result in congestion of other global cell boundaries.

In Fig. 5, there is actually a solution involving rerouting three signal trees for nets $S4$, $S6$, and $S7$. What must be accomplished here is to defer or avoid sequential selection of nets and to give all nets equal consideration during selection and rerouting phases.

The above considerations then lead us to the next phase of global routing.

2. Iterative Rerouting

This phase is an iterative process that is repeated until no positive overflow exists (all demands are satisfied), or no more improvement is possible (local congestion, if any, is identified). During each iteration, a group of nets is chosen to be rerouted. If the new routes achieve a reduction in the overflow count, then the new routes replace the old ones. Otherwise, the itera-

tion is considered a failure, and no change in the routing pattern takes place.

There are two major steps in the Iterative Rerouting phase:
1) selection of nets to reroute:
   a) selecting boundaries,
   b) selecting nets;
2) rerouting of selected nets.

i. Selection Step

In order to reduce the boundary overflow count, it is necessary to identify and reroute some signal nets away from overflow boundaries. The selection step finds a set of signal nets whose removal will reduce total overflows. Care must be taken in this process so that the removal process is not overdone (e.g., the overflow boundaries become under utilized.) Otherwise an obvious solution would be removing every net associated with the boundaries.

A formal treatment of the problem at this selection step is as follows. For signal nets associated with the overflow boundaries, construct a bipartite graph G where an arc between a signal net (Si) vertex and overflow boundary vertex (Tj) exists if and only if the signal net crosses the overflow boundary. Starting with the left vertical boundary of cell (2,2) clockwise in Fig. 5, each overflow boundary is denoted as $T1$, $T2$, and $T3$. Thus the graph associated with the overflow situation in Fig. 5 is depicted in Fig. 6.

Each Ti vertex can have a weight wi indicating the amount of overflow. In the case shown in Fig. 5, the corresponding weight for each vertex Ti of Fig. 6 is 1. In general, the weight associated with vertex Ti is the overflow count number of the associated boundary.

The objective at the selection step is thus:

To find a set of vertices associated with the signal nets to form a cover such that every vertex associated with the overflow boundary is covered ki times where ki is the weight of the ith overflow boundary. The cover should be of minimum cardinality or an obvious solution is all vertices associated with the signal nets.

In the special case where the weight associated with each

boundary is exactly one, the earlier problem is simply a minimum set cover problem [18]:

Instance: Collection C of subsets of a finite set T, positive integer $K <= |C|$.

Question: Dose C contain a cover for T of size K or less?

The preceding problem is NP-complete in the sense of Cook and Karp [19], [20]. The general problem where the weight of each Ti vertex is greater than or equal to one can also be cast into an integer programming problem:

Let Si to be a vertex associated with a signal-net crossing at least one overflow boundary and Tj to be a vertex associated with an overflow boundary where $i = 1, \cdots, m$ and $j = 1, \cdots, n$. m is the total number of Si's and n is the number of overflow boundaries. An incidence matrix A $(n \times m)$ is defined as follows:

$A(i, j) = 1$ if net associated with Sj crosses the overflow boundary associated with Ti, and

$A(i, j) = 0$ otherwise.

The problem then becomes:

$$\text{Min } x1 + x2 + \cdots + xm$$
subject to
$$A \cdot X^T >= W^T, \quad \text{where}$$
$$W = (w1, w2, \cdots, wn),$$
$$X = (x1, \cdots, xm),$$

    wi : the overflow weight associated with boundary i

    xi : 0 or 1.

The preceeding problem belongs to NP-complete class. In practical implementation, it is not necessary to find a minimum solution. Instead, a minimal solution satisfying certain criteria is acceptable.

Due to the iterative nature of the reroute phase, simplicity and efficiency considerations are often of concern. Therefore, the boundary selection process is introduced to restrict the selection step to a subset of signal nets. This biases towards the reduction of overflow counts associated with selected global boundaries. After the boundary selection, the net selection process further selects a subset of nets to reroute.

a) Boundary selection:

One way to select boundaries is to use maximum overflow count as the main criterion. This has the effect of smoothing out the maximum overflow peak. Another method of selecting boundaries is to choose a set of boundaries and pick up one net out of each boundary to reroute. This has the effect of smoothing out all the overflow peaks associated with the selected set uniformly.

Our selection procedure is to choose k largest overflow boundaries as candidates for boundary selection where k is determined by the maximum overflow count. This selection has the effect of smoothing out several overflow peaks simultaneously rather than that of biasing toward a particular peak.

b) Net Selection:

Nets associated with the boundaries selected in subsection a) are then selected for rerouting to eliminate or reduce overflow

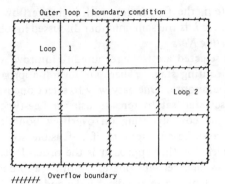

Fig. 7. Identification of overflow boundaries.

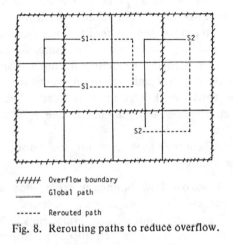

Fig. 8. Rerouting paths to reduce overflow.

counts. A simple procedure is to choose nets with high overflow counts. A later section discusses the procedure to reroute those selected nets to reduce the total overflow count.

The net selection problem is as follows. Fig. 7 shows a set of overflow boundaries, where each overflow boundary is marked. Consider each outside boundary as having a positive overflow count because no wire can be routed across the chip boundary. The following statement can be made.

If there is a loop formed by the overflow boundaries, the total overflow count along the boundary loop will not be reduced by rerouting nets associated with those boundaries unless it is a net crossing this boundary set more than once, and there exists a rerouting path lying entirely within or outside the loop as shown by Fig. 8.

The net selection rule bypassing boundary selection can be stated:

1) Find all loops formed by the overflow boundaries.
2) For each loop find existence of a net (or nets) that cross the boundary more than once in the loop. If such a case exists, the net is chosen.
3) If 1) exists, and 2) does not exist, then no net can be chosen to give improvement to eliminate overflow associated with the loop(s).
4) If 1) does not exist, then use a simple greedy algorithm to select a set of nets discussed earlier with the bipartite graph formulation using all overflow boundaries as the T vertices.

In the system, the first k nets with high overflow counts are selected where k is the same number discussed in subsection a).

ii. Rerouting Nets

The nets selected are to be rerouted to improve wirability in the global routing stage. There are several approaches to accomplish the reroute. One easy way to select one net at a time from the selected set to reroute using a Lee-type algorithm with dynamic cost updating on overflow boundaries. This guarantees a successful reroute if a feasible solution exists. The drawback for this approach is the order-dependency nature in the actual reroute phase. Another approach is to reroute each net of the selected set independently. This necessarily means no dynamic updating of overflow boundaries count during the rerouting process. The drawback, however, is uncertain convergence. Additional overflow may be introduced in the rerouting process resulting in an overall increase of overflow counts (worse results). To insure convergence, the nets can be rerouted independently by not dynamically updating overflow counts and accept an overall reroute pattern only if the total overflow count is reduced. The penalty is increased computational requirements due to a slower convergence rate.

The overall control procedure for the rerouting stage is described as follows.

Step 1) If no overflow boundary exists, stop and report success. Otherwise, select boundaries.

Step 2) Select nets from selected boundaries in Step 1). If there are no more nets to select, stop and report failure.

Step 3) Reroute selected nets in Step 2). If there is improvement in total overflow count, then accept the solution and go back to Step 1). Otherwise, go back to Step 2).

When the algorithm terminates unsuccessfully, it identifies a set of global boundaries which, due to supply limitations, allow no improvements in reducing congestions. This provides information for further analysis by the designer to either perform replacement or repartitioning of logic.

Another feature included in global routing to improve wirability at the detail level is the inroduction of a SLACK factor in the overflow computation:

$$\text{OVERFLOW} = \text{DEMAND} + \text{SLACK} - \text{SUPPLY}.$$

Inclusion of a slack factor in the overflow computation is actually an overestimation of the demand. Therefore, it will always improve the detailed routing task if there is room for this overestimation. Of course, this is not always the case, especially for a tight design.

A uniform slack requirement on all global boundaries can usually give a good routing result when the design is not very tight. This can be achieved by requiring SLACK=0. Then, repeat global routing with SLACK=1, 2, etc. However, as the design gets more crowded and the routing requirements get tighter, the uniform slack procedure will be inadequate. A more selective generation of slack requirements based on the global routing pattern of the nets and more detailed signal net-pin information can help to identify a global boundary-

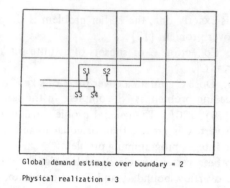

Global demand estimate over boundary = 2
Physical realization = 3

Fig. 9. Realization requirement exceeds demand estimate.

crossing slack matrix. This selection can be used to account for the fact that at certain locations more routing space is required to realize actual routing patterns than is estimated by the global router. Fig. 9 shows such a case where three tracks are required for actual implementation (maximum track requirement occurs in the interior of global cell), and the global routing estimate is only two crossing each boundary. This is a limitation of the global routing approach due to the coarse grid structure. The actual wire demand within each global cell is not considered in this simplfied environment.

V. GENERAL GLOBAL ROUTING PROBLEM FORMULATION

A general formulation for the global routing problem is now introduced.

Consider the cell grid picture in Fig. 10 with the following parameters associated with it.

m_h is the number of cells on the horizontal side,

m_v is the number of cells on the vertical side,

n is the total number of signal nets,

$S = S_1, S_2, \cdots, S_n$ is the set of all signal nets,

$C(i, j, 1)$ is the vertical boundary capacity (for horizontal crossing) of cell (i, j),

$C(i, j, 2)$ is the horizontal boundary capacity (for vertical crossing) of cell (i, j), $i = 1$ to $(m_v - 1)$, $j = 1$ to $(m_h - 1)$.

As boundary conditions, the parameters are

$$C(i, j, 1) = 0 \quad \text{for} \quad i = 0 \quad \text{or} \quad i = m_h$$

$$C(i, j, 2) = 0 \quad \text{for} \quad j = 0 \quad \text{or} \quad j = m_v.$$

$P(i, j) \subseteq S$ is associated with cell (i, j), and
$P(i, j) = \{S_t \in S \mid S_t \in P(k, l) \text{ for some } k, l \text{ such that } (i, j) \neq (k, l)\}$.

In other words, $P(i, j)$ is a subset of signal nets in S having pins in cell (i, j) and there exists $P(k, l)$ such that $P(i, j) \cap P(k, l)$ is nonempty where cell (i, j) is different from cell (k, l).

Thus in Fig. 10(a), the parameters are

$$m_h = 5, \quad m_v = 5; \quad n = 4;$$

$$C(i, j, k) = 1, \quad \text{for all} \quad i, j = 1 \text{ to } 5;$$

$$P(2, 3) = \{S1, S4\}; \quad P(3, 5) = \{S2, S4\}.$$

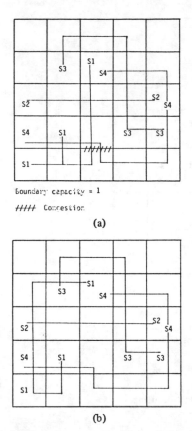

Boundary capacity = 1

//// Congestion

(a)

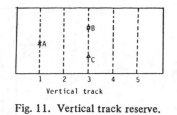

Fig. 11. Vertical track reserve.

subject to

$$|P(i,j)| \leqslant C(i,j,1) + C(i,j,2)$$
$$+ C(i,j-1,1) + C(i-1,j,2).$$

In the most simple case where global cell size is reduced to one routing grid (e.g., $C(i,j,k) = 1$ for all i,j,k, and $|P(i,j)| = 1$ for all i,j), the formulation is reduced to the detail routing problem with a single layer. There is currently no efficient algorithm dealing with this formulation concerning the cell boundary capacity constraint coupled with the inequality constraint of the cell-net relationship. The SLACK(i,j,k) variable is what was discussed earlier in Section IV. Simplified formulations by Lynch [21] and Raghaven et al. [22] have shown the problem to be in the NP-complete class of problems. Careful study and treatment of this formulation using only a small number of variables may yield very useful heuristic algorithms that can be generalized and used for better global routing results.

VI. Concepts of Reserved Tracks for Signals

Before carrying out the global routing stage, each horizontal and vertical track within a global cell is scanned and labelled. A vertical track is *reserved* for a particular signal net if, over the extent of the global cell the track belongs to, there exists pin(s) of the same signal and the second-level metal is free of any obstruction for this signal. A vertical track is said to be *unavailable* for any signal if more than one distinct signal pin occupies the same track or there is an obstruction on the second-level metal track. When two distinct signal pins occupy the same track, the particular track can be used if these two signal nets do not cross each other. If signals do cross each other, assigning the vertical track to any pin will result in an unroutable situation for the other signal net. The track is made unavailable for vertical assignment to facilitate ease of channel routing success. In a tight design case, vertical assignment overflows may exist due to this strategy. However, the routing map is not affected by this assignment phase and the clean-up mazerunner can still have the last chance to finish up those failures. A horizontal track is similarly defined for a signal on the first-level metal (this occurs when macro cell metals extended into routing channel area). Otherwise the track is available for any signal. The global router will take advantage of the reserved track information for a preferential global routing pattern assignment.

Fig. 11 shows an example where routing track 1 is reserved for signal net A, routing track 3 is not available for any signal due to a potential routing conflict, and routing tracks 2, 4, 5 are available for later assignments.

(b)

Fig. 10. (a) A global routing pattern. (b) A solution for case in (a).

Fig. 10(a) shows that at the horizontal cell boundary of cell (3, 4) the capacity is exceeded by one with the particular routing pattern. Fig. 10(b) shows the same case with a solution.

Defining a variable $X(i,j,k)$ similar to that of $C(i,j,k)$ as the crossing count over the global cell boundary, then the objective of the formulation is:

To find the interconnect patterns for the signal set S over the global cells such that

$$X(i,j,k) \leqslant C(i,j,k)$$
$$|P(i,j)| \leqslant C(i,j,1) + C(i,j,2)$$
$$+ C(i,j-1,1) + C(i-1,j,2).$$

The constraint on the cardinality of $P(i,j)$ simply says that there should be enough boundary capacities for all signal nets in $P(i,j)$ associated with cell (i,j) to get out.

More precisely, the signal nets in S over the global grids are to be interconnected to

$$\underset{i,j,k}{\text{Min Max}} \ (X(i,j,k) - C(i,j,k))$$

subject to

$$|P(i,j)| \leqslant C(i,j,1) + C(i,j,2)$$
$$+ C(i,j-1,1) + C(i-1,j,2).$$

Let $\text{SLACK}(i,j,k) = C(i,j,k) - X(i,j,k)$, then the optimization criterion is

$$\underset{i,j,k}{\text{Max Min SLACK}}(i,j,k)$$

159

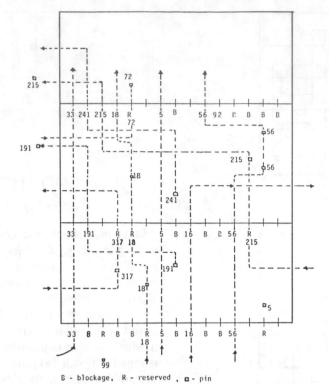

Fig. 12. Example of vertical assignment given global routing.

VII. Vertical Track Assignment

Using the supply–demand data and global routing pattern, vertical assignment selects a vertical track for each net segment crossing each horizontal global cell boundary.

An extension of Munkres' algorithm [23], [24] is used to assign a set of nets (demand) to the vertical tracks (supply) over each cell simultaneously. Each net is analyzed to reflect a weight relative to each track in the vertical channel independent of all others for assignment consideration. The focus of analysis is on:

1) existence of all signal pins inside the current global cell,
2) existence of all signal pins in the global cell above the current global cell,
3) vertical track assignment location at the global cell below, and
4) change of directions for the net under consideration at global cells above.

Fig. 12 shows the result of the vertical assignment process through a few global cells given the global routing patterns and reserved tracks. It is important to point out at this stage that routing density is not known prior to the assignment. A linear scan procedure was used to determine the cost matrix (linear with respect to the problem size, $N \times T$ where N is the number of signal nets crossing the top boundary of current global cell and T is number of vertical tracks available or reserved in current global cell).

Another way of performing this assignment is to use routing density as the criterion. Given a vertical assignment pattern, routing density can be easily determined. A local optimization objective to perform the assignment is then to minimize maximum density by the assignment process.

VIII. Channel Routing

Setting supply greater than or equal to demand in the global routing phase will satisfy the channel routing density requirements as a lower bound. To allow flexibility for macro metal intrusions into the routing area, the channel router is implemented to recognize and avoid those intrusions. Moreover, it takes advantage of all unused active areas to do routing when necessary. In certain areas of the array, internal drivers are provided. When they are selected for use, there are both blockages and pin access requirements *within* the routing area. Fig. 13(a) shows such a possible channel routing environment for our gate array and Fig. 13(b) shows the case where interior channel blockage occurs due to the presence of internal driver circuit. A modified left-edge algorithm was used with design rule checking (e.g., metal spacing, via to metal spacing, etc.). Since the operations are performed on a two-level routing map, the macro intrusion, internal driver metal pattern, and pin locations in the routing area are readily detectable by the software at any point. The channel router also uses "wrong-way jog" to enhance routability. Fig. 14 shows an example of wrong-way jog.

Any remaining unconnected subnets after the channel routing stage are then passed to the mazerunner for completion. To allow user control, several run time options are provided.

1) Route any/all channels using options 2) and 3).
2) Allow dog leg [25].
3) Use left-edge, right-edge, bottom-up or top-down, etc.

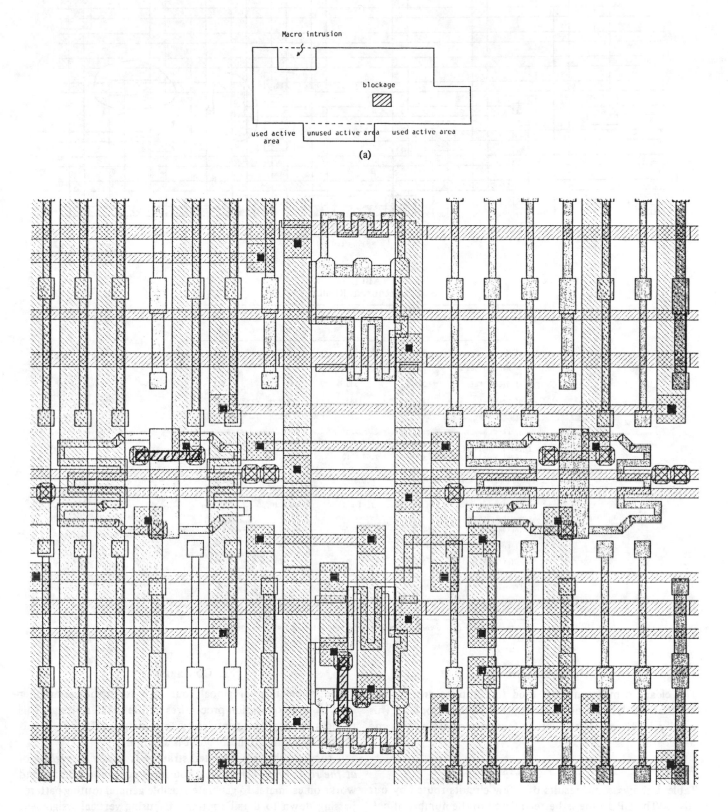

Macro intrusion

blockage

used active area   unused active area   used active area

(a)

 internal channel blockage

(b)

Fig. 13. (a) A channel routing environment. (b) Example where internal channel blockage due to internal driver circuits.

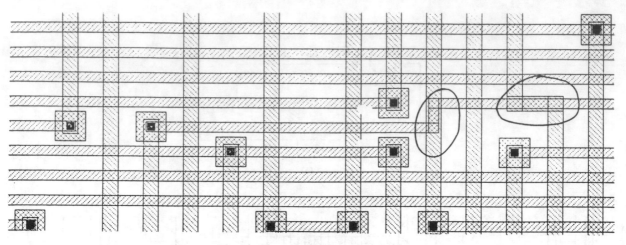

Fig. 14. Example of "wrong-way jog."

TABLE I
SOME ROUTING RESULTS

| | CIRCUIT | | | |
	1	2	3	4
# of Macros	301	317	334	442
# of Macro Types	91	74	58	62
# of Signal Nets	361	535	631	983
# of Net-Pins	1151	1483	1462	3012
Usage of Array	60%	75%	40%	75%
# of Vias	2166	3720	2782	6356
Metal 1 Length (Available)	$1.44 \times 10^6 \mu$	$1.44 \times 10^6 \mu$	$1.44 \times 10^6 \mu$	$1.44 \times 10^6 \mu$
Metal 1 Length (Actual)	327592μ	450422μ	$6.4 \times 10^5 \mu$	$1.06 \times 10^6 \mu$
Metal 2 Length (Available)	$2.44 \times 10^6 \mu$	$2.44 \times 10^6 \mu$	$2.44 \times 10^6 \mu$	$2.44 \times 10^6 \mu$
Metal 2 Length (Actual)	325220μ	590307μ	$6.6 \times 10^5 \mu$	$1.19 \times 10^6 \mu$
# of From-Tos	1557	2007	2728	5044
Completion Rate	100%	100%	100%	99%
# of Manually Routed From-Tos	0	0	0	49
CPU Time (sec.) (IBM 4341-2, 1.4 MIPS)	609*	1040*	1713	3932

* Extrapolated due to computer upgrade from IBM 4341-1 to 4341-2 in 1982.

IX. MAZERUNNER

The clean-up mazerunner is used to complete any connections that are left unconnected by the channel router. This generally involves subnets with a few points within the same channel area.

X. RESULTS

Table I shows some results of a few circuits routed by our system. The CPU time is better related to the number of net-pins and number of signal nets than the percentage utilization. Fig. 15(a) shows the basic power bus structure of our 23 000-transistor gate array. Fig. 15(b) shows the macro metal connections of circuit 4 in Table I. Fig. (15c) and (d) shows the corresponding metal 1 and metal 2 connections of circuit 4.

XI. CONCLUSION

The whole idea of global routing is to introduce another level of interconnection procedures before detail routing. This reduces the problem complexity to a manageable size. This is exactly what hierarchical layout is all about.

A successfully demonstrated strategy is to *use the right tool at the right time*. The global routing stage ignores details and works on estimates to generate feasible general routing patterns passing down to detail routing. By using vertical assignment and channel routing, most nets are routed very efficiently. The Lee-type mazerunner is then used on the few leftover nets to find more complex routing patterns. Further improvement on the channel router using the algorithms in [10] and [26] are under study. A vertical assignment procedure to

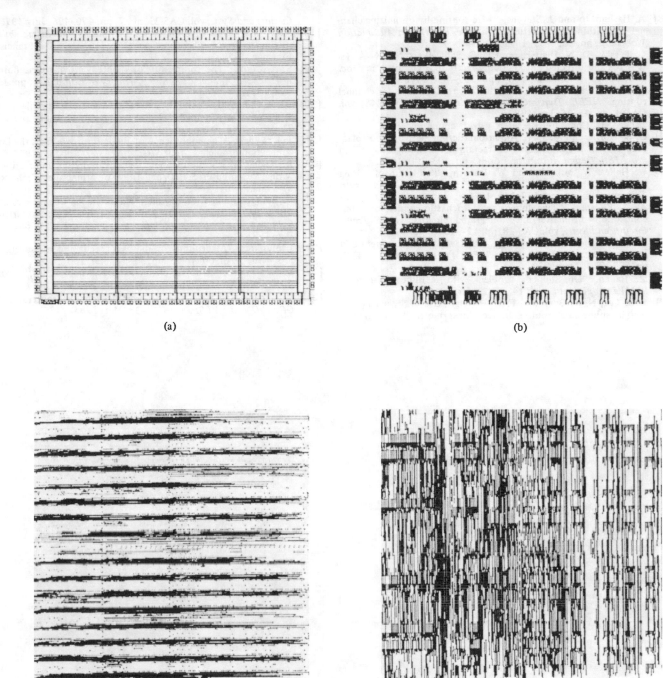

(a)

(b)

(c)

(d)

Fig. 15. (a) Generic power bus structure. (b) Macro metal pattern. (c) First metal routing. (d) Second metal routing.

minimize density function within each assignment step is being investigated.

REFERENCES

[1] K. A. Chen, M. Feuer, K. H. Khokhani, N. Nan, and S. Schmidt, "The chip layout problem: An automatic wiring procedure," in *Proc. 14th Design Automation Conf.*, pp. 298–302, 1977.

[2] K. H. Khokhani and A. M. Patel, "The chip layout problem: A placement procedure for LSI," in *Proc. 14th Design Automation Conf.*, pp. 291–297, 1977.

[3] J. Soukup, "Global router," in *Proc. 16th Design Automation Conf.*, pp. 481–484, 1979.

[4] B. T. Preas and W. M. vanCleemput, "Routing algorithms for hierarchical IC layout," in *Proc. 1979 ISCAS*, pp. 482–485, 1979.

[5] E. F. Moore, "The shortest path through a maze," Bell Tel. Lab. Rep., 1959.

[6] C. Y. Lee, "An algorithm for path connections and its applications," *IRE Trans. Electron. Comput.*, vol. EC-10, pp. 346–365, Sept. 1961.

[7] D. W. Hightower, "A solution to line-routing problems on the continuous plane," in *Proc. 6th Design Automation Workshop*, pp. 1–24, 1969.

[8] A. Hashimoto and J. Stevens, "Wire routing by optimizing channel cell within large apertures," in *Proc. 8th Design Automation Workshop*, pp. 155-169, 1971.

[9] G. Persky, D. N. Deutsch, and D. G. Schweikert, "LTX—A system for the directed automatic design of LSI circuits," in *Proc. 13th Design Automation Conf.*, pp. 399-407, 1976.

[10] T. Yoshimura and E. S. Kuh, "Efficient algorithms for channel routing," *IEEE Trans. Computer-Aided Design of ICAS*, vol. CAD-1, pp. 25-35, Jan. 1982.

[11] J. Lee, STC Computer Research, private communication.

[12] C. S. Horng and M. Lie, "An automatic/interactive layout planning system for arbitrary-sized rectangular building blocks," in *Proc. 18th Design Automation Conf.*, pp. 293-300, 1981.

[13] W. Heller, W. Mikhail, and W. Donath, "Prediction of wiring space requirements for LSI," *Design Automation and Fault-Tolerant Comput.*, pp. 117-144, 1978.

[14] A. A. ElGamal, "Two-dimensional stochastic model for interconnections in master slice integrated circuits," *IEEE Trans. Circuits and Syst.*, vol. CAS-28, pp. 127-137, Feb. 1981.

[15] M. Hanan and J. M. Kurtzberg, "Review of the placement and quadratic assignment problems," *SIAM Rev.*, vol. 14, no. 2, pp. 324-342, Apr. 1972.

[16] M. Hanan, "On Steiner's problem with rectilinear distance," *SIAM J. Appl. Math.*, vol. 14, no. 2, Mar. 1966.

[17] J. H. Lee, N. K. Bose, and F. K. Hwang, "Use of Steiner's problem in suboptimal routing in rectilinear metric," *IEEE Trans. Circuits and Syst.*, vol. CAS-23, no. 7, pp. 470-476, July 1976.

[18] M. R. Gary and D. S. Johnson, *Computers and Intractability: A Guide To the Theory of NP-Completeness.* San Francisco: Freeman, 1979.

[19] S. A. Cook, "The complexity of theorem proving procedures," in *Proc. 3rd Annual ACM Symp. Theory of Computing*, pp. 151-158, 1971.

[20] R. M. Karp, "Reducibility among combinatorial problems," in *Complexity of Computer Computations*, R. E. Miller and J. W. Thatcher, Eds. New York: Plenum Press, 1972.

[21] J. F. Lynch, "The equivalence of theorem proving and the interconnection problem," *ACM SIGDA Newslett.*, no. 53, 1975.

[22] R. Raghavan, J. Cohoon, and S. Sahni, "Mahattan and rectilinear wiring," Computer Sci. Dept., Univ. of Minnesota, Tech. Rep. 81-5, Mar. 1981.

[23] F. Bourgeois and J.-C. Lassalle, "An extension of munkres algorithm for the assignment problem to rectangular matrices," *Commun. Assoc. Comput. Mach.*, vol. 14, no. 12, pp. 802-805, Dec. 1971.

[24] J. Munkres, "Algorithms for the assignment and transportation problems," *J. SIAM*, vol. 5, pp. 32-38, Mar. 1957.

[25] D. N. Deutsch, "A dogleg channel router," in *Proc. 13th Design Automation Conf.*, 1976.

[26] M. Burstein and R. Pelavin, "Hierarchical channel router," *INTEGRATION, the VLSI J.*, pp. 21-38, Mar. 1983.

WIRE ROUTING BY OPTIMIZING CHANNEL ASSIGNMENT WITHIN LARGE APERTURES

Akihiro Hashimoto and James Stevens

Center for Advanced Computation
University of Illinois, Urbana, Illinois

ABSTRACT

The purpose of this paper is to introduce a new wire routing method for two layer printed circuit boards. This technique has been developed at the University of Illinois Center for Advanced Computation and has been programmed in ALGOL for a B5500 computer. The routing method is based on the newly developed channel assignment algorithm and requires many via holes. The primary goals of the method are short execution time and high wireability. Actual design specifications for ILLIAC IV Control Unit boards have been used to test the feasibility of the routing technique. Tests have shown that this algorithm is very fast and can handle large boards.

INTRODUCTION

In the past few years, several new wire routing algorithms have been introduced. Until that time Lee's [1] algorithm had been the only widely known technique. The primary advantages that these new algorithms have over Lee's method are that they require much less computer time and much less storage space. The most noted of these new algoritms are the line search technique of Hightower [2] or Mikami [3], the cellular routing method of Hitchcock [4] and the stepping aperture technique of Lass [5]. The channel routing technique being introduced in this paper attempts to combine the advantages of each of the above algorithms. The major objectives of this new method are to increase the speed of wiring and to increase the flexibility so that a large percentages of connections can be made with a minimum wire length.

In the channel routing algorithm all connections are represented as collections of wire segments as is done by Hightower and Lass. These line segments however are free to move sideways within the boundaries of a given space which may correspond to a cell used by Hitchcock but is many times larger than the cells he uses. Finally wire segments can be moved from one space to another in a manner similar to the method used by Lass. The final assignment of the wire segments to positions within a space is accomplished in an optimum manner by a simple and efficient procedure. While the channel routing algorithm allows great flexibility in the placing and moving of wire segments, it also is very fast and easy to implement.

The channel routing algorithm was designed to be used in routing two layer printed circuit boards which have the following properties. The type of package employed is a dual inline integrated circuit package arranged on the board in straight rows and columns (Fig. 1). Also there must be a capability for a large number of "floating" via holes. The positions of these vias are determined by the channel routing program. The reason for this is that the program will assign all horizonital wire segments to one side of the board and all vertical wire segments to the other. (A modification of this rule will be discussed in Appendix A.) All connections between horizontal and vertical wires are made by means of plated through via holes. In general, each connection made on the board will cause four vias to be assigned. The dependance of this algorithm on inexpensive and reliable via holes is considered the overriding factor determining its usefulness.

The description of the channel routing technique depends heavily on the definition of a space and a channel. First of all a space is an area on a board which extends from one edge to the other and may be of any desired width. For the purposes of this description the width of a space may be defined to be equal to the distance between rows or columns of packages on the board (Fig. 2). Within a given space it will be possible in general for several wires to run parallel to each other. The maximum number of wires which can run parallel is determined by the minimum spacing between wires and the width of the space. This number will be referred to as the number of channels available in that space. As a subdivision of a space, a channel also runs from one edge of the board to the other.

The algorithm itself is composed of two stages, space assignment and channel assignment. Space assignment is the process of

Reprinted with permission from *Proc. 8th Annual Design Automation Workshop Under Joint Sponsorship SHARE, ACM, and IEEE*, 1971, pp. 155–169.

breaking each connection into a set of horizontal and vertical wire segments and assigning these wire segments to spaces on the board. The details of how each connection is originally made will be discussed in a later section. The assigning of wire segments to spaces takes no account of available channels within the spaces; they are simply assigned as if each space had an infinite capacity. After all wire segments have been assigned to spaces, the channel assignment stage of the algorithm finalizes the positions of the wire segments. More than one wire segment can be assigned to one channel as long as there is no overlap between wire segments in that channel (Fig. 3). The algorithm used to assign channels places all of the wire segments in a given space into the minimum possible number of channels. In this way it can be determined whether all of the wire segments can be placed within the physical dimensions imposed by the board layout. If a given space is overcrowded, i.e. requires too many channels, wire segments can be moved from it to a nearby space which is not overcrowded. If all of the spaces on the board become filled or overcrowded through shifting of wire segments then this algorithm would not be able to wire one hundred percent of the connections.

The remainder of the paper is devoted to explaining the details involved in implementing the channel routing program. Also, a discussion is given of some of the possible extensions and modifications that can be applied to the algorithm. The specific example of wiring an ILLIAC IV Control Unit board is used to demonstrate the application. It should be noted that ILLIAC IV CU boards are large, 165 components, multilayer boards which were layed out and wired with much difficulty due to transmission line effects. For the purposes of this paper it will be assumed that these effects due to extra high speed can be ignored so that all of the signal wires can be connected on two layers using minimum distance wiring wherever possible.

BOARD LAYOUT

The package arrangement on an ILLIAC IV CU boards (Fig. 4) is composed of eleven rows of packages with fifteen packages per row giving a maximum of 165 packages per board. Each component is a dual inline integrated circuit package with sixteen pins. The arrangement of holes for mounting these packages is very orderly with all of the sixteen holes lined up in two rows. The orderly arrangement of packages on the board gives rise to large open spaces which run between rows and columns of packages. It is the

wiring of these spaces which is of major interest in this algorithm. The board being considered has two sides available for wiring, one for horizontal wire segments only and one for vertical wire segments only. Since wire segments are to be assigned to spaces on the board, the side used for horizontal wire segments will be divided into horizontal spaces and the other side into vertical spaces.

On the horizontal side of the board there are ten identical horizontal spaces between rows of packages. There is also another type of horizontal space, the space beneath a row of packages. Because of the arrangement of pins the only difference between these two types of horizontal spaces is their width (Fig. 5). On the vertical side of the board there are also two types of spaces, the spaces between columns of packages and the spaces beneath columns of packages. Since the vertical spaces beneath columns of packages contain the package pins they must be handled slightly differently than any other spaces. For the specific case of ILLIAC IV CU boards there are 19 horizontal channels and 16 vertical channels available per package. This can be broken down to the channel capacity for each type of space on the board. Horizontal spaces between packages have a 14 channel capacity while horizontal spaces beneath packages have 5 channels. Vertical spaces between packages have a 9 channel capacity and vertical spaces beneath packages have 7 channels (one wire channel between each of 7 pairs of pins).

Plated through communicating holes, called vias are assigned by the algorithm at the end points of each wire segment. These vias allow wire segments on one side of the board to be connected to wire segments on the other side so that a completed connection can be made. In general many of these vias will be assigned during the course of wiring one board. For this reason the vias must not have fixed positions and at the same time they must be reliable. In the process of performing the routing the information concerning vias can be used to create drilling instructions for producing the necessary holes. At this point it should also be mentioned that great care must be taken to allign the end points of the two wire segments to be connected so that the proper connection will be made and no shorts will occur.

SPACE ASSIGNMENT

Each connection made on the board is specified by a starting pin and an object pin. The connection itself is in general broken up into five wire segments (Fig. 6). At each of the two pins a short vertical wire segment (A,E) is constructed which initially extends to the center of one of the adjoining horizontal spaces. A horizontal wire segment (B) is next started from the end of the short vertical wire (A) at the starting pin and extended to the center of a vertical space which adjoins the package containing the object pin. From that point a vertical wire (C) is created which extends to the center of the horizontal space adjacent to the object pin. The final wire is a horizontal wire (D) which completes the connection. Many alternatives to the above description exist all of which produce a minimal or close to minimal connections. In an attempt to distribute wire segments evenly over the horizontal and vertical spaces, an arbitrary means of selecting one of the possible minimal paths has been employed.

The space assignment method described above is a simplified technique which may not allow complete wiring of the board. Basically, incomplete wiring will result from one of two conditions. First, it may not be possible to make all of the connections required within the physical restriction imposed by the board dimensions. This problem may be solved by using a different placement of components on the board or by decreasing the number of components assigned to the board. Secondly, incomplete wiring may result from the overcrowding of one area of the board. It is possible that the wire segments assigned to one space may overflow the channel capacity of that space whereas all other spaces may have very few channels used. Some steps can be taken to avoid such a situation or to alleviate it after it occurs. At the time of assigning a wire segment to a space it is possible to determine how many wires have already been assigned to that space and to the other space on the board. This information can be used to place a wire segment in the least crowded of the spaces to which it can be assigned. On the other hand, after all channel assignments have been made it is also possible to reassign wire segments to new spaces so that a more even distribution of required channels can be realized. Since each wire segment can be specified by its end points, it is an easy matter to reassign a wire segment to a new space and to make the necessary adjustments to the end points of wire segments it connects to.

In this initial assignment of wire segments to spaces, all wire segments are assumed to travel down the center of their assigned spaces. This assignment creates three kinds of conflicts which must be resolved at

a later time. The conflict of many wires overlapping in the center channel of each space is resolved by the second stage of the algorithm. Via conflicts which arise from having many wire segments end at the same point are also resolved by the second stage. One other type of conflict may result from this method of space assignment. Since the connection to each package pin is made by a short vertical wire segment which does not use up any channel area, (vertical channels go between package pins) it is assumed that these connections can always be made and therefore no information must be kept about them. However two of these short wires may be caused to overlap when channels are assigned to horizontal wires (Fig. 7(a)). This type of conflict can be handled in several ways and is presently resolved by an extra stage in the algorithm which reroutes one of the connections. If space is available in an adjacent channel this is quite easily done (Fig. 7(b)).

CHANNEL ASSIGNMENT

After space assignment is completed each space will contain a set of wire segments. These wire segments are regarded as a set of intervals having an upper bound and a lower bound (Fig. 8). The object of channel assignment is to position all wire segments in the fewest possible number of channels without any wire segments overlapping. The first step in the procedure is to search the list of intervals for the element which has the greatest upper bound. This element is assigned to the first channel and eliminated from the list. The list is then searched for the interval which has the greatest upper bound which is less than the lower bound of the previously chosen interval. This element is also assigned to the first channel and eliminated from the list. The search is repeated until no element fulfills the requirements, at which time the entire process is repeated for the next channel. When all of the intervals have been eliminated from the list the channel assignment is complete. This algorithm always uses the minimum possible number of channels to finalize the positions of all of the wire segments in a given space. The proof of minimality will be given in the next section along with an interesting extension.

When the channel assignments are being made, precautions must be taken to ensure that the wire segments which must be joined to form a connection have end points which coincide. At the conclusion of the space assignment stage many conflicts may exist since all wire segments are assumed to travel down the center of the space. Assume that two horizontal wire segments 1 and 2 (Fig. 9(a)) have the same end point. If vertical wire segment 3 is to connect to 2 and vertical wire segment 4 is to connect to 1 then the two connections are sharing one via hole. Channel assignment is now performed on all vertical wire segments. Since the upper bound of wire 4 is not less than the lower bound of wire 3, these two wire segments may not be assigned to the same channel. For this reason there can no longer be a via conflict between connections 1-4 and 2-3, but wire 1 may be in conflict with wire 2 (Fig. 9(b)). If such a conflict exists then wire 1 and wire 2 cannot be placed in the same channel when horizontal channel assignment is performed. When all channel assignment is complete there is no conflict between connection 1-4 and connection 2-3 (Fig. 9(c)). Other combinations of conflicting wire segments can be formed, but the channel assignment procedes in such a way as to always eliminate wire overlap and via conflicts.

PROOF OF OPTIMAL CHANNEL ASSIGNMENT

The set of wire segments assigned to a given space can be regarded as a set of intervals as previously stated (Fig. 8). This set of intervals can be represented by a directed graph (Fig. 10(a)). The nodes represent intervals and an arrow is directed from one node to another if and only if the lower bound of the first interval is greater than the upper bound of the other interval. Assigning wire segments to channels corresponds to covering the directed graph which represents the set of wire segments with a set of directed paths. A directed path is composed of a set of nodes which are connected by arrows such that all arrows point in the direction of the path (Fig. 10(a)). Selecting a path corresponds to assigning the wire segments represented by the nodes on the path to one channel without overlap. A path cover of a graph is a collection of paths which includes each node of the graph exactly once. Two nodes in a directed graph are said to be incomparable if no path exists which includes both nodes. It has been shown that the greatest number of mutually incomparable nodes which can be found in the a-cyclic graph is equal to the minimum number of paths which can be used to cover the graph. This property was first proven by Dilworth and the corresponding theorem bears his name. [6] The largest collection of mutually incomparable nodes is referred to as a maximal incomparable set and in general several distinct incomparable sets from one graph may be maximal.

To show that the channel assignment algorithm is optimal it must be shown that the number of channels used is equal to the minimum number of paths which can cover the corresponding directed graph. First of all a method of constructing paths must be found which corresponds to the method used to assign wire segments to channels. Such a path can be formed by choosing the node with the greatest upper bound then choosing the arrow emanating from that node which leads to the node with the greatest upper bound. The path is complete when a node is reached which has no outgoing arrows. The first step in the proof is to show that such a path which we will call a max chain contains exactly one member from each maximal incomparable set of the graph. By definition no path can contain more than one element of a maximal comparable set of nodes. Assume on the other hand that a max chain can be found which contains no members of a given maximal incomparable set. If this assumption leads to a contradiction then the first step is demonstrated. Since none of the nodes in a path $N = (N_1, ..., N_p)$ are in a given maximal incomparable set $M = (M_1, ..., M_q)$ they must all be comparable with a member of that set (Fig. 11.) There is a node N_j which is the highest node with an incoming arrow which leads from a member M_k of M. The next preceding node N_i must have an outgoing arrow which leads to a member M_ℓ of M. From the comparisons described above it follows that the lower bound of M_k is greater than the upper bound of N_j and the upper bound of M_ℓ is greater than the lower bound of M_k. Therefore the upper bound of M_ℓ is greater than the upper bound of N_j. This proves that N is not a max chain since the arrow from N_i to M_ℓ was not chosen. Therefore every max chain contains exactly one member from each maximal incomparable set. Eliminating the nodes along a max chain from the directed graph reduces the size of each maximal incomparable set by one, reducing the number of paths required to cover the graph by one. Therefore the number of max chains which is equal to the number of channels used is always the minimum. The authors discussed the relation between this channel assignment algorithm and the other graph theoretical topics [7].

This algorithm assumes that each wire segment is an indivisible entity. The question arrises as to whether fewer channels would be required if a wire segment could be split into independent subsections connected by perpendicular wire segments. To show that such a modification would not improve the number of channels required, a proof will now be given that the size of the maximal incomparable sets is not decreased. Assume that a wire segment which is a member of a given maximal incomparable set can be broken into several nodes so that none are incomparable with the maximal incomparable nodes. The lower bound of any of these nodes would be equal to the upper bound of the following node to preserve continuity. Figure 11 can again be used to show that a contradiction is generated. Node M_k is shown comparable with node M_ℓ ($L(M_k) > U(N_j) = U(N_i) > U(M_\ell)$). Therefore such a division of one wire segment can not reduce the size of a maximal incomparable set and thus cannot reduce the number of channels required to place a set of wire segments.

If channel assignment is performed on all the wire segments on a board without regard for the interconnections they have, a lower limit can be found for the number of horizontal and vertical channels which must be available for wiring the board. The way this is done is by assigning two wire segments, one horizontal and one vertical, for each connection to be made such that all wire segments are of minimum length. All of the horizontal wire segments are assigned to one horizontal space and all vertical wire segments to on vertical space. Finally channels are assigned within these two spaces without any attempt to preserve the connections. The result is a lower limit on the number of horizontal and vertical channels which must be available on the board in order to wire all of the connections. This result can be used as an optimal limit against which the effectiveness of a given wire routing algorithm can be measured. Also such a result can be used in the early stages of a system design to help determine the physical dimensions of a board once its logic has been specified.

PROGRAM AND RESULTS

The channel routing algorithm has been programmed in ALGOL and run on a Burroughs 5500 computer. The program is divided into two stages just as the algorithm is divided. The first stage completes all connections to be made on the board divides them into horizontal and vertical wire segments and assigns all wire segments to spaces. The second stage of the program finalizes the positions of all wire segments and determines whether the board has been successfully wired. The input to the program is a list of connections each represented only by the coordinates of its two end points. The output of the program is a complete specification of the wiring of the board.

Each wire segment created in the first stage of the program must be associated with information giving its position and how it is to be connected. The position of a wire segment is specified by the coordinates of its

end points. For each end point a pointer is kept which gives the memory location of the wire segment that connects to that end point. If one end of a wire segment connects to a pin, a special pointer is inserted. The final position of a wire segment is also recorded by storing the number of the channel to which it is assigned. A tag is also associated with each wire segment to indicate whether a final assignment has been made for that wire segment. Since partial word fields can be conveniently manipulated in ALGOL, only one word of storage is needed to specify each wire segment completely (Fig. 12).

During the space assignment stage of the program one word of information is created to represent each wire segment used. These words are then stored in one of two arrays. One array for horizontal wire segments and the other for vertical wire segments. The first dimension of each of these arrays tells which horizontal or vertical space the wire segment is assigned to. All of the wire segments assigned to a given space are stored in the array row corresponding to that space. During the second stage of the program the assignment of channels within spaces is done for each space independently. For this reason only one array row must reside in core at a time. In a multiprogramming environment this is quite convenient, especially since all array rows are the same size. In the present program the array rows have 256 elements so several of them can fit easily on a small core machine even during multiprogramming. (B5500 has 32K words of memory.)

Several test runs have been performed using wirelists for ILLIAC IV CU boards. The maximum number of components on a board is 165 and for each component there are 16 vertical channels and 19 horizontal channels available. Some boards with up to 103 components and 500 connections were wired completely in less than 30 seconds processor time. The program was tested using 40 boards with 135 or fewer components. Of these, 23 were wired with less than 13 channels total overflow (i.e. the sum of extra channels required from all spaces which overflowed). The wiring of these 23 boards should be easily completed by reassigning a few wire segments to new spaces. The remaining 17 boards could probably be completely wired on two layers after considerable shifting of the wire segments which have overlapped. All of the CU boards with more than 135 components have at least 151 components and most of these are not considered wireable on two layers. Channel assignments were attempted for these boards without regard for space capacity and up to 1200 connections were assigned in less than one minute processor time.

POSSIBLE IMPROVEMENTS

The short run time and high wireability of the developed routing program have proven the usefulness of the channel routing algorithm. Because of the simple nature of the algorithm a channel routing program is very inexpensive to implement. In order to improve the applicability of the program, some new features listed below are being considered for use in future versions of the routing program.

(a) Parallel running wires

In a high speed logic circuit, any two signal wires are not allowed to run side by side for more than a certain predetermined distance. Because of the data structure employed the detection of wires which run parallel for too long a distance is a simple matter. Also, such a problem can be easily resolved by changing channel numbers. Suppose two wires in channels 1 and 2 of a given space run parallel for too long a distance. The problem can be eliminated by exchanging channel 2 with channel 3.

(b) Better space assignment

The current version of the channel routing program assigns a unique vertical or horizontal space to each wire segment even though there are a few alternatives. The selection of a particular space is made in a fixed way. This assignment scheme may cause some local congestion which could have been avoided if the wire segments were distributed carefully. A more flexible space assignment algorithm is now being developed in order to eliminate the unnecessary local congestion which may decrease the wireability.

(c) Better package placement

The channel routing program generally divides a wire into five wire segments with four vias. If a wire connects the two pins which are either on the same horizontal row of pins or two adjacent horizontal rows of pins, the wire will be divided into three wire segments with two vias. Accordingly, a better routing will be obtained if the channel routing program is combined with a placement algorithm which considers the above fact instead of or in conjunction with the total wire length. Taking advantage of the short run time of the channel routing program, an iterative operation between routing and placement programs seems possible.

(d) Reduction of vias

The approach described above will reduce the number of vias before the routing is made. It is also possible to reduce the number of vias after routing is made. The authors have developed an algorithm which, when applied on a routed board, reassigns the wire segments to sides of the board in such a way that a minimum number of vias are used. The details of the algorithm will be described in Appendix A.

ACKNOWLEDGEMENT

The authors wish to thank Prof. D. L. Slotnick for his valuable advice and encouragement. Thanks are also due to Mr. David Pearson of Automation Technology Inc., for his discussions.

REFERENCES

1. C. Y. Lee, "An algorithm for path connections and its applications", IRE Transactions on Electronic Computers, (September 1961), 346-365.

2. D. W. Hightower, "A solution to line-routing problems on the continuous plane", Proc.1969 Design Automation Workshop,1-24.

3. K. Mikami and K. Tabuchi, "A Computer Program for Optimal Routing of Printed Circuit Conductors." Proceedings of the IFIP Congress, 1968 Vol. 2 pp. 1475.

4. R. B. Hitchcock, "Cellular wiring and the cellular modeling technique", Proc. 1969 Design Automation Workshop, 25-41.

5. S. E. Lass, "Automated printed circuit routing with a stepping aperture", Comm. of ACM, 12, No. 5, (May 1969), 262-265.

6. R P. Dilworth, "A Decomposition Theorem for Partially Ordered Sets," Ann. Math., (2) 51 (1950), 161-166.

7. A. Hashimoto and J.E. Stevens, "Path Cover of Acyclic Graphs", ILLIAC IV Document No. 239, December 24, 1970.

8. U. R. Kodres, "Formulation and solution of circuit card design problems through use of graph methods", in Advances in Electronics Circuit Packaging, Vol. 2, G. A. Walker, Ed., New York: Plenum 1962, 121-142.

ALGORITHM FOR REDUCTION OF VIAS

After accomplishing the wire routing by the channel routing technique described in this paper, we may be able to reduce the number of vias by moving some wire segments from one side of the board to the other. For example, the routing shown in Figure A-1 requires five vias, providing that all the vertical wire segments have been assigned to one side of the board and all the horizontal ones to the other. Small circles in the figure denote the vias. Only two vias remain necessary if the wire segments are assigned as shown in Figure A-2.

Now we will show that the problem of minimizing the number of vias can be solved by finding a maximum bipartite subgraph of the graph derived from the routing. The two sides of a board will be identified as side A and side B for convenience.

Let $V^h = \{v_i\}$ and $V^v = \{v_j\}$ denote the sets of horizontal and vertical wire segments, respectively, and define $V = V^h \cup V^v$.

Let B^c be the set of all the unordered pairs (v_i, v_j) such that v_i and v_j intersect if placed on the same side of the board. Similarly, B^v is defined to be the set of all the unordered pairs (v_i, v_j) such that v_i and v_j must be electrically connected. Let $B = B^c \cup B^v$. Note that one component of a pair in B belongs to V^h and the other to V^v.

Regarding V and B as the sets of nodes and branches, respectively, we obtain a bipartite graph $G = (V, B)$. A bipartite graph is a graph without a circuit of odd length. Graph G of Figure A-3 is derived from Figure A-1. The solid lines and the dotted lines correspond to the branches in B^c and B^v, respectively.

If $V = V^A \cup V^B$ and $\emptyset = V^A \cap V^B$, then the pair $\{V^A, V^B\}$ is called a partition of V. A partition $\{V^A, V^B\}$ is feasible if every branch of B^c connects a node in V^A and a node in V^B. A feasible partition $\{V^A, V^B\}$ defines an assignment of the wire segments to the sides of the board, that is, the wire segments in V^A and V^B can be placed on the sides A and B, respectively, without a conflict. For a given feasible partition $\{V^A, V^B\}$, a branch in B^v requires a via if, and only if, it connects a node in V^A and a node in V^B. Thus, our problem is reduced to that of finding a feasible partition $\{V^A, V^B\}$ which has a minimum number of branches of B^v connecting a node in V^A and a node in V^B.

A subgraph of G, defined by $G^c = (V, B^c)$, consists, in general, of several connected components $G_1^c, G_2^c, \ldots, G_n^c$, $n \geq 1$. Let $G_i^c = (V_i, B_i^c)$ and define $V_i^h = V_i \cap V^h$ and $V_i^v = V_i \cap V^v$ (Fig. A-4).

Proposition 1. No partition other than $\{V_i^h, V_i^v\}$ is feasible for each V_i, $i = 1, 2, \ldots, n$.

This proposition is derived from the definition of the feasibility and the fact that G_i^c is a connected bipartite graph.

If a branch in B^v connects a pair of nodes in V_i for some i, then the branch (or the corresponding via) is called essential. Let B^e be the set of such branches. For example, in Figure A-4 $B^e = \{(5,6)\}$.

Proposition 2. An essential via can not be eliminated.

A branch in B^e must connect a node in V_i^h and a node in V_i^v for some i, but the two nodes can not belong to the same subset of a feasible partition due to Proposition 1.

Accordingly, only those vias which correspond to the branches in $B^v - B^e$ are the candidates for elimination. In our example, $V^v - B^e = \{(1,2), (3,4), (7,8), (9,10)\}$.

Proposition 3. Let $\{V^A, V^B\}$ be a feasible partition. If $V_i^h \subset V^A$ and $V_j^v \subset V^A$ (consequently, $V_i^v \subset V^B$ and $V_j^h \subset V^B$) then all the vias corresponding to the branches connecting V_i and V_j are eliminated. Otherwise, all of the vias must remain.

This proposition follows from the fact that there is no branch between V_i^h and V_j^h, or between V_i^v and V_j^v. For example, in Figure A-4 if $V_1^h, V_2^v \subset V^A$ and $V_1^v, V_2^h \subset V^B$ then the two vias corresponding to the branches (1,2) and (9,10) are removed. On the other hand, if $V_1^h, V_2^h \subset V^A$ and $V_1^v, V_2^v \subset V^B$ then both vias are not removed.

Now, replace each V_i of G by a node v_i^* and eliminate all self-loops, then a graph $G^* = (V^*, B^v - B^e)$, $V^* = \{v_1^*, v_2^*, \ldots, v_n^*\}$ will be obtained (Fig. A-5).

Assign 0 or 1 to each of the node of G^* arbitrarily, and let $V_i^h \subset V^A$, $V_i^v \subset V^B$ if v_i^* is assigned 0 and $V_i^h \subset V^B$, $V_i^v \subset V^A$ if v_i^* is assigned 1, then $\{V^A, V^B\}$ is a feasible partition of V. Conversely, any feasible partition of V has a 0-1 assignment of G^* due to Proposition 1. For a given partition derived from a 0-1 assignment, the number of remaining non-essential vias is equal to the number of branches in G^* which connect two nodes having the same value. The following proposition will be easily derived from the definition of the bipartite graph.

Proposition 4. All the non-essential vias can be eliminated if, and only if, the graph G^* is bipartite.

Thus the via minimization problem is reduced to the following one. "Given a graph G^*, find a bipartite subgraph of G^* with a maximum number of branches."

The graph G^* of Figure A-5 is not bipartite and either the branch (v_1^*, v_3^*) or (v_2^*, v_3^*) must be removed in order to obtain a bipartite subgraph. If (v_1^*, v_3^*) is selected, then the corresponding partition will be as follows:

$$V^A = V_1^h \cup V_2^v \cup V_3^h = \{1, 2, 3, 5\}$$

$$V^B = V_1^v \cup V_2^h \cup V_3^v = \{4, 6, 7, 8, 9, 10\}$$

The layout of the wire segments, specified by this partition, is shown in Figure A-2. Two vias remained; one for the essential branch $(5,6)$ and the other for the non-essential branch $(3,4)$ which was removed in the bipartization process.

The problem of finding a maximum bipartite subgraph of a given graph may be solved by either the minimum cover algorithm or the integer linear program [8]; however, no efficient algorithm is known. But the authors expect rather large connected components, G_i^c's, and a small graph G^*, and hence simplicity in solving the problem.

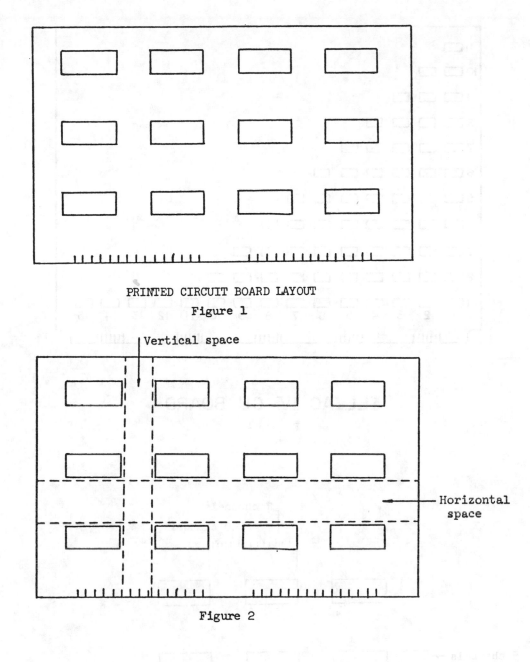

PRINTED CIRCUIT BOARD LAYOUT
Figure 1

Vertical space

Horizontal space

Figure 2

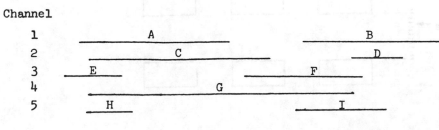

Figure 3

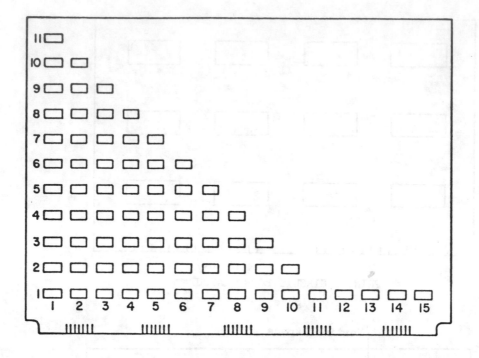

ILLIAC IV CU BOARD
Figure 4

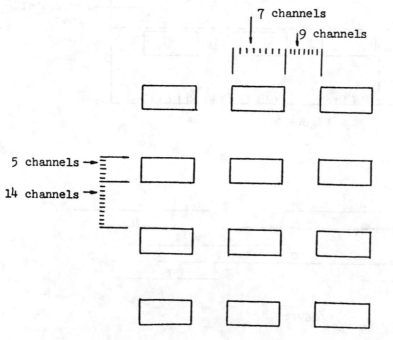

Figure 5

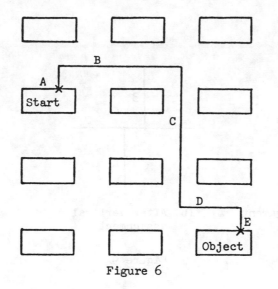

Figure 6

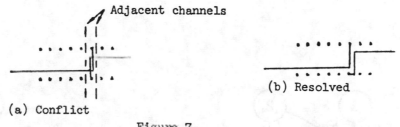

Adjacent channels

(a) Conflict

(b) Resolved

Figure 7

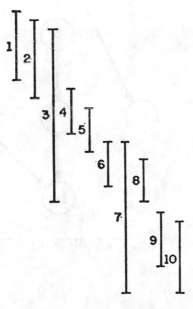

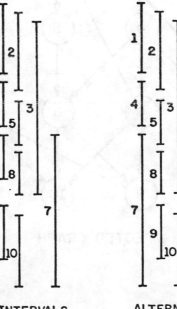

SET OF INTERVALS

INTERVALS ASSIGNED TO FOUR CHANNELS

ALTERNATE MINIMAL SOLUTION

Figure 8

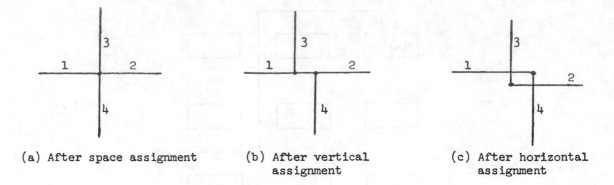

(a) After space assignment

(b) After vertical assignment

(c) After horizontal assignment

Figure 9

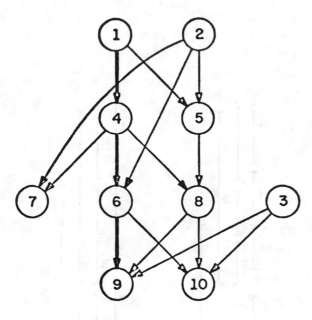

(a) DIRECTED GRAPH

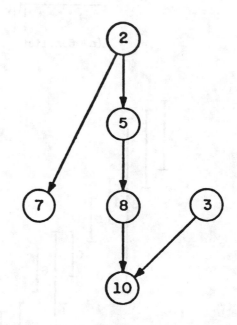

(b) REDUCED GRAPH

Figure 10

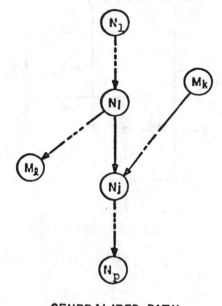

GENERALIZED PATH

Figure 11

Channel	T	Left Pointer	Right Pointer	Left end	Right end

HORIZONTAL WIRE SEGMENT

Figure 12

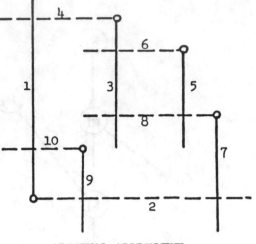

ORIGINAL ASSIGNMENT

Figure A-1

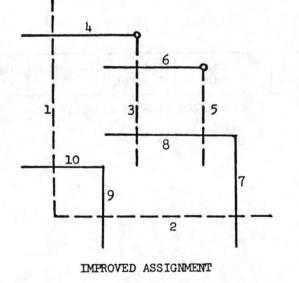

IMPROVED ASSIGNMENT

Figure A-2

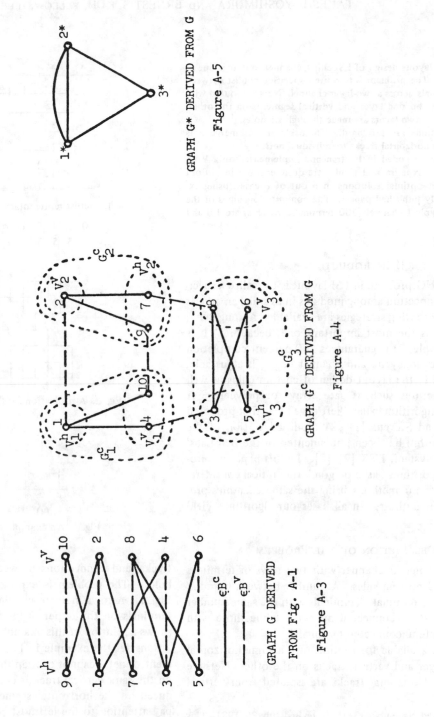

GRAPH G* DERIVED FROM G

Figure A-5

GRAPH G^c DERIVED FROM G

Figure A-4

GRAPH G DERIVED FROM Fig. A-1

Figure A-3

TAKESHI YOSHIMURA and ERNEST S. KUH Member, IEEE

Efficient Algorithms for Channel Routing

TAKESHI YOSHIMURA AND ERNEST S. KUH, FELLOW, IEEE

Abstract—In the layout design of LSI chips, channel routing is one of the key problems. The problem is to route a specified net list between two rows of terminals across a two-layer channel. Nets are routed with horizontal segments on one layer and vertical segments on the other. Connections between two layers are made through via holes.

Two new algorithms are proposed. These algorithms merge nets instead of assigning horizontal tracks to individual nets.

The algorithms were coded in Fortran and implemented on a VAX 11/780 computer. Experimental results are quite encouraging. Both programs generated optimal solutions in 6 out of 8 cases, using examples in previously published papers. The computation times of the algorithms for a typical channel (300 terminals, 70 nets) are 1.0 and 2.1 s, respectively.

I. INTRODUCTION

THE ROUTING problem in LSI layout is to realize a specified interconnection among modules in as small an area as possible. Several routing strategies are available. Among them, channel routing is the most important one; because 1) it is efficient and simple, 2) it guarantees 100-percent completion if constraints are noncyclic and channel height is adjustable, and 3) it is used in the layout design of custom chips as well as uniform structures such as gate arrays or polycells. A channel routing algorithm called "left edge" was first proposed by Hashimoto and Stevens [1]. A modified version of the "left edge" algorithm has been implemented in the Bell Labs' Polycell Layout System, LTX [2]-[5]. In this paper we propose two new algorithms based on graph theoretical considerations. We tested our methods using the same examples provided by previous authors. In all cases our algorithms yield better results.

II. DESCRIPTION OF THE PROBLEM

Consider a rectangular channel with two rows of terminals along its top and bottom sides. A number between 0 and N is assigned to each terminal. Terminals with the same number i $(1 \leqslant i \leqslant N)$ must be connected by net i, while those with number 0 designate unconnected terminals.

Two layers are available for routing. We assume horizontal tracks on one layer and vertical tracks on the other. Nets are laid on tracks. Horizontal tracks are isolated from vertical

Manuscript received November 13, 1980; revised July 20, 1981. This work was supported by the National Science Foundation under Grant ENG-78-24425 and the Alexander von Humboldt Foundation.

T. Yoshimura is with the Department of Electrical Engineering and Computer Sciences, and the Electronics Research Laboratory, University of California, Berkeley, CA 94720, on leave from Nippon Electric Company, Ltd., Japan.

E. S. Kuh is with the Department of Electrical Engineering and Computer Sciences, and the Electronics Research Laboratory, University of California, Berkeley, CA 94720.

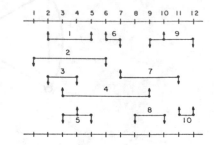

Fig. 1. Netlist representation for routing requirement.

```
0 1 4 5 1 6 7 0 4 9 10 10
2 3 5 3 5 2 6 8 9 8 7  9
```

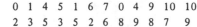

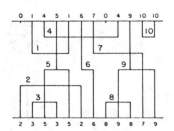

Fig. 2. A realization for the requirement in Fig. 1.

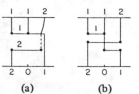

(a)　　　　(b)

Fig. 3. An example with cyclic conflict.

tracks, and connections between them are made through via holes. The problem is expressed by a net list, as shown in Fig. 1. Arrows indicate whether nets are to be connected to terminals on the upper or lower sides of the channel. Fig. 2 shows a solution of this example.

Consider the example in Fig. 3(a). Because the vertical segment of net 1 cannot overlap that of net 2 (shown dotted line) on the same vertical track, a constraint relation has been introduced on the horizontal segments of net 1 and net 2. If we pay attention to the leftmost column, the horizontal segment of net 1 must be placed above that of net 2. By the same reasoning, net 2 must be placed above net 1 if the rightmost column is considered. In other words, this routing specification cannot be realized without splitting some net into more than one horizontal segments as shown in Fig. 3(b). However, this kind of conflicting situation can often be avoided by rearranging the placement. In this paper, we assume that routing

Reprinted from *IEEE Trans. Computer-Aided Design of Integrated Circuits and Systems*, vol. CAD-1, pp. 25-35, Jan. 1982.

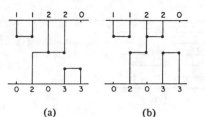

Fig. 4. An example illustrating the advantage of using dogleg. (a) No dogleg. (b) Dogleg.

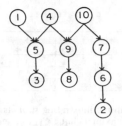

Fig. 5. Vertical constraint graph G_v for the netlist in Fig. 1.

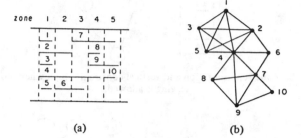

(a)   (b)

Fig. 6. Zone representation and interval graph. Maximal cliques are 12345, 246, 467, 4798, and 7910.

specification is always realizable in the sense that there exists no cyclic conflict in the net list.

The major objective of the problem is to minimize the number of horizontal tracks used to realize the routing requirement.

Consider the example in Fig. 4(a) where an optimal solution is given if splitting of horizontal segments is not allowed. However, the same example can be realized with only two tracks by horizontal splitting as shown in Fig. 4(b).

The splitting of horizontal segments of nets is called "doglegging." This is not only used to avoid the vertical conflict as mentioned, but also used to minimize the number of horizontal tracks. The latter is perhaps more important than the former. In the case of doglegging, we assume that the horizontal splitting of a net is allowed at the terminal positions only, which implies that no additional vertical track is allowed.

III. DEFINITIONS

A. Vertical Constraint Graph

As mentioned earlier, any two nets must not overlap at a vertical column. If we assume that there is only one horizontal segment per net, then it is clear that the horizontal segment of a net connected to the upper terminal at a given column must be placed above the horizontal segment of another net connected to the lower terminal at that column.

This relation can be represented by a directed graph G_v, where each node corresponds to a net and a directed edge from net a to net b means that net a must be placed above net b (Fig. 5). Therefore, if there is a cycle, the routing requirement cannot be realized without dividing some nets. (For example, a cycle $a \rightarrow b \rightarrow c \rightarrow a$ means that net a must be placed above itself.) However, it should be noted here that if the vertical constraint graph is acyclic the routing specification is always realizable.

B. Ancester and Descendent

Node i is said to be ancester of node j (node j is descendent of node i), if there is a directed path from i to j in the vertical constraint graph.

C. Zone Representation of Horizontal Segments

The horizontal segment of a net is determined by its leftmost and rightmost terminal connections. Let $S(i)$ be the set of nets whose horizontal segments intersect column i. Since horizontal segments of distinct nets must not overlap, the horizontal segments of any two nets in $S(i)$ must not be placed on the same horizontal track. This condition must be satisfied at every column. However, it is easy to see that we only have to consider those $S(i)$ which are not subsets of another set.

TABLE I

Column	S(i)	Zone
1	2	
2	1 2 3	
3	1 2 3 4 5	1
4	1 2 3 4 5	
5	1 2 4 5	
6	2 4 6	2
7	4 6 7	3
8	4 7 8	
9	4 7 8 9	4
10	7 8 9	
11	7 9 10	5
12	9 10	

Therefore, we assign zones the sequential number to the columns at which $S(i)$ are maximal. These columns define zone 1, zone 2, etc., as shown in Table I, for the example in Fig. 1. The number of elements in $S(i)$ is called local density, and the maximum among them is called maximum density. Clearly, we need not consider $S(1)$ or $S(2)$ because the horizontal constraints related to these sets are included in that of $S(3)$. The zone representation for this example is shown in Fig. 6(a).

Zones can be defined more clearly by using an interval graph defined by the horizontal segments of nets. A graph $G(\bar{V}, E)$ is an interval graph corresponding to a set of nets, where a node $v_i \in \bar{V}$ represents net n_i and an edge $(v_i, v_j) \in E$ iff $n_i \cap n_j \neq 0$ (i.e., net n_i and net n_j have horizontal overlap) [6], [7]. The interval graph of the net list in Fig. 1 is shown in Fig. 6(b). In terms of an interval graph, a zone is defined by a maximal clique and the clique number is the density.

It should be emphasized that the channel routing problem is completely characterized by the vertical constraint graph and the zone representation. Our problem is to determine an optimum ordering of nets such that 1) the vertical constraints as

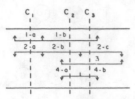

Fig. 7. A four-net example illustrating subnets and cuts corresponding to maximal clique C_1, C_2, and C_3.

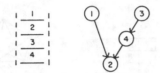

Fig. 8. Zone representation and vertical constraint graph where no doglegging is allowed.

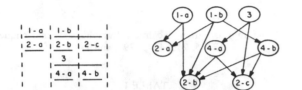

Fig. 9. Zone representation and vertical constraint graph where dogleg is allowed.

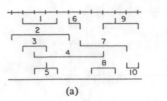

(a)

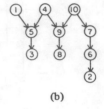

(b)

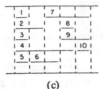

(c)

Fig. 10. Examples of Fig. 1. (a) Netlist. (b) Vertical constraint graph. (c) Zone representation.

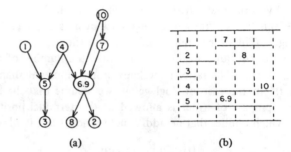

(a)   (b)

Fig. 11. Merging of nets. (a) Updated vertical constraint graph. (b) Updated zone representation.

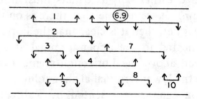

Fig. 12. Netlist after merging of net 6 and net 9 to form net $6 \cdot 9$.

expressed by G_v are satisfied, and 2) the number of tracks needed for realization inherent in the zone representation is minimized. A lower bound of the number of tracks is of course the maximum density.

D. Dogleg

Up to now the vertical constraint graph and the zone representation are defined for the problem where the number of horizontal segments per net is limited to one, i.e., the realization does not allow doglegs. However, these two representations can be extended to the dogleg problem, where the horizontal segment of a net can be divided at its terminal positions. In this case, subnets whose horizontal segments are parts of a net between two consecutive terminals are introduced. Fig. 7 gives a four-net example indicating subnets and cuts corresponding to maximal cliques. Fig. 8 shows the vertical constraint graph and the zone representation of the net list in which dogleg is not allowed. In the dogleg problem, subnets $1-a$, $1-b$, $\cdots$, $4-b$ are considered instead of nets 1, 2, 3, and 4, and the vertical constraint graph and the zone representation are shown in Fig. 9.

IV. A SIMPLE METHOD BASED ON MERGING OF NETS

To find an optimum realization of the channel routing problem is very difficult. The problem has been shown to be NP complete [8]. Therefore, we propose heuristic algorithms which generate optimum or near optimum solutions with a reasonable amount of computation time.

A. "Merging of Nets"

Before describing the algorithm, we will define the following operation.

Definition: Let i and j be the nets for which

a) there exists no horizontal overlap in the zone representation, and

b) there is no directed path between node i and node j in the vertical constraint graph

(i.e., net i and net j can be placed on the same horizontal track).

Then the operation "merging of net i and net j"

a) modifies the vertical constraint graph by shrinking node i and node j into node $i \cdot j$, and

b) updates the zone representation by replacing net i and net j by net $i \cdot j$ which occupies the consecutive zones including those of net i and net j.

Example: In the example of Fig. 10, net 6 and net 9 are candidates for merging. The operation "merging of net 6 and net 9" modifies the vertical constraint graph and the zone representation, as shown in Fig. 11. The updated vertical constraint graph and the zone representation correspond to the net list in Fig. 12, where net 6 and net 9 are replaced by net $6 \cdot 9$. By merging, we mean that net 6 and net 9 are placed on the same horizontal track, although the position of the track is not yet decided.

The following lemma is obvious.

Lemma: If the original vertical constraint graph contains no cycle, the updated vertical constraint graph does not have cycles either.

B. Algorithm Description

Since it is assumed that there is no cycle in the (initial) vertical constraint graph, we can repeat the operation "merging of nets" without generating any cycle in the graph. The following algorithm merges nets systematically according to the zone representation.

Algorithm #1:

	procedure Algorithm #1 (z_s, z_t)
	begin;
a1:	$L = \{\ \}$;
a2:	for $z = z_s$ to z_t do;
	begin;
a3:	$L = L + \{$nets which terminate at zone $z\}$;
a4:	$R = \{$nets which begin at zone $z + 1\}$;
a5:	merge L and R so as to minimize the increase of the longest path length in the vertical constraint graph;
a6:	$L = L - \{n_1, n_2, \cdots\}$, where n_j is a net merged at step a5;
	end;
	end;

If there is a path $n_1 - n_2 - n_3 - \cdots, -n_k$ in the vertical constraint graph, then no two nets among $n_1, n_2, \cdots, n_k$ can be placed on the same track. Therefore, if the longest path length in terms of the number of nodes on the path is k, at least k horizontal tracks are necessary to realize the interconnections. Thus nets are merged so that the longest path length after merging is minimized as much as possible.

Fig. 13 illustrates how the vertical constraint graph is updated by the algorithm. First, net 5 and net 6 are merged, then net 1 and net 7, $\cdots$, and at the fourth iteration net 10 and net 4 are merged. Finally, we have the graph in Fig. 13(e). Then we assign the horizontal tracks to each node of the graph. For example, we can assign track 1 to net $10 \cdot 4$ track 2 to net $1 \cdot 7$, track 3 to net $5 \cdot 6 \cdot 9$, track 4 to net 2 (or net $3 \cdot 8$) and track 5 to net $3 \cdot 8$ (or net 2). Fig. 14 shows the solution corresponding to the graph in Fig. 13(e).

C. Minimizing the Longest Path

The key part of the algorithm is step a5 where two sets of nets are merged. In the following, this process is explained. To make the situation precise, let us introduce several definitions.

(1) $P = \{n1, n2, \cdots, np\}$ and $Q = \{m1, m2, \cdots, mq\}$ $(p \geqslant q)$ are the two sets of nets to be merged. Obviously, elements of P are on separate vertical paths from that of Q.

(2) A modified vertical constraint graph is defined, as shown in Fig. 15, where two fictitious nodes s and t are added, corresponding to a source and a sink, respectively.

(3) $u(n)$, $n \in P \cup Q$: the length of the longest path from s to n

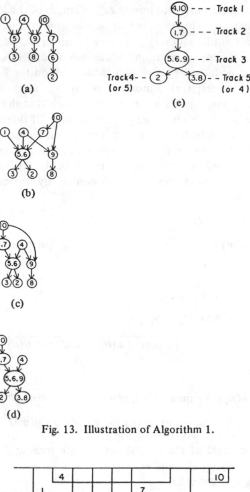

(a)

(b)

(c)

(d)

(e)

Fig. 13. Illustration of Algorithm 1.

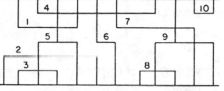

Fig. 14. A solution corresponding to Fig. 13(e).

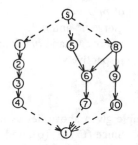

Fig. 15. A modified vertical constraint graph.

(4) $d(n)$, $n \in P \cup Q$: the length of the longest path from n to t

Example:

$$Q = \{6, 7\}$$

$$P = \{1, 3, 4\}$$

$$u(1) = 1, u(3) = 3, u(6) = 2, \cdots$$

$$d(1) = 4, d(3) = 2, d(6) = 2, \cdots.$$

The purpose here is to minimize the length of the longest path after merger. However, it will be too time consuming to find an exact minimum merger, hence a heuristic merging algorithm will be given. Let us introduce some basic intuitive ideas. First, a node $m \in Q$ is chosen, which lies on the longest path before merger; furthermore, it is farthest away from either s or t. Next, a node $n \in P$ is chosen such that the increase of the longest path after merger is minimum. If there are two or more nodes which will result in a minimum increase we choose n such that $u(n) + d(n)$ is maximum or nearly maximum and that the condition $u(m)/d(m) = u(n)/d(n)$ is satisfied or nearly satisfied. These can be implemented by introducing the following:

(1) for $m \in Q$

$$f(m) = C_\infty * \{u(m) + d(m)\} + \max \{u(m), d(m)\},$$
$$C_\infty \gg 1$$

(2) for $n \in P$, $m \in Q$

$$g(n, m) = C_\infty * h(n, m)$$
$$- \{\sqrt{u(m) * u(n)} + \sqrt{d(m) * d(n)}\}$$

where

$$h(n, m) = \max \{u(n), u(m)\} + \max \{d(n), d(m)\}$$
$$- \max \{u(n) + d(n), u(m) + d(m)\}$$

—the increase of the longest path length passing through n or m, by merging of n and m.

Merging Algorithm

given P, Q;
    begin;
a1:   while Q is not empty do;
          begin;
a2:   among Q, find m^* which maximizes $f(m)$;
a3:   among P, find n^* which minimizes $g(n, m^*)$,
                and which is neither ancestor nor descendent of m^*;
a4:   merge n^* and m^*;
a5:   remove n^* and m^* from P and Q,
                respectively;
          end;
    end;

For the example given above, let us pick $C_\infty = 100$, we obtain Table II(a). Since $f(7) > f(6)$, node 7 is chosen first from Q. Next, we evaluate $g(n, 7)$ using $C_\infty = 100$ and obtain Table II(b). Since $g(4, 7)$ is the smallest, we merge node 4 with node 7.

V. An Improved Algorithm Based on Matching

A. Overview of the Algorithm

In Algorithm #1, nets are merged when they are processed, and it should be noted that a merging may block subsequent mergings. Fig. 16 shows an example. Let us assume that, at zone 1, Algorithm #1 merges net a and net d, net b and net e, respectively, (if we follow the Merging Algorithm of the last

TABLE II(a)

		P			Q	
	1	3	4		6	7
u( )	1	3	4		2	3
d( )	4	2	1		2	1
f(m)					402	403

TABLE II(b)

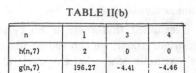

n	1	3	4
h(n,7)	2	0	0
g(n,7)	196.27	-4.41	-4.46

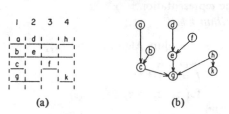

(a)   (b)

Fig. 16. An example to show possible difficulties of Algorithm 1. (a) Zone representation. (b) Vertical constraint graph.

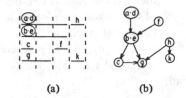

(a)   (b)

Fig. 17. Updated zone representation and vertical constraint graph.

section, these mergings will not occur, but they are assumed for illustration only). The vertical constraint graph and the zone representation are modified as shown in Fig. 17. The merged vertical constraint graph indicates that net f cannot be merged with either net c or net g because a cycle would be created. On the contrary, if net a and net d, net c and net e are merged, respectively, net f can be merged with net b.

To avoid this type of situation as much as possible, or in order to make the algorithm more flexible, we introduce another algorithm. In Algorithm #2 we construct a bipartite graph G_h, where a node represents a net and an edge between net a and net b signifies that net a and net b can be merged. A merging is expressed by a matching on the graph, and it can be updated dynamically. We will explain this idea by using the previous example.

We can see that net d (as well as net e) can be merged with any of three nets a, b, or c in zone 1. So, the algorithm constructs the bipartite graph G_h in Fig. 18, and a temporary merging is feasible, but neither the vertical constraint graph nor the zone representation is updated at this stage. Next, we move to zone 2, where net g terminates at this zone and net f begins at the next zone. So, we add node g to the left side and node f to the right size of the graph G_h, as shown in Fig. 19(a). Since the graph G_v in Fig. 16(b) indicates that net f can be merged with either net a, net b, or net c, three edges are added and matching is also updated as shown by the heavier lines in Fig. 19(a). Of course there is no guarantee that the merging

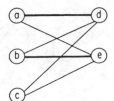

Fig. 18. Graph G_h and a possible matching in processing zone 2.

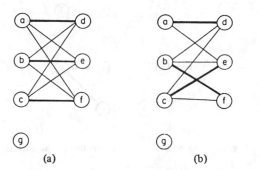

(a)   (b)

Fig. 19. Updating graph G_h in processing zone 3. (a) Updated matching. (b) Modified matching.

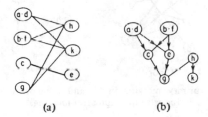

(a)   (b)

Fig. 20. Updated graph G_h and G_v for the processing of zone 4.

which corresponds to the updated matching satisfies the vertical constraint (horizontal constraints are satisfied automatically), so the algorithm checks the constraints and modifies the matching, as shown in Fig. 19(b). This process will be explained later.

At zone 3, net d and net f terminate. This means that, in processing zone 3, node d and node f should be moved to the left side in graph G_h and merged with their partner nets a and b, respectively, as shown in Fig. 20(a). Net c and net e have not been merged yet, since e has not terminated. The vertical constraint graph is also updated, as shown in Fig. 20(b). A matching is next sought for the updated G_h. The procedure will continue until all zones have been processed.

B. Algorithm #2

The general flow of Algorithm #2 is as follows.

    procedure Algorithm #2 (z_s, z_t);
    begin;
Step 1:   for each net n_1 terminating at zone z_s, add node
           n_1 to the left side of graph G_h;
           for $z_i = z_s$ to z_t do;
           begin;
Step 2:   for each net n_r which begins at zone z_{i+1}, add
           node n_r to the right side of graph G_h, and add

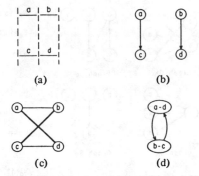

(a)   (b)

Fig. 21. Example to illustrate a matching which violates the vertical constraint. (a) Zone representation. (b) Vertical constraint graph. (c) Graph G_h and matching M. (d) Vertical constraint graph after merging corresponding to the matching M.

           edges between n_r and a node on the left side, if they
           can be merged; then, find a maximum matching;
Step 3:   check if the merging based on the current matching satisfies the vertical constraints. If not, modify
           the matching and graph G_h;
Step 4:   for each net n_1 terminating at zone z_{i+1}, merge n_1
           with the net n_x specified by the matching on graph
           G_h. Then put the merged net $n_1 \cdot n_x$ to the left
           side of G_h;
       end;
    end;

1) Step 1: This step is for initialization. The algorithm simply adds nodes corresponding to the nets terminating at zone z_s. There is no edge in graph G_h at this stage.

2) Step 2: A node corresponding to a net originating at zone z_{i+1} is added to the right side of graph G_h. If the net can be merged with the net corresponding to a node on the left, the edge between them is added. Then, a maximum matching on graph G_h is obtained.

To reduce the CPU time and memory requirement, the number of edges per node is limited by a parameter. (Its value is fixed at 3 in the program.) Edges are selected according to the same intuitive ideas as in the Merging Algorithm of the last section.

3) Step 3: It is obvious that the solution corresponding to any matching on graph G_h does not have horizontal overlaps because G_h is constructed based on the horizontal constraints. However, vertical constraints are not totally considered, so there may be vertical conflict. Fig. 21 shows an example, where the realization corresponding to the matching in Fig. 21(c) is not feasible. Hence, it is necessary to 1) check the vertical conflict for the given matching, and 2) modify the matching when the conflict is found. In order to do these two operations, we use the following algorithm which derives the conditions under which any matching in G_h will lead to feasible solutions.

Algorithm A: Given graph $G_h = (N, E_h)$ and graph $G_v = (N, E_v)$;

    begin
a1: $E_x = 0$;
       while N is not empty do;
       begin;

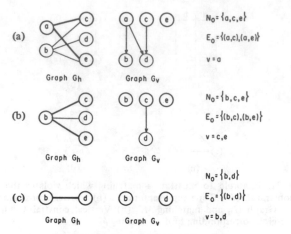

$N_0 = \{a, c, e\}$

$E_0 = \{(a,c),(a,e)\}$

$v = a$

$N_0 = \{b, c, e\}$

$E_0 = \{(b,c),(b,e)\}$

$v = c, e$

$N_0 = \{b, d\}$

$E_0 = \{(b,d)\}$

$v = b, d$

Fig. 22. Illustration of Algorithm A.

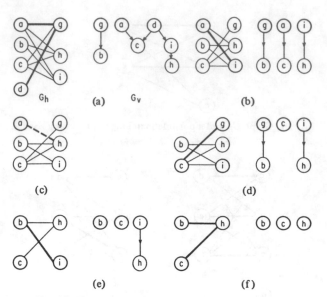

Fig. 23. Example to illustrate the use of the corollary.

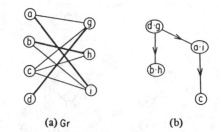

Fig. 24. An arbitrary merging in G_r and the corresponding vertical constraint graph after merging.

a2:   let N_0 be the set of nodes which do not have ancestors in G_v;

a3:   remove the set of edges E_0 from graph G_h, where $E_0 = \{(i,j) \,|\, i, j \in N_0, (i,j) \in E_h\}$;

a4:   if there is a node whose degree is equal to zero in graph G_h, then let it be v, go to a6;

a5:   otherwise, (a) choose one of the nodes $v \in N_0$, which has the smallest number of edges incident to it and (b) let $E_x = E_x + E_y$, where E_y is the set of edges connecting to node v;

a6:   remove node v and connecting edges from graph G_h and graph G_v;

        end;

   end;

Fig. 22 shows how the algorithm works. In the first iteration, $N_0 = \{a, c, e\}$ and two edges (a, c) and (a, e) are removed from G_h. Then node a is removed from G_h and G_v because the degree of node a in G_h is equal to zero etc., $\cdots$.

Theorem: The mergings corresponding to any matching on graph G_h is feasible if and only if E_x is empty at the termination of Algorithm A.

Proof:

if: We first prove that if E_x is empty, any matching of G_h is feasible. Let N_0 be the set of nodes with no ancestors in G_v. Obviously, those edges in G_h which connect nodes in N_0 and which are removed in a3 in the first go-around are feasible edges for matching. Let $N_0^* \subset N_0$ be the set of isolated nodes in G_h after the removal of the edges above. These nodes are connected in G_h only to nodes in N_0 and are obvious candidates for merger, that is, the mergings corresponding to any matching on graph G_h will not create cycles containing nodes of N_0 in graph G_v. Therefore, once these nodes are deleted, they can be forgotten as far as G_v or subsequent merging is concerned. Let N_1 be the second set of nodes with no ancestors. This set consists of new nodes in $\bar{N}_1 \subset N_1$, while the rest belong to $N_0 - N_0^*$. Clearly, nodes in $\bar{N}_1$ are descendents of N_0^* and they are not descendents of $N_0 - N_0^*$. The edges in G_h removed at this step are, therefore, similarly feasible for matching. We next delete $N_1^* \subset N_1$ which are the isolated nodes. Since $E_x = 0$ the process continues until all edges are removed.

only if: We next prove that if E_x is not empty, there exists a matching which is not feasible. We start as in the previous paragraph, but assume that there exists no isolated node after the removal of the initial set of edges. Let $n_0^{(1)} \in N_0$ be a node in N_0. Then there is at least one edge in G_h which connects node $n_0^{(1)}$ and a node, say $n_d^{(1)}$ which is not in N_0. If node $n_d^{(1)}$ is a descendent of node $n_0^{(1)}$, then merger between them is clearly unfeasible. Thus we assume $n_d^{(1)}$ is a descendent of node $n_0^{(2)}$ in $N_0 - \{n_0^{(1)}\}$. In the meantime, node $n_0^{(2)}$ is also not isolated after the initial removal of edges. So, it is connected to a node $n_d^{(2)}$ which, by a similar reasoning, is a descendent of node $n_0^{(3)}$ in $N_0 - \{n_0^{(2)}\}$. Since the number of nodes is finite, there must be a "cycle" of nodes in the sequence $n_0^{(1)}, n_d^{(1)}, n_0^{(2)}, n_d^{(2)}, n_0^{(3)}, n_d^{(3)}, \cdots$. Without loss of generality, we can assume that the cycle is $n_0^{(1)}, n_d^{(1)}, n_0^{(2)}, n_d^{(2)}, \cdots, n_0^{(k)}, n_d^{(k)}, n_0^{(1)}$. Consider the merger according to the set of edges $E_s = \{(n_0^{(i)}, n_d^{(i)}) \,|\, i = 1, \cdots, k\}$, and let $n_{0d}^{(i)}$ be the node produced by the merger of $n_0^{(i)}$ and $n_d^{(i)}$. Since $n_d^{(i)}$ is a descendent of $n_0^{(i+1)}$, $n_{0d}^{(i)}$ is a descendent of $n_{0d}^{(i+1)}$, for $i = 1, 2, 3, \cdots, k-1$. By the same reasoning, node $n_{0d}^{(k)}$ is a descendent of node $n_{0d}^{(1)}$. Hence, the merger creates a cycle in the merged G_v. This completes the proof of the theorem.

Corollary: The merging corresponding to any matching on graph $G_r = (N, E_h - E_x)$ is feasible.

The example shown in Figs. 23 and 24 illustrates the use of the corollary. Fig. 23(a) gives the graphs G_h and G_v where the

TABLE III
No Dogleg

Problem data(#nets)	opt	Solution left edge	Alg.1(cpu in sec.)	Alg.2(cpu in sec.)
ex.1(21)	12	14	12(0.05)	12(0.07)
ex.2(30)	15	18	15(0.07)	15(0.43)
ex.3b(47)	17	20	17(0.17)	17(0.57)
ex.3c(54)	18	19	18(0.18)	18(0.72)
ex.4b(57)	17	23	17(0.23)	17(0.77)
ex.5(62)	20	22	20(0.22)	20(0.88)
dif.ex(72)	28*	39	30(0.40)	28 (1.60)

*Obtained in [5] by means of the method of branch and bound after four hours of computation.

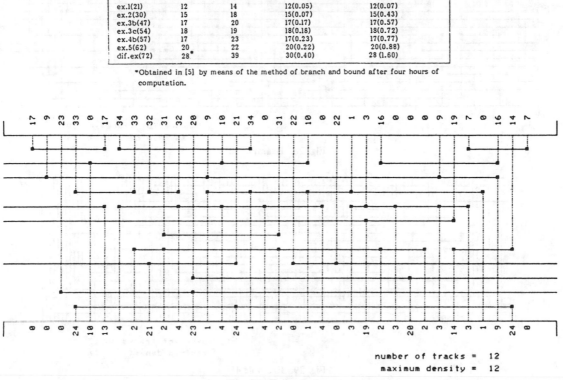

number of tracks = 12
maximum density = 12

Fig. 25. Example 1.

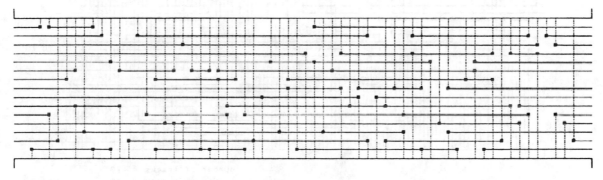

number of tracks = 15
maximum density = 15

Fig. 26. Example 3a.

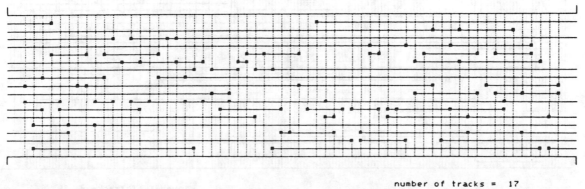

number of tracks = 17
maximum density = 17

Fig. 27. Example 3b.

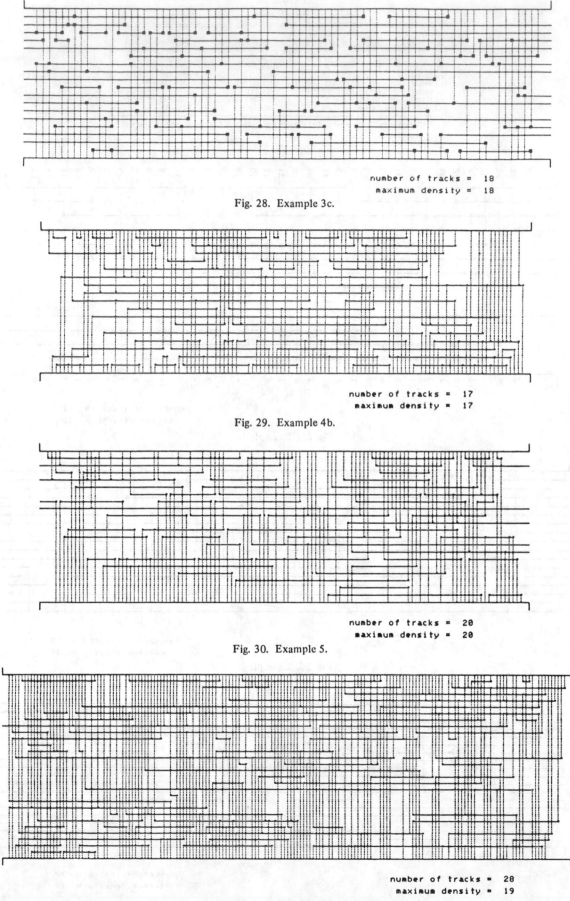

number of tracks = 18
maximum density = 18

Fig. 28. Example 3c.

number of tracks = 17
maximum density = 17

Fig. 29. Example 4b.

number of tracks = 20
maximum density = 20

Fig. 30. Example 5.

number of tracks = 28
maximum density = 19

Fig. 31. Difficult example without dogleg.

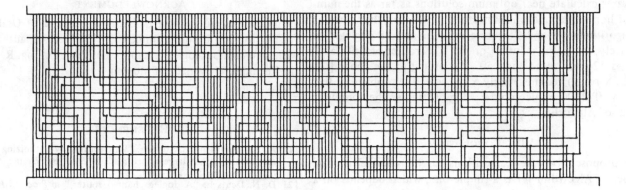

number of tracks = 20
maximum density = 19

Fig. 32. Difficult example with dogleg.

initial nodes with no ancestors and the connecting edges are specially marked. With the removal of the heavier edges, node d is isolated and removed, together with edges (d, g) and (a, g). In Fig. 23(b), the new nodes and edges to be considered next are marked. After the removal of edge (a, i), there exists no isolated node. Let us delete node a and remove edge $(a, h) \in E_x$, as shown in Fig. 23(c). The process then continues with no further problem. In Fig. 24(a) we show the horizontal graph G_r with E_x removed. The heavier edges represent an arbitrary matching. The corresponding vertical graph after merging is shown in Fig. 24(b). In a sense, the bad element (a, h) which could have caused a problem has been deleted before matching.

In the Step 3 of the Algorithm, we check the vertical conflict by applying the Algorithm A to graph $G(N, M)$, where M is the set of matching edges. It is easy to see that the mergings corresponding to the matching are feasible if and only if E_x is empty. If the mergings are not feasible, then we apply Algorithm A to graph $G(N, E_h)$ and recalculate the maximum matching on graph $G_r = (N, E_h - E_x)$. The new matching is feasible because of the corollary.

VI. ADDITIONAL COMMENTS ON THE DOGLEG PROBLEM

So far, we have discussed essentially the problem without doglegs. Of course, the presented algorithms can solve the dogleg problems using the zone representation and the vertical constraint graph introduced in Section III-D. However, additional consideration may be necessary to reduce the number of doglegs and CPU time. Thus we introduce a process "merging of subnets" which merges subnets belonging to the same net to form a net or larger subnet. If we carry this process to the extreme, the problem is reduced to the no-dogleg problem. Hence we impose the following restriction on the subnet merging.

Subnet i and subnet j can be merged only if merging of net i and net j will not increase the length of the longest paths which pass through node i or node j on the vertical constraint graph.

According to the results of the preliminary experiments, this process significantly reduces the CPU time and improves the solutions in the sense that the number of horizontal tracks and the number of doglegs are both reduced.

TABLE IV
DOGLEG

data(#nets)	density	LTX	Alg.1(cpu)	Alg.2(cpu)
dif. ex(72)	19	21	21(1.0)	20(2.1)

VII. COMPUTATIONAL RESULTS

Algorithms #1 and #2 have been coded in Fortran and implemented on the DEC VAX 11/780 computer.

A. Programs

Both programs have a parameter, the "starting column." In the explanation of algorithms, zone processing is carried out from zone 1 to zone n where n is the number of zones. However, we need not process in this order. In fact, we obtained better solution when we processed the zone with the highest density first. The above parameter specifies the starting zone. Programs process zones from this zone toward zone n (or zone 1), then toward zone 1 (or zone n). Usually, the starting zone is chosen among the zones which have the maximum density.

B. No-Dogleg Problem

Here, no dogleg is allowed. Examples 1–5 are the data taken from the existing paper [5], and the "difficult example" was provided by Deustch and Schweikert of Bell Laboratories. Table III compares the number of horizontal tracks of optimum solutions, the results of left edge algorithms, Algorithms #1 and 2. The CPU time for Algorithms #1 and 2 are also listed. Figs. 25–31 show the computer outputs of Algorithm #2.

The result indicates that both Algorithms #1 and 2 reach the optimum solutions for the data in Examples 1–5, and obtain considerably better solutions than the left edge algorithms for all examples.

C. Dogleg Problem

Algorithms were applied for the previous examples and, of course, produced optimum solutions for Examples 1–5. However, some solutions have several doglegs. Table IV shows the results for the "difficult example." We can see that all three programs calculate near optimum solutions as far as the number of horizontal tracks is concerned. However, the solutions of Algorithms #1 and 2 require only 19 and 18 doglegs, respectively, while the solution of the LTX router of Bell Laboratories [2] has more than 50. According to the paper [4], the LTX router requires nearly 5 s to solve the problem of this size on HP-2100 minicomputer. Fig. 32 shows the computer output of Algorithm #2.

VIII. Conclusion

We proposed two new algorithms for the channel routing problem. Algorithm #1 is a simple one, which depends on a process of merging nets instead of assigning horizontal tracks to each net. The basic idea of Algorithm #2 is the same as Algorithm #1, but it uses a matching algorithm to improve the solution. According to the computational results, when doglegs were not allowed, both algorithms produced optimum solutions for five examples taken from a previously published paper, and better solutions than LTX for the "difficult example." As for the dogleg problem, we used the "difficult problem" to make comparison. Algorithm #1 produced a solution which has the same number of horizontal tracks but far fewer doglegs than the solution of LTX. Algorithm #2 generated better solution than Algorithm #1 and LTX in the sense that the number of horizontal tracks and the number of doglegs are both smaller.

Acknowledgment

The authors are grateful to Dr. S. Tsukiyama of Osaka University for his useful comments. The second author also wishes to acknowledge the support of Prof. Dr.-Ing. R. Saal at the Technical University of Munich.

References

[1] A. Hashimoto and S. Stevens, "Wire routing by optimizing channel assignment within large apertures," in *Proc. 8th Design Automation Workshop*, pp. 155–169, 1971.
[2] D. N. Deutsch, "A dogleg channel router," in *Proc. 13th Design Automation Conf.*, 1976.
[3] G. Persky, D. N. Deustch, and D. G. Schweikert, "LTX–A minicomputer-based system for automatic LSI layout," *J. Design Automation and Fault-Tolerant Computing*, pp. 217–255, May 1977.
[4] ——, "LTX–A system for the directed automation design of LSI circuits," in *Proc. 13th Design Automation Conf.*, 1976.
[5] B. W. Kernighan, D./G. Schweikert, and G. Persky, "An optimum channel-routing algorithm for polycell layouts of integrated circuits," in *Proc. 10th Design Automation Workshop*, pp. 50–59, 1973.
[6] T. Ohtsuki, H. Mori, E. S. Kuh, T. Kashiwabara, and T. Fujisawa, "One-dimensional logic gate assignment and interval graphs," *IEEE Trans. Circuits Syst.*, vol. CAS-26, pp. 675–684, Sept. 1979.
[7] C. G. Lekerkerker and J. Ch. Boland, "Representation of a finite graph by a set of intervals on the real line," *Fund. Math.*, vol. 51, pp. 45–64, 1962.
[8] W. Donath, private communication.

Hierarchical Wire Routing

MICHAEL BURSTEIN, MEMBER, IEEE, AND RICHARD PELAVIN

Abstract—We propose a new approach to automatic wire routing of VLSI chips which is applicable to interconnection problem in uniform structures such as gate arrays, switchboxes, channels. Popularity of gate arrays technologies still remains high among VLSI chip manufacturers and, as the scale of integration grows, the interconnection problem becomes increasingly difficult if not intractable. The same is true for problems of switchbox and channel routing, which usually arise in custom designs; the uniformity of wiring substrate unites them with gate array routing problem.

Our approach was initially aimed at gate arrays, but it extends naturally to switchboxes and channels. Uniformity of the wiring substrate is the crucial assumption of the method. It assumes that horizontal and vertical wire segments are realized on different wiring layers and vias are introduced each time a wire changes direction. Any "jogs" ("wrong way" wires) are prohibited. Within these limitations our approach is advantageous over the existing wiring methodologies. Our final layout of wires is independent of both net ordering and ordering of pins within the nets. The wire densities we are able to achieve are often higher than those achieved by other routers. Because of the hierarchical nature of our method it is inherently fast, usually by an order of magnitude faster than the routers based on wave propagation (maze running) technique.

I. INTRODUCTION

WE PROPOSE a new approach to automatic wire routing of VLSI chips which is applicable to interconnection problem in uniform structures such as gate arrays, switchboxes, channels. Popularity of gate arrays technologies still remains high among VLSI chip manufacturers and, as the scale of integration grows, the interconnection problem becomes increasingly difficult if not intractable. The same is true for problems of switchbox and channel routing, which usually arise in custom designs; the uniformity of wiring substrate unites them with gate array routing problem.

Modern automatic wire routing methods can be classified as: (i) "maze-runners" or Lee–Moore type routers [1]–[4], (ii) "line probe" routers [5], (iii) channel routers [6]–[8], (iv) "pattern" routers [9].

Channel routers require a restrictive arrangement of components and do not explore all areas technologically available for wiring. Routers based on "line propagation" are either extremely time consuming or may fail during the search of existing interconnection. That is why "maze runners" are widely utilized for wiring of very dense chips.

All of these routers work in "a net-at-a-time" mode, i.e., all nets are organized in a sequence and a router picks the first not wired net for routing. "A net-at-a-time" router is obviously dependent on the ordering of nets. Moreover, "maze runners"

and "line propagation" routers are also dependent on internal ordering of terminals within a net. Picking different terminals as a source for wave or line propagation may result in different interconnections.

Another disadvantage of "maze runners" is speed. The time for interconnection of one net may be, in the worst case, proportional to the number of grid points of the maze. In order to speed up the maze running the wiring process in often realized in two phases: global wiring and final track allocation. At global wiring phase each net is wired in terms of cells of the gate array [10], [11]. Instead of allocating the wiring track at a final grid resolution, a global route of the net through the cells of gate array is allocated. This limits the area for wave propagation during the final wiring phase to the cells specified by the global route. The idea of global wiring is crucial in the hierarchical approach that we are proposing here.

"Pattern" routers are fairly new and their capabilities are not yet completely clear, but they seem to have the potential of overcoming the abovementioned difficulties of other routing methodologies. The hierarchical routing scheme developed in Section III preserves the flavor of pattern routing and was in a certain sense influenced by the similar ideas.

Our approach should be viewed just as another collection of heuristic tricks for routing. Since most of the wire routing problems (except their trivial particular cases) are *NP*-complete [19] not much more can be expected. The approach was initially aimed at gate arrays, but it extends naturally to switchboxes and channels. Uniformity of the wiring substrate is however the crucial assumption of the method. It assumes that horizontal and vertical wire segments are realized on different wiring layers and vias are introduced each time a wire changes direction. Any "jogs" ("wrong way" wires) are prohibited. But within these limitations our approach is advantageous over the existing wiring methodologies. Our final layout of wires is independent of net ordering and of ordering of pins within the nets. The wire densities we were able to achieve are generally higher than those achieved by other routers. Because of the hierarchical nature of our method it is inherently fast, usually by an order of magnitude faster than the routers based on wave propagation (maze-running) technique.

II. ROUTING HIERARCHY

We will be dealing with a uniform array of cells $G = G(i,j)$, $i = 1, 2, \cdots, p, j = 1, 2, \cdots, q$. Some of the cells carry personalized circuits. Each net is identified with a set of cells of G where interconnection terminals are located. Each cell is surrounded by North, East, South, and West boundaries. Channel capacity of the boundary is the estimated number of wiring tracks that can be driven through this boundary. Each cell

Manuscript received March 3, 1983; revised May 24, 1983.

M. Burstein is with IBM Thomas J. Watson Research Center, Yorktown Heights, NY 10598.

R. Pelavin is with the University of Rochester, Rochester NY.

Reprinted from *IEEE Trans. Computer-Aided Design of Integrated Circuits and Systems*, vol. CAD-2, pp. 223–234, Oct. 1983.

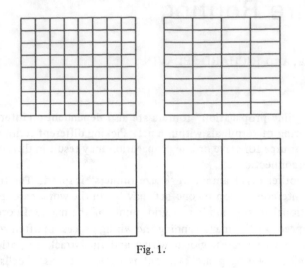

Fig. 1.

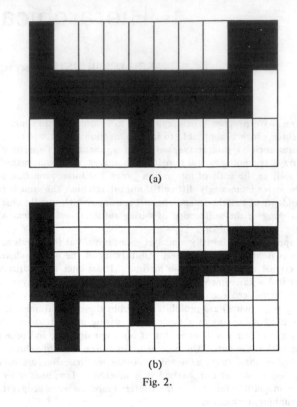

(a)

(b)

Fig. 2.

Cells of the 2 X N Grid

| G(2,1) | G(2,2) | G(2,3) | . . . | G(2,N) |
| G(1,1) | G(1,2) | G(1,3) | . . . | G(1,N) |

Fig. 3.

thus has North, East, South, and West channel capacities associated with it.

Global wire route is a set of cells R with a set of four element Boolean vectors D associated with each cell $r \in R$: $D(r) = (N, E, S, W)$ satisfying natural conditions: if $N = 1$, then the cell to the North of r must belong to R, if $E = 1$, then the East cell of r must belong to R, and so on. In addition to that, global wire route of the net must be connected and contain all terminal cells of this net. The number of cell boundaries the wire route crosses is taken to be the length of the wire route.

The global wiring problem is to construct wire routes for each net in such a way that for any cell-boundary number of wire routes crossing it is less than or equal to the corresponding channel capacity; the number of wires turning within a given cell should be no more that the maximal number of vias that cell can accommodate [12]. We assume here that two interconnection layers are available for routing, one of which is exclusively used for vertical wiring segments, another for horizontal and vias are introduced for each layer change when the wire route makes a turn.

Note that each row or column of cells carries a certain number of wiring tracks, so that the global wiring is not a final solution. But the global wiring problem can be viewed as a generalization of the initial problem, it just accepts the grid of cells with arbitrary capacities and results in the original problem when those capacities are set to 1. Initially the global wiring problem was considered as intermediate stage for gate-array routing [13].

Following the same philosophy that led the separation of the global wiring stage from the final assignment of wiring tracks we may introduce a "super" global routing in terms of pairs of cells, then in terms of quadruples and so on, until the problem vanishes (see Fig. 1). The hierarchical wiring method is, in a sense, a reversed process, where we start with a trivial problem of wiring the grid of two cells and subsequently build the wire routes for every level of chip bisectionings hierarchy. The main question is how do we use the wiring information from the previous hierarchical level in order to obtain it on the current one. Fig. 2 illustrates the answer to this question. Fig. 2(a) presents a wire route on certain level of hierarchy and the grid is dissected for routing on the next level. Obviously the

route has to be located within the shaded area. The latter is naturally partitioned in to the sections of the width $2-2 \times N$ subgrids (Fig. 2(b)). So the problem can be naturally reduced to the problem of global wiring within $2 \times N$ grid. We describe below two different approaches to the $2 \times N$ wiring problem. Our first solution is based on the exact solution of this problem in case of 2×2 grid using an integer linear programming approach and the hierarchical extension of the latter in a "divide and conquer" fashion. The second approach is based on more traditional one net-at-a-time routing within $2 \times N$ grid, imbedding all nets sequentially. It turns out that the Steiner problem with arbitrary costs can be solved in linear time if the width of the grid is only 2. The solution is an extension of dynamic programming procedure described in [14].

III. $2 \times N$ ROUTING—LINEAR PROGRAMMING APPROACH

The problem of Global Wiring within the $2 \times N$ grid can be described as follows. We start with the $2 \times N$ uniform array of cells (Fig. 3). Let $\{e(1), e(2), \cdots, e(m)\}$ denote the set of nets. Each net $e(i)$ is identified with a set of cells of this array, called terminal cells of this net. If cell $G(i,j)$ is a terminal cell for the net $e(k)$ then at least one of the terminals (pins) of

Graph of the 2 X N Grid

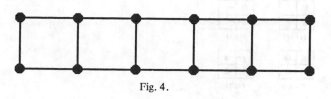

Fig. 4.

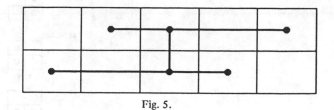

Fig. 5.

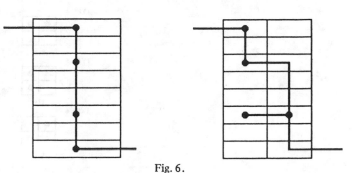

Fig. 6.

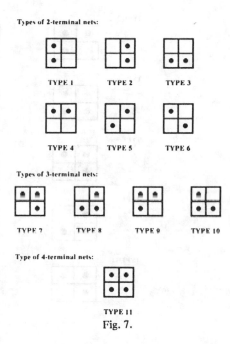

Fig. 7.

this net is located somewhere within this cell.

We also have a $2 \times (N - 1)$ matrix of horizontal channel capacities

$$h(1, 1), h(1, 2), \cdots, h(1, N - 1)$$

$$h(2, 1), h(2, 2), \cdots, h(2, N - 1)$$

and an N-vector of vertical capacities

$$v(1), v(2), \cdots, v(N)$$

associated with our array; $h(i, j)$ denotes the number of available wiring tracks crossing the boundary between the cells $G(i, j)$ and $G(i, j + 1)$; similarly $v(i)$ denotes the number of tracks between $G(1, i)$ and $G(2, i)$.

A graph of our $2 \times N$ grid consists of $2N$ vertices associated with the cells; any two vertices associated with neighbor cells are adjacent. The graph is depicted in Fig. 4.

A route of the net is a subtree of this graph containing all vertices associated with its terminal cells. Our problem is to determine routes for all nets.

3.1. Restrictions

System of net routes (wire layout) has to satisfy two type of restrictions: channel capacity restrictions and via restrictions.

• The number of wires crossing the boundary between the cells $G(i, j)$ and $G(i, j + 1)$ must not exceed $h(i, j)$.

• The number of wires crossing the boundary between the cells $G(1, i)$ and $G(2, i)$ must not exceed $v(i)$.

• For each cell $G(i, j)$, the number of wires bending (changing direction) within this cell for which this cell is not terminal must be within the limits established in [12], i.e., must not exceed the certain value $M(i, j)$ dependent on $v(j)$ and $h(i, j)$.

3.2. Restrictive Wiring

We refer to the above described wiring problem as *general* $2 \times N$ wiring problem, assuming that all possible tree structures are admissible for wire routes. But to be consistent with our general philosophy we also introduce *restrictive* $2 \times N$ wiring problem, where a wire route is not allowed to cross vertical boundary twice, for example, tree structures shown in Fig. 5 are not allowed. We consider this restrictive case because when the current net had been routed on previous levels of hierarchy we accounted for only one crossing of a segment that constitutes both of our boundaries presently. $2 \times N$ wiring problem arises after partitions in the direction "orthogonal" to the direction of the $2 \times N$. Example illustrating this fact is presented in Fig. 6. Suppose on the previous level of hierarchy (for this example) a net was taking the route of Fig. 6(a). All channel capacities of the boundaries the net is crossing might be exhausted. After partitioning in "horizontal" direc-

tion we get our $2 \times N$ wiring problem (Fig. 6(b)). If we now allow crossing of the old boundaries more then once we may run out of wiring tracks for other nets.

3.3. Four Cell Wiring Problem

We start with the consideration of the simplest particular case of our problem: *wiring within* 2×2 *grid*.

In this case it is easy to enumerate all possible types of nets that may occur, depending on the distribution of terminal cells. Clearly, the nets having only one terminal cell are not affecting the wiring problem, so we assume that the number of terminal cells of a net may be 2, 3, or 4. There are only six types of 2-terminal nets, four different types of 3-terminal nets and only one type of 4-terminal nets. All different types of nets are depicted in Fig. 7.

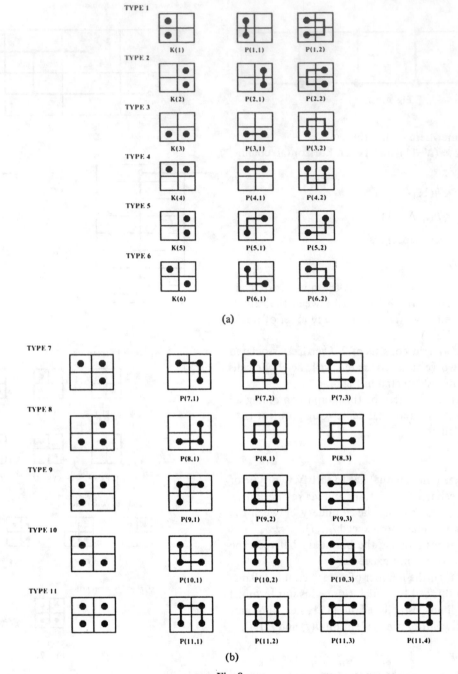

Fig. 8.

Let $K(i)$ be the number of nets of TYPE i, ($i = 1, 2, \cdots, 11$). For each net type there is limited (and small) number of ways it can be routed, and we enumerate all of them. For every net of TYPE 1 there are only two possibilities (denoted $P(1, 1)$ and $P(1, 2)$) presented in Fig. 8. Similarly we enumerate all possibilities for all other types. Fig. 8 contains all possible wire routes for all net types.

The wiring problem is reduced to the question: For $i = 1$, $2, \cdots, 11$, how many nets of TYPE i will be realized in each possibility? Namely, let $x(i, j)$ denote the number of nets of TYPE i realized in the j-th possibility, i.e., the number of nets routed in fashion $P(i, j)$. Determining the values of $x(i, j)$ represents our wiring problem.

3.4. Wiring Within 2×2 Grid

The values $x(i, j)$ can be determined by solution of the *Integer Programming* problem which can be defined as follows.

First of all, the wiring problem may be non solvable, i.e., the solution may result in overflows. Overflowed nets are still classified by types in previous section. Let $x(i)$ (with single index) denote the number of overflowed nets of type $i, i = 1, 2, \cdots, 11$.

We begin with linear restrictions.

$$\forall \, (admissible) \, i, j: \qquad x(i, j) \geqslant 0 \tag{1}$$

$$\forall i = 1, 2, \cdots, 11: \qquad x(i) + \sum_j x(i, j) = K(i). \tag{2}$$

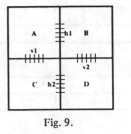

Fig. 9.

Channel capacity restrictions and via restrictions imply another set of linear inequalities. Denote

$H1 = \{(i, j) | P(i, j) \text{ crosses upper vertical boundary}\}.$

$H2, V1, V2$ are defined similarly.

Simple enumeration shows that

$H1 = \{(1, 2); (2, 2); (3, 2); (4, 1); (5, 1); (6, 2); (7, 1); (7, 3);$
$\quad\quad (8, 2); (8, 3); (9, 1); (9, 3); (10, 2); (10, 3); (11, 1);$
$\quad\quad (11, 3); (11, 4)\}.$

$H2 = \{(1, 2); (2, 2); (3, 1); (4, 2); (5, 2); (6, 1); (7, 2); (7, 3);$
$\quad\quad (8, 1); (8, 3); (9, 2); (9, 3); (10, 1); (10, 3); (11, 2);$
$\quad\quad (11, 3); (11, 4)\}.$

$V1 = \{(1, 1); (2, 2); (3, 2); (4, 2); (5, 1); (6, 1); (7, 2); (7, 3);$
$\quad\quad (8, 2); (8, 3); (9, 1); (9, 2); (10, 1); (10, 2); (11, 1);$
$\quad\quad (11, 2); (11, 3)\}.$

$V2 = \{(1, 2); (2, 1); (3, 2); (4, 2); (5, 2); (6, 2); (7, 1); (7, 2);$
$\quad\quad (8, 1); (8, 2); (9, 2); (9, 3); (10, 2); (10, 3); (11, 1);$
$\quad\quad (11, 2); (11, 4)\}.$

Then we have

$$\sum_{(i, j) \in H1} x(i, j) \leqslant h1$$

$$\sum_{(i, j) \in V1} x(i, j) \leqslant v1$$

$$\sum_{(i, j) \in H2} x(i, j) \leqslant h2$$

$$\sum_{(i, j) \in V2} x(i, j) \leqslant v2.$$

Here $v1, v2, h1, h2$ stand for $v(1), v(2), h(2, 1), h(1, 1)$ defined earlier (see Fig. 9).

Similarly, let A be a set of pairs (i, j) such that $P(i, j)$ bends in the upper left cells of our grid. $B, C,$ and D are defined accordingly. We have

$A = \{(2, 2); (3, 2); (5, 1); (8, 2); (8, 3)\}$

$B = \{(1, 2); (3, 2); (6, 2); (10, 2); (10, 3)\}$

$C = \{(2, 2); (4, 2); (6, 1); (7, 2); (7, 3)\}$

$D = \{(1, 2); (4, 2); (5, 2); (9, 2); (9, 3)\}.$

Via restrictions:

$$\sum_{(i, j) \in A} x(i, j) \leqslant M(v1, h1)$$

$$\sum_{(i, j) \in B} x(i, j) \leqslant M(v2, h1)$$

$$\sum_{(i, j) \in C} x(i, j) \leqslant M(v2, h2)$$

$$\sum_{(i, j) \in D} x(i, j) \leqslant M(v1, h2)$$

Values of $M(a, b)$ are presented later.

Denote Y_1, Y_2, Y_3, Y_4 the numbers of leftover wiring tracks on $V1, H1, V2, H2$ boundaries. Then the channel capacity restrictions can be rewritten:

$$\sum_{(i, j) \in V1} x(i, j) + Y_1 = v1$$

$$\sum_{(i, j) \in H1} x(i, j) + Y_2 = h1$$

$$\sum_{(i, j) \in V2} x(i, j) + Y_3 = v2 \quad\quad (3)$$

$$\sum_{(i, j) \in H2} x(i, j) + Y_4 = h2.$$

In the same way we introduce "artificial" variables Z_1, Z_2, Z_3, Z_4 in our via restrictions, corresponding in some sense to "unused vias". We get our last linear equations:

$$\sum_{(i, j) \in A} x(i, j) + Z_1 = M(v1, h1)$$

$$\sum_{(i, j) \in B} x(i, j) + Z_2 = M(v2, h1)$$

$$\sum_{(i, j) \in C} x(i, j) + Z_3 = M(v1, h2) \quad\quad (4)$$

$$\sum_{(i, j) \in D} x(i, j) + Z_4 = M(v2, h2).$$

Now (1), (2), (3), and (4) represent all linear restrictions in our *integer (linear) programming problem*.

We have total of 19 linear equations with 47 variables: 28 of $x(i, j)$ plus $X_i, (i = 1, \cdots, 10, 11)$, plus $Y_i, Z_i (i = 1, 2, 3, 4)$. Here are the linear forms we have to optimize:

(i) $\quad \sum_{i=1}^{11} X_i \rightarrow \min$

(ii) $\quad \sum_{i=1}^{4} Y_i \rightarrow \max$

(iii) $\quad \sum_{i=1}^{4} Z_i \rightarrow \max$

in descending order of priorities. All variables are assumed to be nonnegative integers. Then, (i) reduces the total number of overflowed nets to the possible minimum. In wireable cases this minimum has to be 0. Provided this is achieved, (ii) assures us of the minimal possible wire length, because values of Y_i complement the numbers of wiring tracks that are taken. Finally (iii) attempts to minimize the prospective number of vias used, if there is a choice to make.

The most significant part of this reduction is that the *size of the integer programming problem is always fixed and independent of the number of nets to be wired*. Naturally, we still have to spend $O(m)$ time, where m is the number of nets, just to get the net classification in the previous section. Independent of the method we choose for the solution of the Integer Programming problem, the solution of the 2×2 wiring problem will run in time $C + O(m)$, where only the constant C is dependent on the time spent for solution of the Integer Programming problem.

Remark 1: In the existing implementation we substitute the Integer Programming problem by corresponding Linear Programming problem, invoke standard solution and round off the results.

Remark 2: Solution of the Integer Programming problem indicates for a given net type how many nets of this type are routed which way. We still have to decide finally how the nets will be routed. The choice here is a heuristic attempt to make net lengths shorter. For example, if we know that there are $K(1)$ of type 1 and $x(1, 1)$ of them have to routed straight and $x(1, 2)$ must be detoured (see Fig. 8), then we select the nets whose terminals are located close to the cut-line for detouring and those who have terminals located at the "far left"—for straight connection. In case when we cannot sort out the nets in this fashion, the choice is arbitrary.

3.5. Restrictive Four Cell Wiring

As was mentioned before, we have to consider the *restrictive* 2×2 wiring problem. We achieve that by forcing several variables in our Integer Programming problem of the previous section to be zero. Namely, we forbid the following wiring possibilities (crossing the vertical boundary twice):

$$P(1, 2), P(2, 2), P(7, 3), P(8, 3), P(9, 3), P(10, 3),$$
$$P(11, 3), P(11, 4).$$

So, we set the corresponding $x(i, j)$ to 0. We also eliminate the first two equations from system (2). The rest of the system remains the same, and its solution provides the restrictive 2×2 wiring. The number of variables will be 39; the number of linear restriction will be 17. But we make the following important observation:

If the restrictive 2×2 wiring problem has the solution with no overflows (i.e., $X(i) = 0$, for $i = 1, 2, \cdots, 11$), then this solution is at the same time an optimal solution for the general 2×2 wiring problem.

3.6. $2 \times N$ Wiring Algorithm

Here we describe a heuristic algorithm for the solution of the $2 \times N$ global wiring problem based on 2×2 solutions. It is a recursive procedure based on the "divide and conquer" approach. First of all we partition the $2 \times N$ grid into two

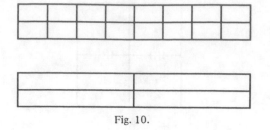

Fig. 10.

parts, generally "in half". Construct a "factorized" 2×2 wiring problem considering only the new cut line and the horizontal line as cell boundaries (see Fig. 10). After this 2×2 grid is routed the problem is reduced to smaller size problems $2 \times N1$ and $2 \times N2$, where $N1 + N2 = N$. The reduction is based on the introduction of new terminal cells of the main grid adjacent to the crossed vertical boundary of the 2×2 grid for every net.

Fig. 11 shows an example of 2×8 routing of one net with 3-terminal cells traced to elementary levels of hierarchy.

Now we have to determine the channel capacities in the factorized 2×2 problem. We, of course, could use the actual channel capacities of the boundaries: $h(1, k)$ and $h(2, k)$ values for horizontal channel capacities (assuming that vertical cut is performed along the kth line) and $\Sigma_{i < k} v(i)$ and $\Sigma_{i > k} v(i)$ for vertical channel capacities.

This definition of channel capacities may be adequate, provided original channel capacity distribution is uniform. It may not be so in the case of nonuniform channel capacity distribution, because during the reduction of general $2 \times N$ problem to the 2×2 factorization, the channel capacities of the vertical lines next to the cut-line disappear from our consideration. It may happen, that the value of $h(1, k)$ is much larger than (say) $h(1, k - 1)$, but the program for the solution of the factorized 2×2 problem will not see the small value of $h(1, k - 1)$; it may force a large number of nets to cross the boundary corresponding to $h(1, k)$, and those nets will overflow on the lower levels of hierarchy because of the small value of $h(1, k - 1)$. In order to avoid this we propose the following heuristic scheme for channel capacity determination.

First of all we compute the quantities $h^*(i, j)$ ($i = 1, 2, j = 1, \cdots, N$).

$$h^*(i, k) = h(i, k);$$

For $j < k$: $h^*(i, j) = \min \{h(i, j) | h^*(i, j + 1)\}, \quad j < k;$

For $j > k$: $h^*(i, j) = \min \{h(i, j) | h^*(i, j - 1)\}, \quad j > k.$

Now the horizontal channel capacities $h1$ and $h2$ are determined as

$$hi = \frac{\sum_j h^*(i, j)}{N}, \quad i = 1, 2.$$

Vertical channel capacities will be defined as $(\Sigma_{i < k} v(i)) \times C(k)$ and $(\Sigma_{i > k} v(i)) \times C(N - k)$ where $C(x)$ are experimentally determined coefficients:

$$C(x) = \frac{x + 1}{2x}.$$

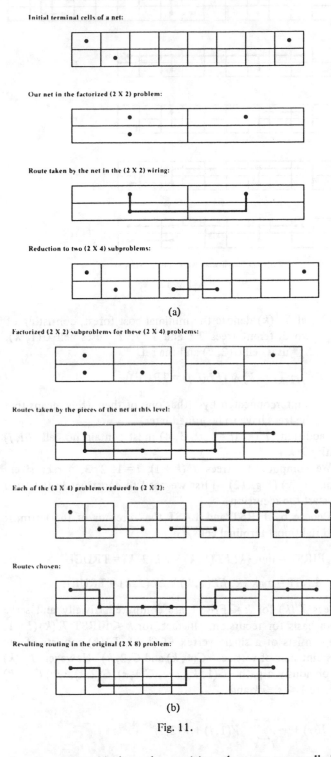

HIERARCHICAL (2 X N) WIRING EXAMPLE

Initial terminal cells of a net:

Our net in the factorized (2 X 2) problem:

Route taken by the net in the (2 X 2) wiring:

Reduction to two (2 X 4) subproblems:

(a)

Factorized (2 X 2) subproblems for these (2 X 4) problems:

Routes taken by the pieces of the net at this level:

Each of the (2 X 4) problems reduced to (2 X 2):

Routes chosen:

Resulting routing in the original (2 X 8) problem:

(b)

Fig. 11.

So, we use actual channel capacities when our supercells become actual cells ($C(1) = 1$), but when the number of cells in a supercell is large the amount of channel capacity we assign in the factorized problem is slightly more then half of the actual channel capacity of the boundary.

3.7. Number of Vias

Here we define the bounds for numbers of vias in a cell for both $2 \times N$ wiring problem and the corresponding factorized 2×2 problem. Let $m = m(k, r)$ is the maximal integer such

that in any cell with k horizontal and r vertical tracks $m(k, r)$ vias may always be realized. These values are calculated in [12]. We use exactly these values for actual bound for the number of vias in a cell of the $2 \times N$ grid. But for supercells of the corresponding 2×2 factorized problem these values have to be reduced, because a change of direction in a supercell may later result in several bends on the lower levels. We use the same heuristic approximation for these bounds in this case. Namely, suppose a supercell contains k horizontal, r vertical tracks and consists of n actual cells. Then the values $M(k, r)$ are determined as

$$M(k, r) = m(k, r) \times C(n)$$

with the same coefficients:

$$C(x) = \frac{x + 1}{2x}.$$

3.8. Concluding Remarks

The essential part of the presented algorithm is the wiring of all the net simultaneously, so that the *resulting layout of routes is order independent.* It is also independent of terminal ordering within the nets. The routing problem is reduced to the solution of some relatively small integer linear programming problems (of the fixed size, independent of the size of the grid and number of nets) in hierarchical fashion, so the time spent on routing on every level of hierarchy is *independent of the number of nets* (although the $O(m)$ time will be spent anyway for net classification, where m is the number of nets). Time for the Integer Programming solution becomes an *additive* constant in the total time for 2×2 wiring. Note also that the hierarchical technique of $2 \times N$ wiring permits parallel processing on a tree structured multiprocessor system, because after the initial partition of the $2 \times N$ grid into two subgrids, the routing problems within the subgrids are independent of each other.

Experiments performed so far indicated that the above described wiring method is powerful in case of $2 \times N$ grid with uniform capacities. But as the capacity deviations grow the wiring quality decreases. But the overflows produced by the solution of Integer Programming problem may not be unavoidable. To handle them we use the Dynamic Programming approach.

IV. $2 \times N$ ROUTING—DYNAMIC PROGRAMMING APPROACH

In this section we describe another, more traditional, way of routing within $2 \times N$ grid. It may be used independently or in combination with the linear programming technique for overflow handling. Denote $G(i, j)$ the cells of our $(2 \times N)$ grid, $i = 1, 2, j = 1, 2, \cdots, n$. Wiring is performed by the algorithm Two-By-N (TBN) which allocates routes for all nets, one at a time.

With each interconnection route of a net we associate a cost, which is comprised of the costs of boundary crossings and costs of vias; the total cost is simply the sum of these costs.

The TBN wiring algorithm is a linear algorithm that finds

the minimal cost tree interconnecting a set of elements located on a $(2 \times N)$ grid.

The input to the algorithm consists of

1) The terminal locations on the $(2 \times N)$ grid: EL is a $(2 \times N)$ Boolean matrix; if $EL(i, j) = $ TRUE then the corresponding cell is a terminal cell.

2) Costs of boundary crossing:

   —HC is an N element vector which stores the costs of crossing the horizontal boundaries; $HC(j)$ indicates the cost that must be added if a wire crosses the boundary between $G(1, j)$ and $G(2, j)$;

   —VC is a $(2 \times (N - 1))$ matrix that stores the costs of vertical boundary crossings; $VC(i, j)$ indicates the cost which must be added if a wire crosses the boundary between $G(i, j)$ and $G(i, j + 1)$, $i = 1, 2$.

   —VIAC is a $(2 \times N)$ matrix that stores the costs of turning within a cell $G(i, j)$; VIAC(i, j) indicates the cost added if a wire crosses two adjacent boundaries of the cell $G(i, j)$.

Note: If the cost is greater than LRG, which is set to a large number, the corresponding boundary is blocked and a wire can not pass through this cell boundary. In other words, cost of crossing a blocked boundary in infinitely high.

Matrices HC, VC, and VIAC represent our cost-functions. They should reflect the boundary capacities and via conditions. Clearly, $VC(i, j)$ and $HC(j)$ must be a decreasing functions of the number of wiring tracks crossing the corresponding boundaries. Selection of the proper cost functions is extremely important because of its influence on the quality of the final routing. We will discuss this point in the next section.

The output of TBN consists of the minimal interconnection tree.

4.1. Minimal Cost Interconnection

The TBN algorithm is a modification of an algorithm developed in [14], which finds a Steiner tree that interconnects a set of elements located on a $(2 \times N)$ grid. But our algorithm solves the problem for an arbitrary cost function associated with crossing a boundary (matrices HC, VC, and VIAC), whereas the algorithm of [14] assumes the rectilinear Steiner tree, which is the case when all the costs HC, VC are equal to 1, and VIAC consists of 0 s.

We need the following definitions.

1) Let $T^1(k)$ denote the minimal cost tree which interconnects the following set of cells:

$$\{G(i, j): (j \leqslant k) \& (EL(i, j) = \text{TRUE})\} \cup \{G(1, k)\}.$$

2) Let $T^2(k)$ denote the minimal cost tree which interconnects the following set of cells:

$$\{G(i, j): (j \leqslant k) \& (EL(i, j) = \text{TRUE})\} \cup \{G(2, k)\}.$$

3) Let $T^3(k)$ denote the minimal cost tree which interconnects the following set of cells:

$$\{G(i, j): (j \leqslant k) \& (EL(i, j) = \text{TRUE})\} \cup \{G(1, k); G(2, k)\}.$$

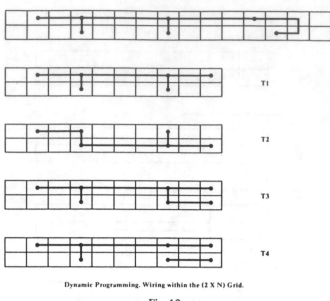

Dynamic Programming. Wiring within the (2 X N) Grid.

Fig. 12.

4) Let $T^4(k)$ denote the minimal cost forest, consisting of two different trees T^* and T^{**}: T^* uses cell $G(1, k)$, T^{**} uses cell $G(2, k)$ and the set

$$\{G(i, j): (j \leqslant k) \& (EL(i, j) = \text{TRUE})\}$$

is interconnected by either one of them (this means that the trees have to be joined later).

In addition, $T^i(k)$ ($i = 1, 2, 3, 4$) must contain no cell $G(i, j)$ with $j > k$.

We compute the trees $T^i(k + 1)$, $i = 1, 2, 3, 4$ recursively from $T^i(k)$ (Fig. 12). First we need to construct initial trees to start the recursion.

Denote by FIRST and LAST the abscissas of the leftmost and rightmost terminal cells, i.e.,

$$\text{FIRST} = \min \{k: EL(1, k) \vee EL(2, k) = \text{TRUE}\}$$

$$\text{LAST} = \max \{k: EL(1, k) \vee EL(2, k) = \text{TRUE}\}.$$

Trees $T^i(k)$ for $k \leqslant $ FIRST are computed trivially and serve as a basis for recursion. In fact, for $k \leqslant $ FIRST $T^l(k)$ ($l = 1, 2$) consists of a single vertex $-G(l, k)$, $T^4(k)$ consists of the disjoint pair of vertices $-G(1, k)$ and $G(2, k)$. However, $T^3(k)$ is obviously a path $-G(1, k), \cdots, G(1, s), G(2, s), \cdots, G(2, k)$ where $1 \leqslant s \leqslant k$ and

$$H(s) + \sum_{i = s}^{i = k - 1} V(1, i) + V(2, i)$$

is *minimal.*

In rectilinear case, when all costs are the same, $T^3(k)$ is the adjacent pair of vertices $-G(1, k)$ and $G(2, k)$. But in general, the cost $H(k)$ might be too high and detouring might result in a cheaper route (see Fig. 13(a)).

Suppose that FIRST $\leqslant k \leqslant $ LAST and the trees $T^i(k)$, $i = 1, 2, 3, 4$ are constructed. To construct $T^i(k + 1)$ we enumerate all possible extensions from $T^i(k)$, $i = 1, 2, 3, 4$, and select the

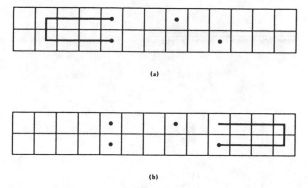

(a)

(b)

Fig. 13.

cheapest one. For example, to construct $T^1(k + 1)$ we select the cheapest way of adding $G(1, k + 1)$ to each of the $T^i(k)$, $i = 1, 2, 3, 4$, and select the cheapest among them. Note, that $T^4(k + 1)$ always becomes either trivial extension of $T^4(k)$ or one of the vertices $G(1, k + 1)$ or $G(2, k + 1)$ is isolated whereas the other component is identical to $T^1(k + 1)$ or $T^2(k + 1)$.

When the value $k =$ LAST is reached our trees are covering all terminal cells. We temporarily select the cheapest among $T^1(\text{LAST})$, $T^2(\text{LAST})$, $T^3(\text{LAST})$ and denote it by T. In rectilinear case it is obviously the interconnection we need. But it is possible that the costs $H(s)$ for $s \leqslant$ LAST are so high that it is cheaper to take a right side detour. In a way symmetric to the construction of $T^3(\text{FIRST})$ we select a path $-G(1, \text{LAST}), \cdots, G(1, s), G(2, s), \cdots, G(2, \text{LAST})$, where LAST $< s \leqslant n$ and

$$H(s) + \sum_{i = \text{LAST}}^{i = s - 1} V(1, i) + V(2, i)$$

is *minimal*.

Then we join the components of $T^4(\text{LAST})$ by this path, (see Fig. 13(b)), compare the cost of resulting interconnection with the cost of T and finally select the cheapest.

Note that at each stage of recursion we make a constant number of comparisons, and the total computation time is $O(n)$ because the initiation stage–computing $T^3(\text{FIRST})$–takes $O(\text{FIRST})$ time, construction of T is performed in time O (LAST - FIRST) and final right side detour is computed in at most $O(n - \text{LAST})$ time.

In case when the interval [FIRST, LAST] is small in comparison with $[1, n]$ we can limit the detours from either side (Fig. 13) in order to eliminate unnecessary computations. We usually limit the detour outside of the [FIRST, LAST] segment by 2 or 3, so in most cases the time spent for $2 \times n$ routing will be $O(\text{LAST} - \text{FIRST})$.

4.2. Cost Functions

Costs of boundary crossings–horizontal and vertical are chosen to be an exponentially increasing functions of boundary capacities. If B is a boundary, $W(B)$ its initial capacity, $C(B)$ the number of already routed nets crossing B, then set

$$VC(HC)(B) = (C1)^{(C(B) - W(B))} + C2.$$

Via costs are chosen similarly. If $M(G)$ is the maximal number of vias that can be realized in the cell $G([12])$ and $T(G)$ is the number of nets already turning in G then

$$\text{VIAC}(G) = (C3)^{(T(G) - M(G))} + C4.$$

The constants $C1$, $C2$, $C3$, $C4$ are defined experimentally during the "fine-tuning" process of algorithm implementation. Sometimes (when the aspect ratio of the grid is large) the coefficients for horizontal costs are chosen differently from those for vertical costs.

4.3. Refinements

Once all the nets are imbedded into the $(2 \times N)$ grid we may remove one of them, update the boundary capacities, compute new boundary costs (all other nets remain imbedded) and again run the TBN procedure (against the background of all the other wires). We perform this rerouting for every net in ramdom order. Very often the nets choose exactly the same route they had before. But several of them in fact get rerouted, improving the distribution of the remaining boundary capacities. We experimentally observed that in repeated rerouting (third and subsequent applications of TBN) nets tend not to choose different routes (which is not the case when maze-type rerouting is applied for a general grid). So, more then one cycle of rerouting appears impractical for the case of the $(2 \times N)$ grid. The small thickness of the grid–2 strips– is probably the reason why the rerouting process converges so fast.

Another trick, which sometimes helps a lot, is rerouting within a sliding $2 \times N$ window; at any level of hierarchy we may select a pair of consecutive rows, usually with largest number of negative boundary capacities, and try to reroute the net segments passing through them.

V. RESULTS

The complexity of the presented wiring routines can be estimated as follows. The time spent for routing of a single net at any level of hierarchy is proportional to the length of horizontal segments of this net (assuming that a constant time is spent for "detouring"). It is observed that the total length of a net at any level of hierarchy is bounded by $O(a + b)$, where $a \times b$ is the grid size. Since there are m nets, and $\log_2 ab$ levels of hierarchy, the upper bound for algorithm complexity can be expressed as

$$O(m(a + b) \log_2 (ab)).$$

This figure should be considered as a worst-case behavior, when almost all nets expand through the whole chip.

We applied this approach to wire several gate array chips, and pieces of custom designs such as switchboxes and channels.

5.1. Gate Arrays

Experiments performed so far indicated enormous speed advantage of the presented method over existing routers. We performed several experiments with global wiring of both

TABLE I

	# circuits	array size	overflow count	wire length
	100	10 X 13		
HGW			1	618
Maze Runner			5	656
	150	11 X 16		
HGW			5	824
Maze Runner			11	890
	200	13 X 17		
HGW			0	1124
Maze Runner			4	1196
	250	15 X 20		
HGW			11	1356
Maze Runner			23	1396
	300	17 X 22		
HGW			2	1956
Maze Runner			4	2188
	350	18 X 24		
HGW			3	2245
Maze Runner			14	2378

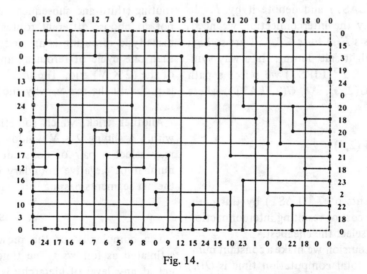

Fig. 14.

actual gate-array chips and hypothetical VLSI chips (artificially created random logic). The global routing of the small array of 400 cells out of which 300 are used with approximately 300 nets takes about 4–4.5 s. (CPU time on IBM 370/3081). The global routing run for hypothetical 200 X 200 array with 500 000 nets took 36 min.

Table I presents some results comparing the work of Hierarchical Global Router (HGR) with simple maze runner.

5.2. Switchbox Routing Problem

Switchbox [15] is a rectangular routing area with no obstuction inside and with signals entering from all four directions. We model it simply by labeling the wiring pins by net numbers; pins having the same positive label have to be interconnected. Several algorithms for this problem were developed recently [15], [16] but with limited success. However, applying the hierarchical approach we were able to wire successfully all small examples presented in the literature. Nevertheless, the

switchbox problem in our opinion still remains extremely difficult. One of our test examples is presented in Fig. 14. In an attempt to get its wiring manually we were able to produce a wiring with 3 overflowed nets—12, 15, and 24. The result of Hierarchical router is significantly better, only one net—24—overflowed (Fig. 14). At this time we do not know whether this example is at all wireable.

5.3. Channel Routing

Channel Routing Problem is a particular case of the switchbox problem when the signals are allowed to enter from two opposite sides of the rectangle. Channel routing is one of the most investigated areas of interconnection art [6]–[8], [17], [18]. The general hierarchical routing technique is applicable to channel routing directly, but in [18] we introduced a specialized version of it; because we are allowed to vary the channel width, the algorithm had to be slightly modified in order to incorporate this flexibility. We additionally observed that

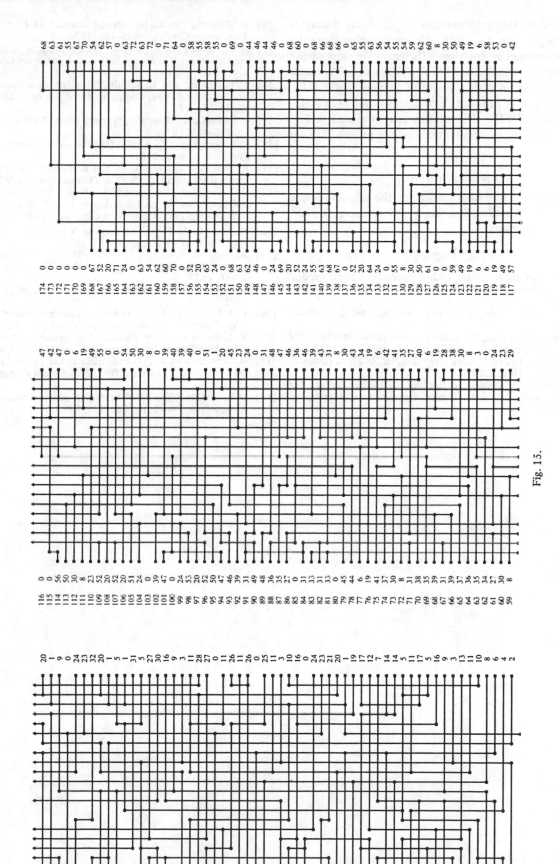

Fig. 15.

IEEE TRANSACTIONS ON COMPUTER-AIDED DESIGN, VOL. CAD-2, NO. 4, OCTOBER 1983

in most channel routing problems the length of the channel is significantly larger than it's width. So, partitioning along the length of the channel is not time efficient. We skipped the vertical (along the length) partitioning stage at all levels of hierarchy and came up with a powerful channel router [18]. This router was able to produce an optimal routing of Deutch's "difficult example" [7]. The routing is reproduced in Fig. 15. All other existing routers required one or more additional wiring tracks.

ACKNOWLEDGMENT

The authors would like to acknowledge the valuable advice and comments of W. Heller, S. J. Hong, J. Kurtzberg, R. Nair, and anonymous reviewers.

REFERENCES

[1] C. Y. Lee, "An algorithm for Path connections and its applications," *IRE Trans. Electronic Comput.*, vol. EC-10, pp. 346–365, Sept. 1961.

[2] E. F. Moore, "Shortest path through a maze," *Bell Syst. Mono.* no. 3523.

[3] S. J. Hong and R. Nair, "Wire-routing machines–New tools for VLSI physical design," *Proc. IEEE*, vol. 71, pp. 57–65, 1983.

[4] F. Rubin, "The Lee connection algorithm," *IEEE Trans. Comput.*, vol. C-23, pp. 907–914, 1974.

[5] D. Hightower, "A solution to the line routing problem on the continuous plane," in *Proc. Design Automation Workshop*, pp. 1–24, 1969.

[6] D. Deutsch, "A dogleg channel router," in *Proc. 13th Design Automation Conf.*, pp. 425–433, 1976.

[7] A. Hashimoto and J. Stevens, "Wire routing by optimizing channel assignment," in *Proc. 8th Design Automation Conf.*, pp. 214–224, 1971.

[8] T. Yoshimura and E. S. Kuh, "Efficient algorithms for channel routing," *IEEE Trans. Computer-Aided Des. of ICAS*, vol. CAD-1, pp. 25–35, 1982.

[9] J. Soukup and J. Fournier, "Pattern router," in *Proc. 1978 ISCAS*, pp. 486–489, 1979.

[10] J. Soukup and J. C. Royle, "On hierarchical routing," *J. Digital Syst.*, vol. 5, no. 3, 1981.

[11] R. Nair, S. J. Hong, S. Liles, and R. Villani, "Global wiring on a wire routing machine," in *Proc. 19th DA Conf.*, 1982.

[12] D. T. Lee, S. J. Hong, and C. K. Wong, "Number of vias: A control parameter for global wiring of high-density chips," *IBM J. Res. Develop.*, vol. 25, no. 4, pp. 261–271, 1981.

[13] N. Nan and M. Feuer, "A method for automatic wiring of LSI chip," in *Proc. ISCAS*, pp. 11–15, 1978.

[14] A. Aho, M. R. Garey, and F. K. Hwang, "Rectilinear Steiner trees: Efficient special case algorithms," *Networks*, vol. 7, pp. 37–58, 1977.

[15] J. Soukup, "Circuit Layout," *Proc. IEEE*, vol. 69, pp. 1281–1304, 1981.

[16] K. J. Supowit, "A minimum-impact routing algorithm," in *Proc. 19th Design Automation Conf.*, pp. 104–112, 1982.

[17] R. L. Rivest and C. M. Fiduccia, "A 'Greedy' channel router," in *Proc. 19th Design Automation Conf.*, pp. 418–424, 1982.

[18] M. Burstein and R. Pelavin, "Hierarchical channel router," *Integration*, vol. 1, 1983 (also *Proc. 20th Design Automation Conf.*, 1983).

[19] D. S. Johnson, "The NP-completeness column: An ongoing guide," *J. Algorithms*, vol. 3, pp. 381–395, 1983.

GENERAL RIVER ROUTING ALGORITHM*

Chi-Ping Hsu

Department of Electrical Engineering and Computer Sciences
and the Electronics Research Laboratory
University of California, Berkeley, California 94720

ABSTRACT

A general and practical river routing algorithm is described. It is assumed that there is one layer for routing and terminals are on the boundaries of an arbitrarily shaped rectilinear routing region. All nets are two-terminal nets with pre-assigned (may be different) widths and no crossover between nets is allowed. The minimum separation between the edges of two adjacent wires is input as the design rule. This algorithm assumes no grid on the plane and will always generate a solution if a solution exists. The number of corners is reduced by flipping of corners. An analysis to determine the minimum space required for a strait-type river routing problem is included.

Let B be the number of boundary segments and T be the total number of terminals. The time complexity is of $O(T(B+T)^2)$ and the storage required is $O((B+T)^2)$. This algorithm is implemented as part of the design station under development at the University of California, Berkeley.

1. Introduction

With the advent of VLSI technologies, the complexity of the circuits on a single chip has increased drastically. Microprocessors and many digital signal processors are now built on a single chip. These processors usually have sets of data busses which interconnect different circuit blocks on the chip. At the chip planning stage, the designer can, in general, determine the sequence of the input and output busses of each block, so that every block has the output busses ordered in the same sequence as that of the input busses of the receiving circuit blocks. Since the input and output busses of the blocks have the same sequence, it is possible to make the interconnections between blocks on a single layer. The problem of interconnecting pairs of pins in two rows with the same sequence on a single layer is usually referred to as the "river routing problem"[1,2,3,4].

This paper presents a routing algorithm which handles a much more general and practical river routing problem. This algorithm can handle arbitrarily shaped rectilinear routing regions with nets for which wire widths can be defined independently.

* This paper is supported in part by Air Force Office of Scientific Research, contract number F49620-79-C-0178, Joint Services Electronics Program, National Science Foundation Grant ECS-8201580, and Bell Laboratories at Murray Hill, New Jersey.

Also, it is a gridless routing algorithm. The minimum separation between the edges of two wires is input as a parameter. This algorithm guarantees that a solution can be found if one exists. The design rule check and an analysis to determine the minimum space needed for a strait-type river routing problem are included.

In section 2, we introduce some basic terminologies and define the river routing problems. In section 3, the algorithm for routing is presented. In section 4, the calculation of the minimum width for the strait-type problem is described.

2. Terminology and Problem Formulation

A *routing region* is a continuous area between circuit blocks that can be used for routing. It is specified by the boundaries of each available layer. (Note that each layer may have different boundaries). A *terminal* is either an input or output pin on the boundaries of a routing region. A terminal is characterized by its location, width, and the layer it is on. A *signal net* (or simply *net*) is a set of terminals to be interconnected by wires. A *routing segment* is a horizontal or vertical wire segment on a specified layer which implements all or part of a signal net. It is represented by its starting and ending points, the width of the segment, its associated layer, and the signal net it belongs to. A *contact* is an area where a set of layers are electrically interconnected. It has a geometry representing its bounding box and a set of associated layers.

A general *routing problem* consists of a routing region, a set of signal net definitions, and a set of design rule parameters. A *solution* to a routing problem is a set of routing segments and contacts inside the routing region, which implements the set of signal nets definitions without design rule violation.

A general *river routing problem* is a special case of the general routing problem where

(a) Only one layer is available for routing in the routing region.

(b) All terminals are on the same layer.

Reprinted with permission from *ACM/IEEE 20th Design Automation Conf. Proceedings*, 1983, pp. 578–583.

(c) Every signal net consists of exactly two terminals.

(d) Terminals are located in such a way that no crossover between signal nets is necessary for a solution to exist.

(e) The routing region has no internal blockage.

Fig. 1 shows an example of a general river routing problem.

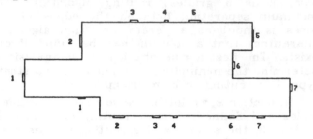

Fig.1   An example of the general river routing problem

A *strait-type river routing problem* is a special case of river routing problem. It is a river routing problem, where

(a) The routing region can be specified by two opposite boundary segment lists that are both monotonic in either X or Y direction.

(b) All terminals are on horizontal(vertical) boundary segments if the two boundary segment lists are monotonic in the X(Y) direction.

(c) Every signal net has one terminal on each of the two boundary segment lists.

A strait-type river routing problem is horizontal if the two boundaries are monotonic in X direction and vice versa. The separation between the two boundaries is usually refered to as the *width* of the strait. Fig. 2 is a horizontal strait-type river routing problem.

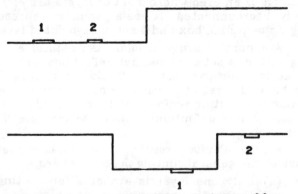

Fig.2   A strait-type river routing problem

3. General River Routing Algorithm

An ordered list of segments is *continuous* if the starting point of every segment, except for the first segment, is the ending point of the previous one. A *path* is a continuous list of alternating horizontal and vertical routing segments. A terminal connected with the first segment of a path will be called a *starting terminal* and the terminal connected with the last segment is an *ending terminal*. We assume that the widths of the routing segments of a signal net are the same and are predefined by the user.

This algorithm routes one net at a time. It starts with the assignment of the starting terminal for each net and the net order is determined by the sequence of terminals on the boundaries. Path searching is done by routing each net in turn as close to the boundaries as possible. Unnecessary corners are then removed by flipping of corners.

3.1. Starting Terminal Assignment

For a river routing problem, all nets are two-terminal nets. Without loss of generality, we assume that every path is counter-clockwise along the boundary. Every net has two possible paths *along the boundary*. Clearly, the two possible paths then correspond to the two possible choices of starting terminal for the net. Fig. 3 shows two possible paths along the boundary for a net {T1,T2}. Path P1 has T1 as its starting terminal and path P2 has T2 as its starting terminal.

The starting terminal for a net is chosen, independently of all other nets, such that the shorter path is selected. This is done by calculating the total length of the boundary segments counter-clockwise between the starting and ending terminals and compare it with half of the total length of the boundaries of the routing region. In fig. 3, path P1 is shorter than path P2. So, terminal T1 is assigned to be the starting terminal for the net.

3.2. Net Ordering

Since every path is counter-clockwise and begins at the starting terminal, there are constraints on the order of the nets to be routed. A net is routed only after all nets, which have one or more terminals on the counter-clockwise portion of the boundaries between the starting and ending terminals of the net, have been routed.

To determine the net order, we first generate a circular list of all terminals ordered in counter-clockwise direction according to their positions on the boundaries. A planarity check is performed to see whether there exist crossovers by using a stack. If so, then the input problem is not a river routing problem. Otherwise, we order the nets as follows:

NET-ORDERING

(1) stack ST = empty;
     i = 1;
     N = the total number of signal nets;
     T = any terminal in the circular list;
     Every starting terminal is marked as NOT PUSHED.
     Every ending terminal is marked as NOT

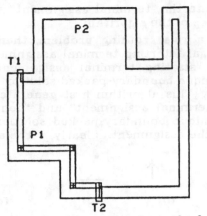

Fig.3 Two possible paths along the boundary, path P1 is shorter, so terminal T1 is assigned as the starting terminal.

MATCHED.

(2) while i <= N
  begin
    if T is a starting terminal that is NOT PUSHED,
    begin
      push T on ST;
      mark T as PUSHED;
    end
    else
    begin
      if T is an ending terminal that is NOT MATCHED,
      begin
        if T and the top element of ST belong to the same net,
        begin
          mark T as MATCHED;
          pop the top element from ST;
          assign the number i to the net;
          increment i by one;
        end;
      end;
    end;
    T = next terminal in the circular list;
  end;

Assert: The nets are ordered in increasing order of the assigned number.

Fig. 4 shows the starting terminal for each net and the net ordering is {1,7,6,5,4,2,3}.

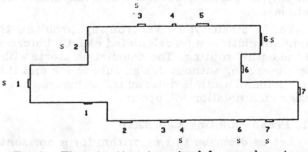
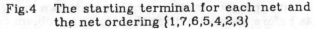

Fig.4 The starting terminal for each net and the net ordering {1,7,6,5,4,2,3}

3.3. Path Searching

For each net in turn, two continuous constraint segment lists are created by tracing through the boundary segments and the existing routing segments that are *exposed* to the current net. A path is then formed by routing counterclockwise as close to one of the constraint list as possible, beginning at the starting terminal and stopping at the ending terminal of the net. The path is then checked against the other constraint list for design rule violation. If a violation occurs, it means that there is not enough space for any solution to exist, and from the topology of the current routing, the user can easily determine where the space should be added.

Fig 5 shows constraint lists 'abcd' and 'ihgf' for net A and the path created by routing as close to list 'abcd' as possible. The path is then checked against the list 'ihgf' for design rule violations.

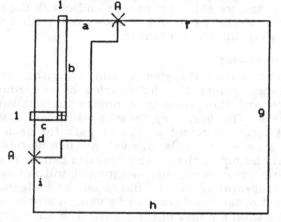

Fig.5 Constraint lists 'abcd' and 'fghi' for net A and the path created by routing as close to one of the constraint list as possible.

3.4. Corner Minimization

After we have done the path searching for all nets without design rule violation, we already have a solution. Every net has a unique path associated with it. All paths are pushed outward against the boundaries and the excess space remains in the center of the routing region. Fig 6 shows an example of the routing after path searching.

The corner minimization is done in a systematic way of flipping corners toward the inside of the routing region. We minimize the corners of one net at a time. The order of the nets for this operation is just the reverse of the order determined by the previous net ordering step. The net that is routed last is minimized first. Equivalently, we are minimizing the corners of the paths from inside nets toward the outside nets of the routing region.

Every corner of a path belongs to one of the eight possible cases as shown in Fig.7. Since every path is routed in the counter-clockwise direction, four of the cases can have their corners *flipped* toward the inside of the routing region. In fig.7, cases a,b,c,d can be transformed to cases e,f,g,h by

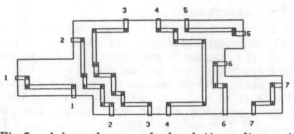

Fig.6 A boundary-packed solution after path searching

flipping toward the inside, where the inside is indicated by a black dot. A constraint segment list is generated by the same way as in the path searching step. Before we flip a corner, we first check whether the flipping will generate any design rule violation against the constraint list. If it does not have any design rule violation, we will flip the corner and thus eliminates two corners of the path. Otherwise, we skip this corner and check the next corner of the path. Fig 8 shows one path before and after flipping of corners.

3.5. Summary

In summary, this river routing algorithm routes all nets against the boundaries of the routing region and then tries to minimize the number of corners. The first step "starting terminal assignment" tries to select a shorter path for each net and spread the nets against all boundaries. It should be noted that if the problem can be routed without crossovers, this assignment will not cause any crossover to occur. Based on the assignment, the net order is determined by using a stack. Let T be the total number of terminals. The while loop in NET-ORDERING will be executed at most 2T times since at most two cycles around the circular terminal list is necessary if the problem is a river routing problem. Because of the assumption that all paths are counter-clockwise against the boundary, the paths are generated easily by going along the constraint list. Finally, the corner minimization is done in reverse net order by flipping the corners toward the inside of the routing region. This step efficiently generates a very desirable final layout due to the fact that all paths are distributed around the whole routing region and the flipping operation is very efficient.

3.6. Existence of A Solution

A solution of a river routing problem is *boundary-packed* if and only if no path can be replaced by another path which is no farther from the boundary anywhere. Given a routing solution with n nets. The routing region is divided into n+1 subregions. If we imagine that we stand inside one of the subregions, a unique boundary-packed solution can be obtained by pushing all the nets against the boundaries. For a legal "starting terminal assignment", it uniquely determines one subregion where we stand. The solution we get after path searching and before corner minimization is *the* unique boundary-packed solution associated

with the "starting terminal assignment" we get in the first step of the algorithm.

Given a river routing problem, there always exists a legal "starting terminal assignment". For any legal "starting terminal assignment", there exists a unique boundary-packed solution if a solution exists. This algorithm first generates a legal "starting terminal assignment" and then tries to find the unique boundary-packed solution associated with the assignment. Clearly, if the algorithm

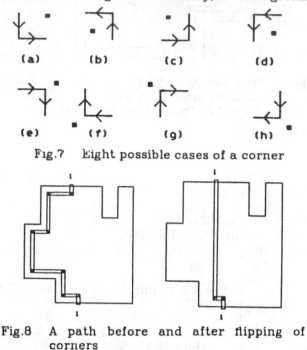

Fig.7 Eight possible cases of a corner

Fig.8 A path before and after flipping of corners

can not find the unique boundary-packed solution, it implies that no boundary-packed solution exists, which in turn implies that there is no solution.

4. Calculation of The Minimum Width For A Strait-type River Routing Problem

In the algorithm described above, we have implicitly assumed that the given routing region has a fixed topology. The algorithm will generate a solution if a solution exists, and returns a partially routed layout if there is no solution for the problem. If a solution exists, the excess space can be obtained by using the boundary-packed layout. However, if a solution does not exist, the information about the minimum additional space needed can only be partially obtained by using the incomplete layout.

For a strait-type river routing problem, the minimum width can be calculated exactly before we do the actual routing. The calculation starts with a pseudo-routing without design rule check and then the minimum width is determined to insure that no design rule violation will occur.

4.1. Pseudo-Routing Algorithm

Let us describe this algorithm for a horizontal strait-type river routing problem as shown in Fig. 2. As before a signal net for this type of problem consists of two terminals. There, one is on the upper boundary and the other is on the lower boundary. Let Xui be the x coordinate of the terminal of signal net i on the upper boundary and Xli be the x coordinate of the terminal of signal net i on the lower boundary. Signal net i will be called a *falling net* if Xui < Xli, a *trivial net* if Xui = Xli, and a *rising net* if Xui > Xli.

Algorithm: Pseudo-routing

This algorithm proceeds net by net, in the order from left to right, generating a set of routing segments for each net in turn. Further, the set of routing segments for each net is generated by the simple procedure:

(1) If the net under consideration is a trivial net, we simply make a straight vertical routing segment connecting the two terminals and proceed to the next net. If the net is not a trivial net, we generate a *continuous* constraint segment list. This list of segments consists of the routing segments of the previous net just routed (or the left boundary segments if the net is the leftmost net) and portions of either

the problem. So this algorithm will always generate a solution if one exists.

the upper or lower (for rising or falling net) boundary segments between the previous net and terminal of the net.

(2) Generate a continuous list of segments with current net width by *licking* along the right edge of the constraint segments with the separation equal to the minimum spacing between adjacent wire edges. Note that boundary segments have width equal to zero.

(3) Generate two vertical routing segments from the two terminals of the net to the segment list obtained in step (2). Delete the segments to the left of the vertical cutline through the left terminal. The resulting continuous segment list is the routing segments of this net.

Fig. 9 shows the routing operation for net 2. The constraint segment list consists of the routing segments of net 1 and portions of the lower boundary between the two terminals of net 1 and net 2. A continuous list is generated by licking along the constraint list and the the resulting routing segment list is obtained for net 2.

Informally, we route one net at a time in the order from left to right. If the net under consideration is a falling net, we go downward from the left terminal of the net as far as we can, and then lick along the upper edge of the constraint segment list as close as possible until we reach the right terminal of the net. If the net is a rising net, we do the similar operation except we go upward instead of downward. No design rule check is per-

formed against the opposite boundaries. If the net is trivial, we simply connect the two terminals directly by a straight vertical routing segment.

The intuition behind this algorithm is that no space to the left of the routing segments of a net can be used by the nets to its right. Since we route the nets from left to right, we would like to route a net in such a way that we leave the maximum available space to the right of the routing segments. Also, we have implicitly assumed that both the left and right boundaries consists of a single vertical segment. It is easy to see that, for each net, no routing segment needs to appear to the left of its left terminal. So, we route the nets by stacking all

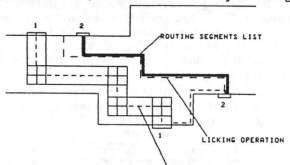

Fig.9   Constraint segment list, licking operation, and the resulting routing segments of a pseudo-routing for net 2

rising nets against the upper boundary, and all falling nets against the lower boundary. Fig. 10 shows the result of a test example.

Note that this algorithm can handle arbitrarily shaped rectilinear routing region as long as the upper and lower boundaries are monotonic in the X direction.

If B is the number of boundary segments and T is the total number of terminals. The time complexity for this algorithm is $O(T(B+T)^2)$ and the storage required is $O((B+T)^2)$.

4.2. Calculation of The Minimum Width

The basic constraint on the width of the strait-type routing problem is that all routing segments must be inside the routing region, ie. between upper and lower boundaries. Since the pseudo-routing algorithm proceeds in such a way that all rising nets are stacked upward against the upper boundary and all falling nets are stacked downward against lower boundary. We can imagine that the upper boundary segments, and the rising

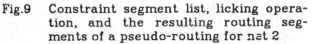

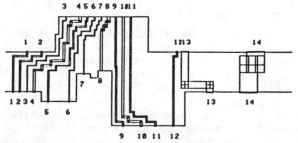

Fig.10 Pseudo-routing for a strait-type river routing problem

net segments (except for those vertical routing segments that connect to the lower boundary), forms a rigid body. Similarly, the lower boundary and the falling net segments (except for the routing segments that connect to the upper boundary), forms another rigid body. The concept is shown in Fig. 11. Now, we can put the two rigid bodies as close together as possible without design rule violation and calculate the *excess space* or the *minimum additional space* needed to have a solution. With the calculated minimum width, the result of the pseudo-routing *is* a solution to the problem. Since the general river routing algorithm guarantees that a solution can be found if one exists. We now can use the calculated width and do the actual routing by using the general river routing algorithm.

Let B be the number of boundary segments and T be the total number of terminals, the time complexity for this analysis is $O(B(B+T)^2)$.

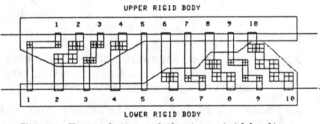

Fig.11 Formulation of the two rigid bodies

5. Experimental results

The two algorithms are implemented in the C language on a VAX-11/780 running the Berkeley Unix operating system, as part of the automatic placement and routing package of the design station under development at the University of California, Berkeley.

Fig. 12 shows the result of an example routed by the general river routing algorithm. There are 12 boundary segments and 14 terminals and the time required is 0.3 CPU seconds. Fig. 13 is a strait-type river routing problem with 22 boundary segments and 28 terminals. The analysis for the minimum width takes 0.2 seconds and the actual routing takes 1.0 seconds. Fig. 14 is a practical problem with 126 terminals and 12 boundary segments. The excess space has been detected and removed by using the boundary-packed solution. The time required is 7 seconds.

6. Conclusions

A general river routing algorithm is described. This algorithm can route signal nets with different widths inside an arbitrarily shaped routing region with one layer for routing. It guarantees that a solution can be found if one exists, and the solution is very close to the manual layout. The analysis to find the minimum width for a strait-type river routing problem guarantees that an optimal solution can be found. A simple generalization of the algorithm can handle a river routing problem with multiple-terminal nets.

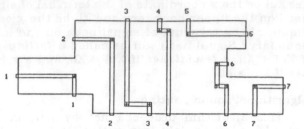

Fig.12 An example routed by the general river routing algorithm

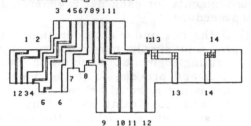

Fig.13 A strait-type river routing problem analyzed and routed

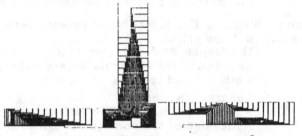

Fig. 14 A large and practical example

7. Acknowledgement

The author would like to thank professor E. S. Kuh for strong support and helpful advising, K. Keller who provides the convenient interactive graphics editor KIC, M. Takahashi for programming the interface with KIC, and the friendly users of professor R. Brodersen. Professor R. Newton and Professor A. Sangiovanni-Vincentelli have expedited this project.

8. References

[1] Baratz, A. E., "Algorithms for Integrated Circuit signal Routing" (Ph.D. dissertation), Dept. of Electrical Engineering and Computer Science, M.I.T., August 1981.

[2] Tompa, M., "An Optimal Solution to a Wire-Routing Problem", Proceedings of the twelfth Annual ACM Symposium on Theory of Computing, 1980, p161-176.

[3] Leiserson, C. E. ; Pinter, R. Y., "Optimal Placement for river Routing", Proceedings of the CMU Conference on VLSI systems and computations, October 1981.

[4] Pinter, R. Y., "River Routing: Methodology and Analysis", the third Caltech Conference on VLSI, March 21-23, 1983.

On Optimum Single-Row Routing

ERNEST S. KUH, FELLOW, IEEE, TOSHINOBU KASHIWABARA, AND TOSHIO FUJISAWA,

FELLOW, IEEE

Abstract—The problem of single-row routing represents the backbone of the problem of general routing of multilayer printed circuit boards. In this paper, the necessary and sufficient condition for optimum single-row routing is obtained. By optimum routing we mean minimum street congestion. A novel formulation is introduced. Examples are given to illustrate how optimum routings are derived. A graph theory interpretation of the condition is also given.

I. INTRODUCTION

RECENT ADVANCES in microelectronics have drastically changed the tasks and design philosophy of circuit designers. One of the primary concerns nowadays is the efficient layout of chips and circuit modules which may contain thousands of interconnected devices and units. While CAD packages for layout are frequently used for various purposes in industry, the general problem of circuit layout is far from solved. As a matter of fact, basic study in the field from a theoretical point of view is lacking. One main problem seems to be the difficulty in formulating explicitly stated problems which are relevant to practical circuit layout.

In this paper we deal with a crucial problem, the problem of single-row routing. It arises in the layout design of multilayer printed circuit boards and backplanes. It is a simple problem, it can be unambiguously stated, and it represents the backbone of the general routing problem. The problem was first introduced by So of the Bell Laboratories [1]. Subsequently, algorithms and sufficient conditions for routing to minimize the tracks needed have been proposed [2]. In the present paper, we introduce a novel formulation of the same problem. With the new formulation, it becomes possible to understand the intricacies of the problem, thus we have been able to obtain a complete set of necessary and sufficient conditions for optimum routing. A graph theory interpretation is also given. Although an efficient algorithm has yet to be worked out to employ these conditions for general routing, examples are given to illustrate how the conditions are used to obtain optimum routing.

Manuscript received May 23, 1978. This research was sponsored in part by the National Science Foundation under Grants ENG 76-84522 and INT 77-10234, and by the Ministry of Education of Japan under Grant in Aid for Scientific Research D-265294, 1977.

E. S. Kuh is with the Department of Electrical Engineering and Computer Sciences and the Electronics Research Laboratory, University of California, Berkeley, CA 94720.

T. Kashiwabara and T. Fujisawa are with the Department of Information and Computer Sciences, Osaka University, Toyonaka, Osaka 560, Japan.

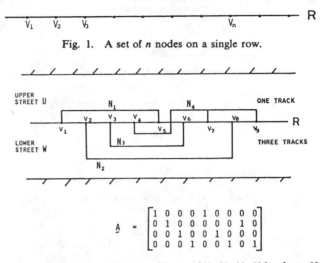

Fig. 1. A set of n nodes on a single row.

Fig. 2. A realization of the net list $L = \{N_1, N_2, N_3, N_4\}$ where $N_1 = \{v_1, v_5\}$, $N_2 = \{v_2, v_8\}$, $N_3 = \{v_3, v_6\}$, and $N_4 = \{v_4, v_7, v_9\}$. Upper-street congestion = 1 track, lower-street congestion = 3 tracks in this realization.

II. FORMULATION OF THE PROBLEM

Given a set of n nodes evenly spaced on a row which is located on the real line R as shown in Fig. 1. A net list $L = \{N_1, N_2, \cdots, N_m\}$ is given which prescribes the connection pattern of the m nets to the n nodes. The specification can be expressed in terms of an $m \times n$ 0-1 matrix $A = [a_{ij}]$, where $a_{ij} = 1$ if net N_i is to connect node v_j, and $a_{ij} = 0$, otherwise. It should be stressed that a node is to be connected to one and only one net, thus there exists one and only one 1's in each column of the matrix.

A net list is to be *realized* with a set of m nonintersecting nets which consist of only horizontal and vertical paths connecting the nodes according to specification. An example depicting a realization of a given net list together with the matrix specification and some pertinent terminology is shown in Fig. 2. The space above the real line R is referred to as the upper street and the space below R is referred to as the lower street. The number of horizontal tracks needed in the realization in the upper street, is called the upper-street congestion. Similarly, we define the lower-street congestion. In the realization, we allow a net to switch from upper street to lower street, and conversely, as shown by the net N_4 in Fig. 2.

Previously, it has been shown that given a net list, a realization always exists [2]. An optimum realization is one which minimizes the street congestion in both streets.

Reprinted from *IEEE Trans. Circuits Syst.*, vol. CAS-26, pp. 361–368, June 1979.

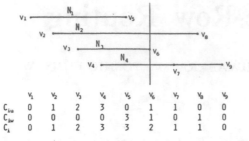

	v_1	v_2	v_3	v_4	v_5	v_6	v_7	v_8	v_9
c_{iu}	0	1	2	3	0	1	1	0	0
c_{iw}	0	0	0	0	3	1	0	1	0
c_i	0	1	2	3	3	2	1	1	0

Fig. 3. Interval graphical representation of the matrix A in Fig. 2 and the node cut numbers.

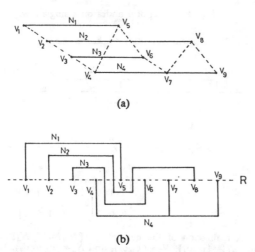

(a)

(b)

Fig. 4. (a) Interval graphical representation together with the reference line. (b) Net list realization which corresponds to the interval graphical representation of Fig. 4(a) and the matrix A of Fig. 2.

Sufficient conditions for realization to minimize street congestion together with a routing algorithm were proposed in [2]. Unfortunately, the algorithm has been found to be incomplete and it could fail. A counter example is given in the Appendix. In this paper we give the necessary and sufficient condition for optimum realization. We introduce a new formulation of the problem as follows.

Consider the set of m nets given by the net list in terms of the matrix A. Let us draw a set of m horizontal intervals representing the m nets in the order given by the matrix from top down. This is illustrated in Fig. 3 for the problem given by the matrix A in Fig. 2. Note that each horizontal line corresponds to the interval specified by a row of A between the extreme left and right 1's. Each node is appropriately marked on the lines as shown. We call this the interval graphical representation of the matrix A. Since there are $m!$ row permutations, there exist a total of $m!$ interval graphical representations for a given net list. In the following we will first demonstrate that, for each representation, there corresponds a *unique* realization.

The interval graphical representation of the example in Fig. 3 is redrawn in Fig. 4(a) together with a set of line segments in broken lines connecting the nodes. Let us define a *reference line* as the continuous line segments which connect the nodes in succession from left to right. The m interval lines together with the reference line form

a graph. The crux of our proposed realization lies in a topological mapping of the graph so constructed. Let us stretch out the reference line and set it on top of the real line R. The m horizontal interval lines are mapped topologically into vertical and horizontal paths. Nets and portions of nets which lie above the reference line are mapped into paths in the upper street. Similarly, nets and portions of nets below the reference line are mapped into paths in the lower street. This process defines a unique realization. For the example in Fig. 4(a), This mapping results in the realization as shown in Fig. 4(b). It becomes obvious that the problem of finding an optimum realization is reduced to that of finding a matrix A which represents an optimum ordering of the m nets in the form of horizontal intervals. In order to pursue further we need to understand the property of street congestions in terms of the new formulation.

First, let us define the *cut number* of a node v_i, denoted by c_i, as in [2]. Let us draw a vertical line at v_i superimposed on the interval graphical representation as shown in Fig. 3 for example by the line at v_6. The cut number c_i is defined as the number of nets cut by the vertical line, disregarding the net to which v_i belongs. The nets cut by the vertical line are called the nets which *cover* the node v_i. Thus at v_6 in Fig. 3 $c_6 = 2$, and nets N_2 and N_4 are said to cover v_6. We next introduce the *upper cut number* at v_i, c_{iu}, as the number of nets cut by the vertical line above v_i. Similarly, we define the *lower cut number* at v_i, c_{iw}, as the number of nets cut by the vertical line below v_i. Obviously, for all i, $c_i = c_{iu} + c_{iw}$. These cut numbers of all the nodes for the example are listed in the table in Fig. 3. We similarly, define a net *covering* the node v_i *from the above* as the net which intersects with the vertical line above v_i, and a net *covering* the node v_i *from below* as the net which intersects with the vertical line below v_i. Thus in Fig. 3, N_2 covers the node v_6 from the above and N_4 covers the node v_6 from below. Let

$$C_u = \max_i c_{iu}$$

and

$$C_w = \max_i c_{iw}. \qquad (1)$$

From the topological mapping just introduced, it becomes clear that C_u gives the track number in the upper street and is equal to the upper-street congestion for the realization. Similarly, C_w is equal to the lower-street congestion. Therefore, our problem of finding an optimum realization or routing[1] amounts to finding an ordering of the m nets among the $m!$ permutations for which the max $\{C_u, C_w\}$ is a minimum.

It would be hopeless to generate all $m!$ realizations in order to obtain an optimum one. In the next section, we will first study the characteristics of an optimum realization to gain some insight. The necessary and sufficient conditions for an optimum realization will be given in Section IV.

---

[1]We sometimes use the term routing to mean realization.

III. OPTIMUM ROUTING—A PREAMBLE

Before we discuss optimum routing, it is necessary to introduce the term, *cut number of a net* N_j, denoted by q_j as was done in [2]. We define q_j as the maximum of the cut numbers of the nodes which belong to the net N_j. For the example in Fig. 3, we have $q_1 = 3$, $q_2 = 1$, $q_3 = 2$, and $q_4 = 3$. It is clear that the cut number of a net is an important property in determining the net ordering for a optimum routing. For example, if the first net from the top has a cut number q then C_w is at least q, because at one of the nodes which belong to the first net, the lower cut number is q. Similarly, if the last net chosen has a cut number q, then C_u is at least q, because at one of the nodes which belong to the last net, the upper cut number is q. Thus it makes sense to select those nets with least cut numbers as the outer nets for optimum routing.

In an optimum realization, let Q_0 be the street congestion, thus $Q_0 = \max \{C_u, C_w\}$. Let us further denote by

$$q_m = \min_j q_j$$

$$q_M = \max_j q_j \qquad (2)$$

then we can state the following.

Proposition 1: $Q_0 \geq \max \{q_m, q_t\}$ where $q_t = \lceil q_M/2 \rceil$ and $\lceil x \rceil$ is the smallest integer not smaller than x.

Proof: $Q_0 \geq q_m$ has already been shown by the argument above. $Q_0 \geq q_t = \lceil q_M/2 \rceil$ is proven by first assuming that we assign the net with q_M in one of the middle rows. For example, if q_M is even, at the node v_i where $c_i = q_M$, the best we could do is to choose $q_M/2$ nets covering the node from the above and $q_M/2$ nets covering the node from below. Thus $c_{iu} = c_{iw} = q_M/2$. If q_M is odd, at the node where $c_i = q_M$, the best we could do is to choose $(q_M + 1)/2$ nets on one side of v_i and $(q_M - 1)/2$ nets on the other. Thus either $c_{iu} = (q_M + 1)/2$ and $c_{iw} = (q_M - 1)/2$, or $c_{iu} = (q_M - 1)/2$ and $c_{iw} = (q_M + 1)/2$. It is also clear that with any other ordering we cannot do better than this. Therefore, we have shown that $Q_0 = \max \{C_u, C_w\} \geq \lceil q_M/2 \rceil$. Q.E.D.

From the above, we see that the strategy to obtain an optimum routing is to choose those nets with the lowest cut number as the outer rows and to divide up c_i properly between c_{iu} and c_{iw} at those nodes where the cut number is larger than q_m. Although detailed specifications are to be worked out for an optimum routing, it seems that an increasing ordering of nets based on the cut number of a net from the outer rows to the center rows is perhaps the right strategy for an optimum routing. Before we give the necessary and sufficient conditions for optimum realization, let us consider an example.

Example 1

Given the net list as represented by the interval graphical representation of Fig. 5. The net list consists of 16 nodes and 8 nets. The cut numbers of the nodes and nets are marked on the figure. From that, we see $q_m = 2$ and $q_M = 3$; thus $Q_0 \geq 2$. Our question is whether $Q_0 = 2$ can be

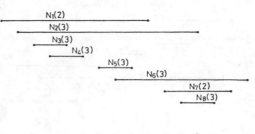

$$C_i \quad 0 \quad 1 \quad 2 \quad 3 \quad 3 \quad 2 \quad 2 \quad 3 \quad 3 \quad 2 \quad 2 \quad 3 \quad 3 \quad 2 \quad 1 \quad 0$$
$$\quad v_1 \quad v_2 \quad v_3 \quad v_4 \quad v_5 \quad v_6 \quad v_7 \quad v_8 \quad v_9 \quad v_{10} \quad v_{11} \quad v_{12} \quad v_{13} \quad v_{14} \quad v_{15} \quad v_{16}$$

Fig. 5. Example 1, an arbitrary ordering together with cut numbers.

TABLE I
PERTINENT INFORMATION FOR OPTIMUM ROUTING IN EXAMPLE 1

node	associated net	nets cut
v_4	N_4	N_1 N_2 N_3
v_5	N_3	N_1 N_2 N_4
v_8	N_6	N_1 N_2 N_5
v_9	N_5	N_1 N_2 $.N_6$
v_{12}	N_8	N_2 N_6 N_7
v_{13}	N_2	N_6 N_7 N_8

realized. For comparison, the problem does not satisfy the sufficient condition given in [2]. From our discussion so far, it is clear that we should select N_1 and N_7 as the outer rows because they have the lowest cut number; but how about the rest?

First we consider all the nodes with cut number less than 3. These are v_1, v_2, v_3, v_6, v_7, v_{10}, v_{11}, v_{14}, v_{15}, and v_{16}. Clearly, at any of these nodes, since the cut number is less than 3, c_{iu} and c_{iw} at these nodes will not cause trouble. This means that we only need to concern ourselves with the remaining six nodes with cut number equal to three: v_4, v_5, v_8, v_9, v_{12}, and v_{13}. At these nodes, we must make sure that in assigning the nets, the cut number is divided up between the lower street and the upper street. A 2–1 division or a 1–2 division of the cut number is fine. But a 3–0 or a 0–3 division is not. In Table I we list all necessary information at these nodes.

The first column gives all the nodes with cut number larger than 2. The second column indicates the net to which the node belongs. The third column gives the nets which cover the node or nets cut by a vertical line drawn at the node. From the table, we can decide an ordering of the eight nets such that nets in the second column will have at least one net above it and one net below it among the nets specified in the third column. For this example, there exist many such orderings, for example,

$$N_1 \ N_3 \ N_4 \ N_5 \ N_6 \ N_2 \ N_8 \ N_7.$$

The interval graphical representation and its corresponding realization are shown in Fig. 6. Thus $Q_0 = 2$ has been realized.

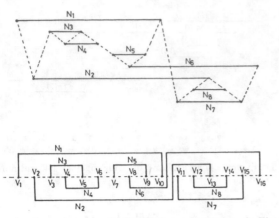

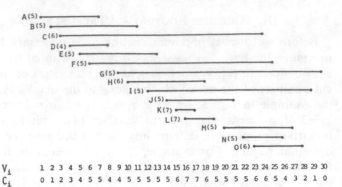

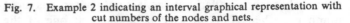

Fig. 7. Example 2 indicating an interval graphical representation with cut numbers of the nodes and nets.

Fig. 6. An optimum realization of the example given in Fig. 5.

IV. THE NECESSARY AND SUFFICIENT CONDITIONS

In general, the task of obtaining an optimum routing is more involved. However, the concept is the same. We must check those nodes at which the cut number is large. A table similar to that of Example 1 needs to be constructed. We must test whether an optimum division of the cut numbers between the lower street and the upper street is possible. Although the determination of a feasible order may not be simple, a set of necessary and sufficient conditions can always be stated.

First we need to define some useful terms. Let

$$l \triangleq q_M - q_t. \qquad (3)$$

A net x is said to *cover* a net y at a node v which belongs to y if x covers v. Similarly, net x is said to *cover* net y at w from *the above* if net x *covers* node v from *the above*, and net x is said to *cover* net y at v from *below* if x covers v *from below*.

Theorem 1

There exists an optimum realization with street congestion $Q_0 = q_t$ if and only if there exists an ordering such that for each v_i with $c_i = q_t + k (k = 1, \cdots, l)$ the net associated with v_i is covered at v_i from the above and below by at least k nets.

Proof: Since at each v_i with $c_i \leqslant q_t$, c_{iu} and $c_{iw} \leqslant q_t$, we only need to concern ourselves with those v_i where $c_i > q_t$. At these nodes, with cut number $q_t + k$, $k > 0$, if there are at least k nets above the node and k nets below the node, the maximum cut number, C_w and C_u are at most q_t. Since from Proposition 1 $Q_0 \geqslant q_t$ thus the optimum $Q_0 = q_t$ is realized. This proves the sufficiency. To prove the necessity, we assume that at those nodes v_i with cut number $q_t + k$, there are less than k nets covering v_i from the above or below. Then, since $c_{ui} + c_{wi} = c_i = q_t + k$, either c_{ui} or c_{wi} must be greater than q_t. Therefore, the street congestion is larger than q_t.   Q.E.D.

To deal with the general situation we need to introduce the definition of *p-excess property*.

By *p-excess property* we mean that there exists an ordering such that for each v_i with $c_i = q_t + k$ ($k = p + 1, \cdots, l$) the net associated with v_i is covered by at least $k - p$ nets from the above and from the below.

Theorem 2

There exists an optimum realization with street congestion $Q_0 = q_t + p$, if and only if p is the least nonnegative integer for which the p-excess property holds.

The proof of this theorem is exactly the same as that of Theorem 1 and is therefore omitted.

Remark: Theorem 1 is a special case of Theorem 2 when $p = 0$.

Example 2

In this example there are 30 nodes and 15 nets. An interval graphical representation is shown in Fig. 7 together with the cut number of the nodes and of the nets. For convenience we use alphabets in capital letters to designate nets. It is seen that $q_m = 4$, $q_M = 7$, $q_t = 4$, and $l = 3$. In Table II, we give those nodes with cut number larger than 4, the associated nets, and the nets that cover the pertinent nodes. The nodes are grouped into three parts according to their cut numbers. First, we must check whether the conditions in Theorem 1 are satisfied. To determine the net ordering, it is useful to note that there are two nets which have cut numbers less than 5. They are net D and net G, and they are assigned right away to the outer rows. As to the others, we will start from inside out by considering the first part of Table II. Both nets K and L have cut number 7, we need to assign them in the middle. We next consider nets in the second part, namely: C, H, and O. Since they have cut number 6, and we will temporarily assign them next to nets K and L. In checking with the nets which cover the nodes with cut number 7, a tentative ordering of

$$F \ C \ H \ L \ K \ J \ I \ G$$

will satisfy the conditions that both L and K have three nets above and below. Similarly, for the second part in Table II a tentative assignment of

$$M \ N \ F \ O \ C \ H \ L \ K \ J \ I \ G$$

is made. This will satisfy the conditions that O, C, and H

TABLE II

EXAMPLE 2: NODES WITH CUT NUMBER LARGER THAN q_t, THE
ASSOCIATED NETS, AND THE NETS WHICH COVER THE NODES

Nodes	Cut number	Associated nets	Nets covering nodes
17	7	K	F C H L J I G
16	7	L	F C H K J I G
24	6	C	F O N M I G
18	6	H	F C L J I G
23	6	O	F C N M I G
15	6	K	F C H J I G
7	5	A	D B E F C
11	5	B	E F C H G
13	5	E	F C H I G
6	5	F	D A B E C
25	5	F	O N M I G
12	5	I	E F C H G
14	5	J	F C H I G
21	5	J	F C M I G
20	5	M	F C J I G
22	5	N	F C M I G
19	5	L	F C J I G
10	5	H	B E F C G

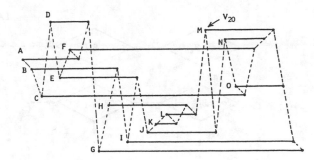

Fig. 8. A specific ordering representing an optimum with
$Q_0 = q_t + 1 = 5$.

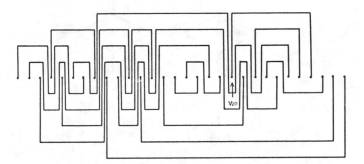

Fig. 9. An optimum realization with $Q_0 = 5$.

have two nets above them and two nets below them
among those which cover the pertinent nodes.

The remaining part of Table II contains nodes with cut
number 5. An ordering must be obtained to satisfy the
remaining conditions with the clue that net D and net G
are assigned in the outer rows. We discover that there are
various constraints among the nets with cut number 5. An
ordering as

$$D M N F A B E O C H L K J I G$$

satisfies all conditions except at node 20 where net M
should have a net from the above among the pertinent
ones. After much cut and try, we are convinced that not
all the conditions can be simultaneously satisfied. This
means that the conditions in Theorem 1 cannot be met,
thus $Q_0 > q_t$. By applying Theorem 2, it becomes clear that
Theorem 2 is satisfied with $p = 1$. Therefore, we conclude
that $Q_0 = q_t + 1 = 5$, and the ordering given above gives
one such solution. The interval graphical representation is
drawn in Fig. 8 together with the reference line. The
realization is shown in Fig. 9. It is seen that at node v_{20},
$C_{20w} = 5$, which is the maximum C_w among the nodes.
Thus an optimum $Q_0 = 5$ has been realized.

To gain further insight of the conditions given in Theorem 1 and Theorem 2, we will classify the nodes into
different sets according to their cut number. Let $\mathcal{S}_k$,
$k = 1, 2, \cdots, l$ be the set of nodes with cut number equal to
$q_t + k$. Then the test to be made can be stated in terms of
net covering to satisfy a specific table generated as in
Example 2. The necessary and sufficient conditions for an
optimum realization can be summarized by the flow chart
as given in Table III. In the flow chart we use the
terminology "cover $\mathcal{S}_k(j)$" to mean nets covering nodes in
the set with cut number $q_t + k$ from the above and below j
times.

V. GRAPH THEORY INTERPRETATIONS

The necessary and sufficient conditions for optimum
routing can be interpreted in terms of graphs. This may
lead to an efficient algorithm which is to be determined.
For a given net list and a specified value of p we first
construct a bipartite graph G as follows.

We assign a node v_i in G to every node v_i with cut
number $c_i > q_t + p$. This forms the first set of nodes of the
bipartite graph G. The second set of nodes consists of
node N_j in G representing all the nets. The edges are
defined according to the covering relation. Thus there
exists an edge between v_i and N_j if, in the net list, net N_j
covers node v_i. Next we choose the orientation of the
edges to satisfy the covering property at each node v_i. An
edge enters v_i from node N_j if net N_j covers v_i from the
above; and an edge leaves from v_i to node N_j if net N_j
covers v_i from below. Thus to satisfy the p-excess property, for each node v_i with cut number $c_i = q_t + k (k = p +
1, \cdots, l)$, both the number of edges entering v_i and leaving
from v_i must be at least $k - p$. For Example 1, with $p = 0$,
the bipartite graph is shown in Fig. 10(a) with an edge
orientation selected according to the ordering in Fig. 6.

Next, we introduce a reduced digraph G' from the
directed bipartite graph G by the following operations.
First identify node v_i in G to its associated net, node N_j by
shorting v_i and N_j. Parallel edges with same directions are
then merged. The reduced digraph G' thus consists of m
nodes representing all the nets. For the present example,
G' is shown in Fig. 10(b). We note that G' defines a
precedence relation among the nets. A proper ordering
can be obtained by observing the precedence relation and

TABLE III

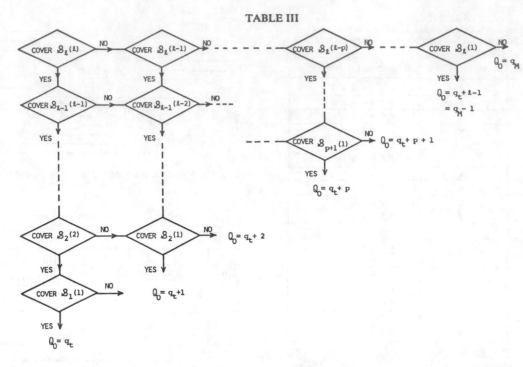

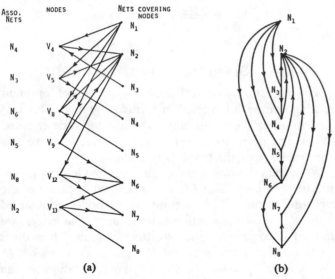

Fig. 10. (a) A bipartite graph G representing net ordering of Example 1. (b) The reduced digraph G'.

tracing through the digraph G'. First, it is important to note that the reduced digraph G' has no cycle. This is almost self-evident because G' gives the precedence relation among nets. We are now in a position to state the following theorem.

Theorem 3

The necessary and sufficient condition for a net list to have the p-excess property is that there exists an orientation for the edges in the bipartite gaph G with the following properties. i) For each node v_i, the number of edges entering v_i and leaving from v_i must be at least $k-p$, and ii) the reduced digraph G' is acyclic.

Proof: i) is obvious and follows directly from the definitions of G and the p-excess property. To prove necessity for ii), we assume that there is a directed cycle in G'

$$[N_1, N_2], [N_2, N_3], \cdots, [N_{k-1}, N_k], [N_k, N_1].$$

Then, reading the list from left to right, we see that N_1 is above N_2, N_2 above N_3, $\cdots$, N_{k-1} above N_k, and N_k above N_1. Therefore, we conclude that N_1 is above N_1, which is a contradiction. To prove sufficiency for ii), we introduce an ordering among nets based on the orientation of the edges in G'. For two distinct nets N and N' we say $N < N'$ if and only if there exists a directed path

$$[N_1, N_2], \cdots, [N_{k-1}, N_k]$$

where $N = N_1$ and $N' = N_k$.[2] Since there exists no directed cycles, one of the following cases must hold for any two distinct nets N and N'.

1) $N < N'$
2) $N' < N$
3) there exists no order relation between N and N'.

Because of the transitive relation that $N < N'$ and $N' < N''$ imply $N < N''$, there exists a partial order $<$ among the nets in G'. Therefore, a proper ordering among the nets can be obtained.   Q.E.D.

In our example in Fig. 10(b) N_1 is the minimal net. By starting with N_1 and tracing through G', we obtain a proper ordering

$$N_1 \ N_3 \ N_4 \ N_5 \ N_6 \ N_2 \ N_8 \ N_7.$$

This is the order which we obtained in Fig. 6. Obviously, there exist many proper orderings because, at each stage,

[2] A net N is said to be the minimal net among a set of nets if there is no net N' in the set of nets with relation $N' < N$.

we may pick up arbitrarily a minimal net among many. For example, N_5 may be picked ahead of N_3 to obtain the following proper orderings:

$$N_1 \ N_5 \ N_6 \ N_3 \ N_4 \ N_2 \ N_8 \ N_7$$

and

$$N_1 \ N_5 \ N_3 \ N_4 \ N_6 \ N_2 \ N_8 \ N_7.$$

VI. Conclusions

In this paper we have demonstrated that by the use of an interval graphical representation of the net list of the single-row routing problem we obtain a set of necessary and sufficient conditions for optimum realization. The conditions are given in terms of feasible orderings of the nets. A graph theory interpretation is also given. So far, an efficient algorithm has not yet been worked out to implement a procedure for properly ordering the nets. However, cut-and-try has been rather easy for modestly complicated problems.

The special case with $Q_0 \leqslant 2$ is of interest for practical printed-circuit-board routing. For this case, a simple algorithm can be derived by the use of directed graphs. As the general problem of multilayer printed-circuit-board routing depends in a crucial way on single-row routing, especially situation with maximum of two rows, the major problem remains to be solved is optimum layering. The problem is to decompose in an optimum way a single-row multilayer net list into a number of single-row single-layer problems [3].

In conclusion, as far as optimum single-row routing is concerned, the present paper gives a complete set of necessary and sufficient conditions. Work is still needed to derive an algorithm for the implementation of net ordering to satisfy these conditions.

Appendix

In [2], Theorem 2 stated that a net list L over R is realizable with $C_u = C_w = M \geqslant \rho' = \lceil \rho/2 \rceil$ if the following holds: for every unit interval (a,b) with density $d(a,b) = I$, $I > M+1$, there exists at least $2(I-M)$ nets covering (a,b) such that each of them has cut number less than I. The theorem gives a sufficient condition for realization (not necessarily the optimum). An Assignment Algorithm for realization was also given. In the Assignment Algorithm, there usually exists leeway as to which nets among several eligible ones should be assigned at each step. No specific information on the ordering was given. In the following example, it is shown that an arbitrary ordering will fail in obtaining a realization.

Example 3

Consider the net list given in Fig. 11 together with the density for all the intervals, the cut number of the nodes and of the nets. Since $\rho = 6$, $M \geqslant 3$, we only need to check first those intervals with density larger than or equal to 3. The conditions given in [2, theorem 2] are satisfied with

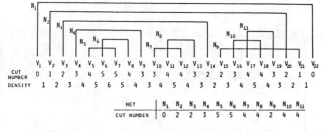

Fig. 11. A counter example to [2, algorithm 2].

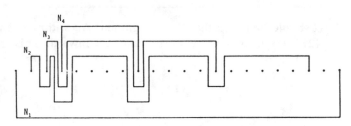

Fig. 12. An ordering which fails.

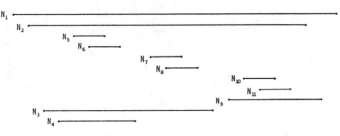

Fig. 13. An optimum ordering for Example 3.

$M = 3$; therefore, a realization with street congestion equal to 3 should exist.

Let us test the Assignment Algorithm. The first interval from the left is $(4,5)$, and we have four nets, N_1, N_2, N_3, and N_4 to choose for U and W. Since the algorithm does not indicate priority, we choose N_4 in U and N_1 in W. The next interval $(8,9)$ can be passed. The next one is $(10,11)$, and we have N_2 and N_3 to choose for U. We assign N_3 in U. The next interval $(12,13)$ can be passed. The next interval is $(16,17)$, and we assign N_2 in U. The last one $(18,19)$ can be passed. We thus completed all intervals with density 4. This is summarized in Fig. 12.

The left-most unit interval with density 5 is $(5,6)$. There are 4 nets, N_1, N_2, N_3, and N_4 two of which must be in U and two remaining must be in W. Thus the algorithm fails.

The same problem can be solved easily with our present method. From the given information, we have $q_m = 0$, $q_M = 5$, $q_t = 3$, and $l = 2$. The information to satisfy Theorem 1 is summarized in Table IV. It is easily checked that an ordering as

$$N_1, \ N_2, \ N_5, \ N_6, \ N_7, \ N_8, \ N_{10}, \ N_{11}, \ N_9, \ N_3, \ N_4$$

satisfies all the conditions. Thus $Q_0 = q_t = 3$, and an optimum ordering is shown in Fig. 13.

TABLE IV
EXAMPLE 3

Nodes	Cut number	Net Asso.	Nets cut
6	5	N_6	N_1 N_2 N_3 N_4 N_5
7	5	N_5	N_1 N_2 N_3 N_4 N_6
5	4	N_5	N_1 N_2 N_3 N_4
8	4	N_6	N_1 N_2 N_3 N_4
11	4	N_8	N_1 N_2 N_3 N_7
12	4	N_7	N_1 N_2 N_3 N_8
16	4	N_{11}	N_1 N_2 N_9 N_{10}
17	4	N_{10}	N_1 N_2 N_9 N_{11}

It should be noted that the Assignment Algorithm given in [2] always assigns nets from outside inward. This is because the way that routing is defined in the upper street and lower street. The interval graphical representation given in the present paper avoids completely the difficulty of sequential routing.

REFERENCES

[1] H. So, "Some theoretical results on the routing of multilayer printed-wiring boards," in *Proc. 1974 IEEE Int. Symp. on Circuits and Systems*, pp. 296–303.
[2] B. S. Ting, E. S. Kuh, and I. Shirakawa, "The multilayer routing problem: Algorithms and necessary and sufficient conditions for the single-row, single-layer case," *IEEE Trans. Circuits Syst.*, vol. CAS-23, pp. 768–778, Dec. 1976.
[3] B. S. Ting and E. S. Kuh, "An approach to the routing of multilayer printed circuit boards," in *Proc. 1978 IEEE Int. Symp. on Circuits and Systems*.

Part V
Layout Systems

LTX — A Minicomputer-Based System For Automated LSI Layout

G. PERSKY, D. N. DEUTSCH, and D. G. SCHWEIKERT

Abstract — LTX is a minicomputer-based design system for large-scale integrated circuit chip layout which offers a flexible set of interactive and automatic procedures for translating a circuit connectivity description into a finished mask design. The system encompasses algorithms for two-dimensional placement, string placement, exploitation of equivalent terminals, decomposition of routing into channels, and channel routing. Circuit connectivity is preserved during interactive procedures. LTX runs on an H-P 2100/21MX-series computer with a disc operating system and 32K words of memory.

In current polycell-style layouts, about a week is typically required for completion of the layout design of an LSI chip containing 500 cells.

I. INTRODUCTION

LTX is a minicomputer-based design system intended for layout of large-scale integrated circuit chips. LTX stands for LSL To XYMASK, where LSL [1] and XYMASK [2] are respectively the Bell Labs languages for describing circuit connectivity and fabrication masks, and implies that the system is meant to translate a circuit description (LSL) into a finished mask design (XYMASK). LTX is highly interactive, but possesses many automatic features. Its strength lies in the implementation of what can be called "Directed Automatic Design", exemplified by the following operational characteristics:

1. The user chooses from a variety of design strategies. The system provides algorithmic assistance in carrying them out.

2. The results of following a particular strategy are available immediately so that the decision to repeat, revise, or proceed, can be made at once.

3. The data base and program sections possess the re-entrant properties necessary for repetition and recycling.

Interactively-controllable design algorithms are provided by LTX for global placement, local placement, and routing. They are backed by a full complement of interactive graphics capabilities allowing extensive human intervention. The LTX algorithms are essentially simple ones, the underlying philosophy being that large complex problems can be modularized to yield good results from a well balanced combination of uncomplicated strategies.

The predecessor of LTX was the GRAFOS system [3]. which utilized PDP-8 hardware and was originally intended to interpret a specialized symbolic routing language. Subsequently, channel routing and polycell manipulation features were added. GRAFOS was then successfully used to design a number of polycell PMOS

The authors are with Bell Laboratories, Murray Hill, New Jersey.

LSI chips. but with sufficient difficulty to stimulate development of the more sophisticated and powerful LTX system.

Thus far, design work with LTX has been limited to integrated circuits using the "standard cell" or "polycell" approach. In this layout style, the basic logic gates are implemented as "cells" of rectangular outline arranged in parallel rows and routed through the intervening channels (see, for example, Fig. 16). Therefore, many of the algorithmic facilities are directed toward this particular layout style, which is so amenable to automation. However, the data base and program modules have been organized to permit easy adaptation of the system to other modular design styles as well. LTX always achieves a layout — in terms of completing all connections and obeying all (local) design rules. Because it performs automatic placement and routing and assumes the burden of guarding the integrity of the circuit, LTX greatly speeds up the design process. As measured from first data input, perfunctory oversize layouts are available literally in minutes. Subsequently. much time may be spent in trying to obtain a more compact layout and in accommodating special design requirements (lead lengths for critical paths, etc.). Since there is currently no known way of estimating the minimum required area. it is, unfortunately. difficult to determine the proper amount of layout time expenditure. Realistic design times for polycell layouts have varied from hours for simple configurations to about a week for very large complex circuits.

The LTX hardware configuration consists of a Hewlett-Packard 2100/21MX-series processor with 32K 16-bit words of main memory. a moving-head disc. mag tape. card reader. and an optional line printer. User interaction and graphic display are carried out on a Tektronix 4014 storage scope terminal. LTX is run under DOS-M. a disc operating system developed by Hewlett-Packard [20]. To facilitate programming and maintenance. the various portions of LTX. i.e. input processor, display, router, etc., have been constructed as independent program modules which interact with the LTX data base stored on disc but which are individually executable under DOS-M.

The LTX disc data base is accessed via the DOS-M disc driver but is unknown to the DOS-M file structure, LTX maintaining its own data file directory. The data base was designed for compactness. maintainability. and execution speed. To reduce search times. all data lists are "pointered". LTX program modules use dynamic core allocation for the data lists they acquire from disc, and are structured such that. in most cases, few additional disc accesses are required once execution is underway. Certain modules are themselves organized into disc overlays in order to maximize the amount of core available for data.

Many of the items in this paper have been presented previously [4-7]. However. this paper is a complete and self-contained description of the current LTX system. Certain system functions, strategies. and results are presented here for the first time. The remainder of the paper is organized into seven sections. Section II introduces the main concepts which underlie LTX. Section III deals with input and output, and Section IV with display and interactive editing. Section V covers

Reprinted from the *J. Design Automation and Fault Tolerant Computing*, vol. 1, no. 3, pp. 217–255, May 1977, with the permission of the publisher Computer Science Press, Inc.. 1803 Research Blvd., Rockville, MD 20850.

program modules which carry out global placement and local placement operations, while Section VI treats the routing problem, from global net topology to channel routing. The last two sections present a detailed example and concluding remarks.

II. LTX CONCEPTS AND DEFINITIONS

From the standpoint of physical design, the principal distinguishing characteristic of large scale integrated circuits is their extraordinarily high degree of complexity. Given finite resources, the most practical approach to dealing with such complexity is to break the design problem down into smaller, more manageable units. These can then be attacked with suitably specialized tools, and subsequently reintegrated back into the original whole. In LTX the necessary modularization is achieved by subdivision of the layout region into geometric entities called "blocks", "domains", and "superdomains", which are easily identified and manipulated. These are illustrated in Fig. 1 for a polycell-style arrangement.

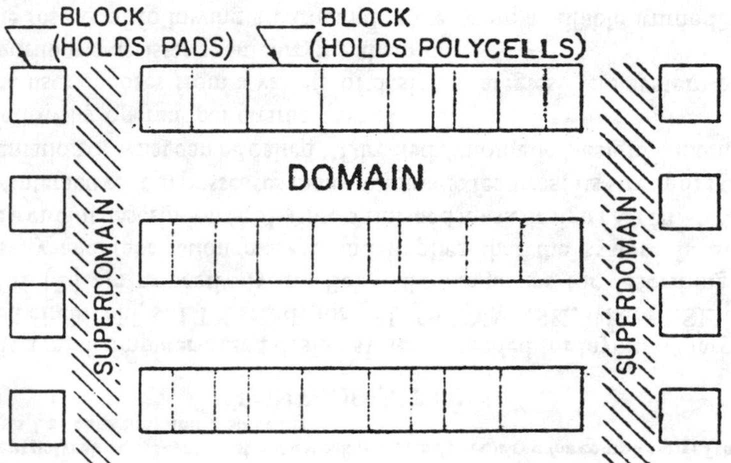

An LTX *domain* is a region, bounded by a closed Manhattan figure which encloses specific cell terminals. In general, a net connecting terminals at various locations on the chip may thus lie entirely within a single domain or interconnect more than one domain. The portion of an interconnection net lying within a domain is called an *intradomain component*, and the portion connecting domains an *interdomain component*. A net terminal not within any domain is regarded as belonging to the interdomain component. The intradomain net components in a given domain are routed by application of a particular algorithm. Therefore, a domain is usually a region of regular routing.

LTX is adapted to design styles possessing a basic row and column layout of blocks, e.g. rows of polycells. In polycell layout, the horizontal wiring channels between blocks correspond to domains, while the regions between columns of blocks, called *superdomains* (see Fig. 1), carry interdomain routing traffic and, for routing purposes, are treated like vertical domains. Thus, the overall layout pattern bears similarity to a street layout, the polycell rows corresponding to the city blocks, the domains to sections of east-west streets, and the superdomains to north-south streets. Within this framework, LTX permits multiple columns of polycell rows, but does not generally permit "dead-end streets".

Modularity in the LTX design style leads to the desired modularity in the design process. Briefly, one arranges the cells among and within blocks, routes the domains, and then routes the superdomains. For polycell circuits, the actual layout stages are:

1. Partitioning of the cells into rows, and crude global cell placement
2. Initial one-dimensional placement
3. Definition of domains, decomposition of nets by domains, and determination of interdomain net topology
4. Optimization of cell positions within rows for "best" channel routing
5. Routing of domains (horizontal channels)
6. Routing of superdomains (vertical channels).

The operations are performed through a combination of automatic and interactive procedures, the particular mix being under the control of the user. An important feature of the system is that the steps can be recycled, beginning anywhere from step 1 to step 6. Therefore, at any time, effort can be focused on the overall placement problem or on the detailed local cell placement, wherever it is needed most. The major LTX program modules which perform or assist with the principal design functions and I/O are described in the next four sections.

III. LTX INPUT AND OUTPUT

III.1. *Input Processor*

The LTX Input Processor accepts six input files. The first is the circuit connectivity description and the second is a library of logic element types. e.g. polycell library. In the circuit connectivity file, i.e. LSL description, the specific cell occurrences comprising the circuit are declared, and the input and output signals for each cell are named. From this file the Input Processor builds the cell occur-

An LTX *block* is defined as a collection of circuit elements, for example polycells, which are organized in a group and normally confined to the interior of a rectangular region identified with the block. For practical purposes, a block may be considered as a bin which holds circuit elements. Various types of blocks can be defined, each characterized by its dimensions and the types of circuit elements which may, or may not, belong to it. Figure 1 shows two types of blocks: "cell blocks", which can contain only polycells and are effectively synonymous with the polycell rows, and "pad blocks", which can contain only chip input and output pads, which are treated by LTX as specially defined one-terminal cells.

BLOCK (HOLDS PADS) — BLOCK (HOLDS POLYCELLS)

DOMAIN

SUPERDOMAIN

Figure 1. Illustration of BLOCKS, DOMAINS, and SUPERDOMAINS

rence and interconnection net portions of the data base. In the cell library, each cell type entry specifies the cell boundary, the positions of the connection points or "terminals", the mask level on which the routing must contact a given terminal, and the sets of electrically and logically equivalent terminals, if any. Electrically equivalent terminals are terminals which correspond to the same electrical node. Logically equivalent terminals are electrically distinct but logically interchangeable (e.g. the 3 inputs of a 3-input NAND).

The remaining input files describe the characteristics and placement of blocks, the assignment of cells to blocks and placement of cells within blocks, and the assignment of layout parameters. General cell placement is specified relative to the origin of a host block, or in an alternative "string placement" mode for polycell rows, by simply naming the cells in the block in left-to-right order. Cell assignment and placement at input are optional since these tasks can also be carried out automatically or interactively. A cell's block assignment or placement can be designated as "HARD", prohibiting automatic routines from moving the cell.

III.2. *XYMASK Output*

Final output from LTX is a hard coordinate description, in the XYMASK language, of the cell positions and routing which is readable by plotting and mask making systems. The routing description consists of path centerlines, path widths, mask level designations, and contact locations. For actual mask generation, the LTX cell library is replaced by the XYMASK cell library describing the "innards" of the cells.

III.3. *Other I/O*

Binary data dumps permit jobs to be saved on tape at any stage of completion, for later reloading onto the LTX system. A disc-based "deep-store" and retrieval facility for cell placements provides rapid recovery from unsuccessful trial placements. In addition, block and cell placements can be output in a format readable by the LTX input processor, thus allowing partial or complete transfer of placement results on those occasions when complete reinput of a job is necessary. Output is available for driving a Calcomp plotter.

A very important output capability is offered by a program module which calculates the net capacitances, outputting them in a format readable by the MOTIS simulation program [8]. It permits circuit simulation to be readily carried out with final net capacitances, thus facilitating discovery of critical nets and timing problems before circuit fabrication.

IV. DISPLAY AND INTERACTION

IV.1. *Display*

LTX provides facilities for inspecting the layout at any stage of the design process, and for performing manual manipulations of the data base. Although LTX emphasizes automatic design, it contains nevertheless an interactive design system which offers a full array of display and manipulative features.

A complete set of windowing commands permits magnification, windowing, centering, and horizontal and vertical stepping. A display window may be stored and recalled. The various element types in the LTX data base — blocks, cells, terminals (including electrical and logical equivalence relations), domains, and nets — can be displayed individually or in combination. Cells and nets can be labeled, and regions of maximum routing congestion can be indicated. Essentially instantaneous search and identification features are provided.

IV.2 *General Interactive Procedures*

Interactive procedures in LTX fall into two distinct categories: those that are of general applicability and those that are limited to polycell-style layout. The more important of the general capabilities are:

1. Domain creation and editing:
   Construction of domains, as closed Manhattan figures, can be performed either by straightforward digitization of alternate corners or, in the case of rectangular domains, by a procedure which automatically chooses the horizontal dimensions by analyzing placement of cells in the appropriate blocks.

2. Reflection and displacement of blocks and cells:
   Groups of cells or blocks (including their member cells) can be selected with the cursor and can be moved or reflected about the x or y axis. Displacement operations can be performed either with or without grid snaps. The LTX grid is both movable and adjustable in spacing; it can be prealigned with any chosen cell or block edge prior to performing grid snap displacements.

3. Editing of layout parameters:
   Various layout parameters, such as horizontal and vertical grid spacings, can be inspected during a job and changed as desired. Other parameters (e.g. cell count) can be inspected, but not changed.

4. Block length editing:
   Block lengths can be altered in order to adjust block capacities.

5. Equivalent terminal exchanges:
   The system permits the user to make certain "legal" edits of net connectivity, as discussed in Section VI.3.

6. Channel routing editing:
   LTX offers routing edits specialized to channel routing. These are described in Section VI.3.

IV.3. *Polycell Procedures*

For polycell layout work, the principal interactive functions are cell insertions and exchanges. A cell selected with the cursor can be inserted into any other position in either its original or a different polycell row, or two selected cells in the same or different rows can be exchanged. The rows involved in these operations are automatically repacked so as to keep their left edges stationary and all cells abutting. The x-axis reflection status of a transferred cell is made to agree with that of its new row.

V. PLACEMENT

V.1. Introduction
V.1.1. Defining the Problem.

In LTX, the items to be placed are cells, each cell having a specified size corresponding to its cell type (see Section III.1). A cell is placed in a "block" which has a location and a capacity; any number of cells can be placed in a block, provided the sum of the cells' sizes is not greater than the block's capacity. For example: in the traditional placement problem of only one item in a block, the cell size and block size are set to unity; in a polycell layout having rows of equal-height cells, the width of a cell serves as a measure of its size and the width of a block then represents its capacity.

Both cells and blocks are classified by "types". A block type definition consists of its capacity and a list of cell types which can or cannot be placed in blocks of that type. This permission/exclusion list allows a variety of items to be handled in a completely parallel fashion.

Cells can be preplaced in blocks, and the placement can be qualified as either "HARD" or "SOFT". "HARD" means that the placement cannot be changed by an algorithm. LTX algorithms. "SOFT" means the placement may be changed by an algorithm if an appropriate figure-of-merit is improved.

An equipotential connection of 2 or more cell terminals is a "net".

The placement problem involves placing cells in blocks, with the goal of minimizing some figure-of-merit approximating the quantity of routing (e.g., length, area), without violating any design constraints.

V.1.2. Design Constraints
V.1.2.1 Net Length

Perhaps the most commonly used layout constraint is a limit on the lengths of individual nets. The limit is often specified to be the same for all nets and is frequently not satisfied. In a manual design, such failure in a trial layout is easily overcome. Nets which exceed the limit are analyzed to determine if the extra length defeats the desired circuit performance; if not, the limit is ignored. A good designer can make this judgement on the basis of a very perfunctory analysis for most of the violations. This manual approach is basically "over-constrained" with the relaxation of constraints dependent on a human's decision-making ability. Of course, relatively few trial layouts can be considered.

In an automated placement procedure it is not appropriate to copy manual procedures. The computer is poor at decision-making; the human intuition that a specific, but unforseen, question may be resolved by a particular analysis is impossible to program. Except for trivial problems, blindly doing all the relevant analysis is usually economically impractical. On the other hand, the computer has a strength which the human does not: it can evaluate thousands of trials. As a result, we look for an "under-constrained" approach and the LTX placement algorithms, like many other placement programs, rely on "global" or integrated figures-of-merit to loosely control individual net lengths.

V.1.2.2 Cell Site Restrictions

A very simple constraint which often stymies the trial use of existing placement programs is the requirement that some cells be excluded from certain cell sites (or blocks in LTX terminology). A simple example is a DIP board which has sockets (blocks) which accommodate either 16 or 24 pin DIPs (cells), but not both. LTX handles these constraints with a permission/exclusion list which excludes the 16-pin cell type from the 24-pin block type and vice versa.

V.1.2.3 Fixed Inputs and Outputs

Electrical circuit layout problems seem to be sharply divided into those with specified input and output (I/O) locations and those where the I/O locations can be chosen to suit the layout. Unfortunately, this seems to have led to a dearth of placement programs which can handle both conditions gracefully. With I/O pad locations generally allowed on all four edges of an integrated circuit, the placement of input and output pads is particularly important in IC layout.

LTX gives each I/O pad the same status as a cell. Then each possible pad location is defined to be a block of a type which permits only I/O-type cells. Conversely, the "normal" interior block is of a type which excludes I/O cells. Layouts with inputs and outputs having specified pad locations are handled simply: and corresponding I/O cells are "HARD" preplaced. For those input and output locations which are discretionary, the corresponding I/O cells will be placed using the same criteria as for any "normal" cell, but will be iteratively exchanged only with each other.

V.1.3. The Multi-Algorithm Approach

LTX attacks the placement problem in two stages: first, using a number of two-dimensional (2-D) algorithms with global, but relatively crude, figures-of-merit; second, for modular styles having a single row of cells within a block, using a number of one-dimensional (1-D) algorithms with local, but relatively refined, figures-of-merit. The algorithms are individually quite simple, never moving more than two cells at a time, and never "looking ahead" more than one move. However, the interweaving and sequencing of these algorithms is believed to overcome the limitations of any individual algorithm, providing high performance with more consistency than would be obtained with one or two more complex algorithms.

V.2. Two-Dimensional Placement

LTX does not use a partitioning algorithm to initially create groups of cells for later placement on the chip, but begins, instead, with placement. There are a number of reasons for omitting the traditional first step of partitioning. The simplest reason is that it is feasible, from a standpoint of both computer time and performance, to do placement with, say, 500 individual cells. Another reason is that minimizing the number of connections between logic blocks within the chip does not have the sharply defined cost function one associates with, say, minimizing the number of chip-to-chip connections. For example, placing two cells on opposite sides of a channel may require less routing area than adjacent placement in the same row.

PLAC is the 2-D placement program for the LTX system. To help distinguish PLAC from the several 1-D placement programs used by LTX, we will refer to it as PLAC(2D).

V.2.1. *Summary of Functions*

There are two major phases: the initial placement phase which sequentially selects unplaced cells according to their connectivity to other cells and places them so as to minimize total routing length, and the iterative improvement phase which performs pairwise exchanges to reduce total routing length. Routing length for a single net is approximated by the half-perimeter of the rectangle enclosing its terminals.

These two phases involve algorithms which are quite simple. However, to reduce blockage of exchanges between unequal-size cells, the initial placement and iterative improvement phases are interlaced: a fraction of the cells are placed, then the placement of the placed cells is iteratively improved, then the next fraction is placed, followed by more iterative improvement (but on a larger number of cells), and so forth. This allows iterative improvement to be attempted while the blocks are still empty enough to generally allow unrestricted cell exchanges.

The initial placement strategy enables PLAC(2D) to be influenced by the cells which are preplaced, or "seeded". This provides an important feature of PLAC(2D): the designer can affect the results and choose his level of involvement. He may choose to preplace nothing, or just a few cells from each major functional cluster. PLAC(2D) allows the human designer to do the overall arrangement, communicated through the "seed", while relieving him of the tedium of the less pattern-oriented decisions. For example, the size of functional clusters is rarely the same as that of the rows in a rectangular layout, requiring the breakup of the clusters which "don't fit" as integral pieces. Here, human intuition is relatively poorer, and PLAC(2D) relatively better, than on the selection of an overall arrangement.

V.2.2. *Figures-of-Merit*

Design algorithms generally depend on various "figures of merit" to compare alternative arrangements or select the most promising candidate for the next operation. The selection of a figure-of-merit is a delicate choice between realism and computer time. The 2-D figures-of-merit are described in this section: the algorithms which use these figures-of-merit are discussed in section V.2.3.

V.2.2.1. *Connectivity – IOC*

During the constructive placement phase, the placement sequence is intended to "grow" strongly-connected clusters of cells before congestion becomes a problem. PLAC(2D) uses a simple "inside-outside" connectivity (IOC), which is computed for unplaced cells as follows: for each candidate cell, set IOC=0; then

for each net connected to the candidate cell, add 1 if the net's other terminals are connected only to placed cells ("inside" connections), or subtract 1 if connected only to unplaced cells ("outside" connections).

V.2.2.2. *Routing Length Lp*

In a typical IC placement problem, the usual choice for the "real" figure-of-merit is actual routing area, provided that the critical nets are not so long as to affect the required circuit performance.

During the phase of initial placement of unplaced cells, there is no obvious means of evaluating the routing area or blocking effects of connections to unplaced cells. Even in the iterative improvement phase, after all cells are placed, a complete routing of every trial placement is economically impractical for two-dimensional placement problems of any reasonable size. Thus, a simpler figure-of-merit is substituted for routing area.

An approximation of total routing length, denoted Lp, is used by PLAC(2D) as the figure-of-merit. The length of each net is individually approximated and denoted lp. Each net terminal is considered to be at the center of the connected cell's block. (For long polycell rows, each row is temporarily decomposed into a specified number of blocks to improve the net length approximation.) PLAC(2D) approximates a net's length (lp) as the half-perimeter of the smallest rectangle which encloses the net's terminals, an approximation recommended in [9]. This approximation is a lower bound to the length of the rectilinear Steiner minimal tree routing [10] (non-terminal "tie" points are allowed) and basically assumes that each net can be routed without being blocked by other nets. For 2- and 3-terminal nets, which usually comprise 75% of the nets, the approximate length is equal to the Steiner length. For 4- and 5-terminal nets the Steiner length is not more than the length of the enclosing rectangle plus twice the width [10], limiting the approximation error and generally avoiding the pitfalls of models based on a "connection matrix" representation of nets with more than 2 terminals [11]. For nets with a large number of terminals (e.g. clock lines), the approximation is poor but since such a line often "goes everywhere" anyway, the crucial difference in net length for any two placements of the net's connected cells is not likely to be substantially in error.

While the PLAC(2D) net length estimate is very simple, any other net length estimate could be used. The PLAC(2D) strategies have been applied to the placement of DIPs on a large wire-wrap board, using the actual wire length calculated by the router as the placement figure-of-merit. However, limited experiments [12] found that essentially the same results were achieved using the much simpler half-perimeter estimate used in the LTX version. The length of the minimum spanning tree (no non-terminal "tie" points) would be an attractive choice if nets typically have more than 3 terminals. The spanning tree length is a better bound than lp, always within 3/2 of the Steiner length [13], and much faster to calculate than the Steiner length.

V.2.2.3. Combined Lp and Row Length

In doing placement on polycell layouts, the Lp figure-of-merit tends to completely fill the blocks representing the middle rows, making the chip wider, while leaving the top and bottom rows empty. While final row length uniformity can be obtained by specifying block capacities totaling an amount exactly equal to the sum of the cell sizes, this creates a "box-filling" problem on initial placement and severely restricts pairwise exchanges, if cells are of unequal sizes. Even if these constraints are not significant, such uniformity may actually cost more routing length than it's worth.

Assume that delta W is the change in width of the "widest" row due to a pairwise exchange of two cells, i.e.,

$$delta\ W = delta\ [\ max\ over\ i\ (sum\ of\ cell\ widths\ in\ row\ i)\ ].$$

Then the change in chip area, due to this reduction in width, is (delta W * H), where H is the height of the chip. On the other hand, a change in routing length, (delta Lp), could result in a change in chip area of as much as (delta Lp * h), where h is the track spacing. Thus, in the iterative improvement stage, if all cells are placed, we attempt to minimize chip area by using

$$Lp + beta\ *\ [\ max\ over\ i\ (sum\ of\ cell\ widths\ in\ row\ i)\ ]$$

as a figure-of-merit, where beta is a user specified constant with a recommended value on the order of H/h.

Occasionally two or more rows may be equal to the maximum cell row width. In this case, a single pairwise exchange cannot reduce the maximum row width. Where neither the Lp nor the maximum width term is changed, the count of widest rows is used as a figure-of-merit in hopes that a reduction in this count will eventually lead to moves which make a meaningful reduction in the widest row width.

V.2.3. Placement Algorithms

PLAC(2D) uses two individual algorithms: constructive placement of unplaced cells (Section V.2.3.1) and iterative exchange of placed cells (Section V.2.3.2). While these two algorithms involve careful choices of level of complexity and figures-of-merit, they are basically straightforward adaptations of known techniques. However, PLAC(2D) uses a third algorithm, the interlace algorithm, which is thought to be original: instead of the customary strategy of constructively placing all the cells before iterating, PLAC(2D) permits iterative improvement to occur with specified frequency during constructive placement. The motivation for doing this is discussed in Section V.2.3.3.

V.2.3.1 Constructive Initial Placement

The PLAC(2D) constructive placement algorithm sequentially places unplaced cells. The procedure consists of the selection of the next cell to be placed, and then the selection of the block in which it will be placed.

Candidate Selection:

The inside-outside connectivity, IOC, is calculated for all unplaced cells. The cell with the highest IOC, regardless of cell type, is selected as the next cell to be placed. In case of an IOC tie, the cell with the larger size is selected.

Initial Placement:

The candidate cell is given a trial placement in each block having the necessary remaining capacity and cell type permission. The block which results in the least approximate total net length, Lp, is selected. In case of a tie, the block with the most remaining capacity is chosen. In evaluating Lp for these placements, nets not connected to the candidate cell do not change length and are not actually calculated for each trial placement. During this phase, some terminals on the calculated nets belong to unplaced cells. In calculating lp, terminals with undefined locations are ignored, i.e. assumed to be within the rectangle enclosing the "placed" terminals.

V.2.3.2 Iterative Exchange

Consider the cells numbered in the order of their appearance in the cell connectivity list. One "pass" through the iterative exchange consists of the following: Take each cell i (in numerical order) and exchange it with each cell j, for j>i, if all of the following conditions are met — the cells are in different blocks, the cell types are permitted in the new blocks, and the total net length is reduced. Nets which are not connected to either cell i or cell j, and nets which are connected to both cell i and cell j, will not change length and these net lengths are not calculated for the trial exchanges. As a user option, the combination of Lp and row length described above (Section V.2.2.3) is used in place of Lp, if all cells are placed.

Note that an exchange between every pair of cells is tried once and only once. The profitable exchange of, say, Cell 17 with Cell 42 does not preclude the possibility of subsequently exchanging Cell 17 and Cell 43. There is no search for the best exchange; profitable exchanges are made immediately without considering the possibility that the current exchange may eliminate a larger profit on a subsequently considered exchange. While the results of this exchange strategy are input-sequence dependent, a trait to be generally avoided in design algorithms, it is believed that doing ALL profitable exchanges achieves approximately the same results with less computing time than doing only the MOST profitable exchange (assuming, in both cases, the strategy is repeated until improvements are exhausted).

V.2.3.3 Construct/Exchange Interlace Algorithm

The usual procedure for combining initial placement and iterative improvement is to do them in sequence. Unfortunately, if the blocks are full, or nearly full, profitable exchanges between cells of unequal sizes may be prevented by size constraints. PLAC(2D) attempts to reduce this problem by interlacing iterative improvement with the initial constructive placement, allowing exchanges to be tried before all the cells are placed and before the blocks are too full. The interlace algorithm is quite simple: assume that N is the number of unplaced cells; the user specifies K, the number of construct/exchange cycles; then PLAC(2D) repeatedly performs initial constructive placement of approximately 1/Kth of the unplaced cells, followed by a single iterative pairwise exchange pass (this does not necessarily exhaust exchange improvements). Following the final construct/exchange cycle, exchange passes are repeated until all iterative improvements are exhausted.

Choices for values of K have been explored experimentally [5]. Generally $K=1$ (all cells constructively placed before any iteration) gives higher Lp than $K>1$. For problems with unequal size cells and nearly full blocks. Lp tends to decrease as K increases. although there is a substantial random component, for $K>1$.

V.3. One-Dimensional Placement

V.3.1. Summary of Functions

In the polycell-style layout of large-scale integrated circuits, rectangular cells are arranged in parallel horizontal rows and the interconnection nets are routed in the intervening "channels". Therefore, a major task of a polycell placement program is the 1-D arrangement of cells so as to minimize the channel heights required to contain the associated routing.

There are three 1-D placement algorithms:

1. For each row the constructive 1-D placement procedure takes each cell, as assigned by the 2-D placement algorithms to the blocks representing that row, and calculates a target x-coordinate for each cell. The cells are then packed left-to-right in order of their target coordinates, starting at the left edge of the left block. A similar strategy is used to constructively place LTX-generated feed-throughcells.

2. The iterative cell insertion takes, in turn, each cell in the specified row and performs a trial insertion between each adjacent pair of cells in the row. The figure of merit is horizontal routing length, totaled for all nets connected to all cells in the row, ignoring net terminals beyond the two channels immediately adjacent to the cell row.

3. The third 1-D algorithm, which will be denoted PRO(1D), standing for Placement for Routing Optimization, is the most heavily used, usually in conjunction with interactive moves. PRO(1D) consists of adjacent pairwise exchange moves interlaced with the reflection of individual cells about their vertical axes. While it restricts its attention to one specified row at a time. PRO(1D) can analyze the routing in the two adjacent channels with much more accuracy and detail than any of the previous algorithms. Three figures-of-merit are involved: FM1) the number of nets which are "unroutable" due to cyclic vertical constraints (see Section V.3.2.4), FM2) the maximum count of nets crossing any terminal column ("channel density") (Section V.3.2.3), and FM3) the number of terminal columns in which the channel density occurs (Section V.3.2.3). These three figures-of-merit are used hierarchically: lack of change in FM1 leads to the use of FM2: lack of change in either FM1 or FM2 leads to the use of FM3.

V.3.2. One-Dimensional Figures-of-Merit

The various figures-of-merit for the 1-D placement are designed to cope with varying levels of missing information and evaluate figures-of-merit from either a full chip or limited two adjacent-channel point of view.

V.3.2.1. Average Midpoint – Constructive 1-D Placement

Used by two constructive 1-D placement algorithms, the essence of this figure-of-merit is easily stated: first, each cell is given a "target" x-coordinate equal to the average x-coordinate of the midpoints of its connected nets. The net midpoint is calculated by averaging the x-coordinates of the net's left-most and right-most terminals, regardless of row. However, before intra-block left-to-right ordering has been done — the typical situation after using the 2-D algorithms discussed above — the exact cell terminal positions are unknown, and the net terminals are assumed to be located at the cells' block centers.

Feed-through cells, which provide interchannel net connections and are created with fixed row membership (see Section VI.1.2), are placed within their rows after the left-to-right placement of functional cells. Thus, cell terminal x-coordinates are known, and are used to calculate midpoints of the feed-through's connected net, separately, for the two channels adjacent to the feed-through's row. The target x-coordinate is the average of these midpoints.

V.3.2.2. Total Horizontal Routing Length – Single Cell Insertion

For a specified cell row, with specified left-ro-right cell order, we calculate, separately for the adjacent channels. the horizontal routing length for each net connected to any cell in the row, using the actual terminal coordinates. Only net terminals in the 2 channels adjacent to the specified row are considered.

V.3.2.3. Channel Density, Span – PRO(1D)

We define the "local density" at a given longitudinal coordinate in a channel as the number of nets that intersect the perpendicular to the channel at that coordinate. Clearly, the local density is a discontinuous function, which can change value only at coordinates corresponding to terminal positions (where nets either begin or end). In the usual case in which the cell terminals are aligned with a grid, we are interested in the local density only at "on-grid" positions, i.e. at terminal columns. The "channel density" is defined to be the maximum value attained by the local density anywhere along the channel. It is evident that, if the wiring parallel to the channel is limited to one mask level, then the channel density constitutes a lower bound on the number of routing tracks required, whatever the routing style. Only if net connections are confined to one channel edge (one-sided channel) is it guaranteed [14] that the routing can be realized at this lower bound. However, the sophisticated channel router used by LTX (Section VI.2) can usually route the typical two-sided channel "at density" or near it. making the channel density a very good approximation to the routing track count. The total number of terminal columns for which the local density equals the channel density is called the "span".

A channel density calculation procedure which does not require net terminal positions to be on grid is used by the router. On the other hand, in PRO(1D) where a new density calculation must be done for each cell reflection or exchange, speed is an overriding consideration. The channel density calculation used by PRO(1D) "snaps" off-grid terminals to a grid. This algorithm executes more rapidly but incurs possible minor inaccuracies. The detailed procedure, which is quite straightforward, is presented in [6].

V.3.2.4. Unroutable Net Count – PRO(1D)

One of the difficulties inherent in routing two-sided channels is that, on occasion, the routing can be blocked by "cyclic vertical constraint loops". When two distinct nets have connections to terminals on opposite sides of the channel but in the same, or nearly the same, vertical column they are said to form a "vertical constraint pair" because the net having the top terminal connection must be routed with its horizontal segment topmost to avoid an illegal overlap of the vertical segments. A cyclic constraint occurs whenever a chain of constraint pairs closes in a loop, because then it is impossible to perform the routing without at least one vertical overlap. Figure 2A shows a simple loop involving two nets, each of which has one horizontal path segment and two terminals. The figure is drawn with the constraint at the left edge honored and the one on the right violated. Honoring the constraint on the right would necessarily entail violating the one on the left. In more complicated cases, many nets may be involved in a constraint loop. PRO-(1D) attempts to eliminate such conflicts by rearranging the polycells in the row in order to reduce the "unroutable net count", which is defined as the number of vertical constraint pairs involved in cyclic constraint loops. Figure 2B shows the constraint loop of Figure 2A resolved by reflection of the upper left hand cell. Figure 2C shows another method of resolving a constraint loop, i.e. by inserting a break into one of the horizontal paths. The LTX channel router, which is capable of dividing, or "dogleging", a horizontal trunk, often can route an otherwise unrout-

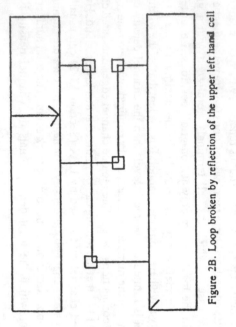

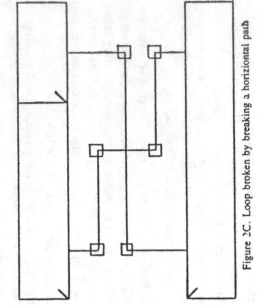

Figure 2B. Loop broken by reflection of the upper left hand cell

Figure 2C. Loop broken by breaking a horizontal path

able channel in this manner. Nevertheless, the "straight trunk" routability criterion in PRO(1D), which predates dogleg routing in LTX, remains important because satisfying it (a) guarantees routability, albeit more strongly than necessary for the dogleg router. (b) can almost always be done without an increase in density, and (c) seems to increase the chances of the dogleg router routing the channel "in density"

There is, unfortunately, no shortcut to determining routability. One must, in effect, attempt to route the channel, and if the task cannot be completed, report some quantitative measure of the incomplete portion. The routability algorithm used by PRO(1D) includes many of the steps of a "straight trunk" channel router (e.g. [15]), but does not involve the assignment of nets to specific routing tracks. It consists of three parts.

First, a "constraint table" is prepared. This consists of vertical constraint pairs.

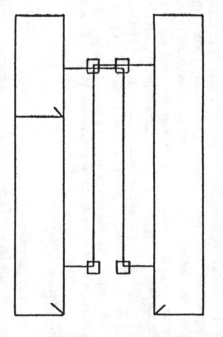

Figure 2A. A cyclic vertical constraint loop

wherein the top net in each pair has a connection to a terminal lying directly above (or within a horizontal tolerance limit of) a terminal connection of the bottom net. (Mask level "self-constraints" are included by PRO(1D). See [6] for details.) Second, a top-down "pseudo-route" is performed in order to determine routability. Since unconstrained nets are of no interest in this calculation, the candidates for "routing" are chosen directly from the constraint table itself. The top pair members are scanned for the first net that is not also a bottom pair member. This net is free to be "placed" topmost. It is marked "routed", and all constraint pairs of which it is a member are eliminated. The next net that is similarly free is marked "routing" and its constraint pairs eliminated. In this fashion, the constraint list is repeatedly scanned until it is either empty or constraint pairs remain that cannot be eliminated. If the former, the channel is routable. If the latter, it is not routable. The remaining pairs in the constraint table are again scanned for placement candidates, but this time they are chosen from bottom pair members and are checked against blockage from above. The bottom up pseudo-route cannot eliminate all the remaining pairs. It removes constraint chains that might be hanging downward from cyclic constraint loops. Those pairs that still survive must themselves be members of loops or be caught between them. The number of pairs remaining is reported as the unroutable net count. This count may exceed the number of distinct unrouted nets because a net may belong to more than one constraint loop, and because it may include nets belonging to constraint chains that happen to be caught between constraint loops. The unroutable net count is, therefore, a somewhat crude figure-of-merit, although PRO(1D) has utilized it quite successfully in achieving routability.

A formal presentation of the procedure outlined above is given in [6].

V.3.3. One-Dimensional Placement Algorithms

V.3.3.1. Constructive Algorithms

For each row the constructive 1-D placement procedure takes each cell, as assigned by the 2-D placement algorithms to the blocks representing that row, and calculates a target x-coordinate for each cell equal to the average x-coordinate of the midpoints of its connected nets. The cells are then packed left-to-right in order of their target coordinates, starting at the left edge of the row.

Basically the same procedure is used to place feed-through cells. The initial feed-through placement procedure takes the rows in bottom-up sequence. In calculating the target x-coordinate for a given feed-through, its connection to an unplaced feed-through cell in the row above, if any, is ignored.

Once the feed-through cells have been initially placed, an individual row can be selected and its feed-through cells' target x-coordinates calculated, treating the adjacent channels equally. The target x-coordinate for a non-feed-through cell is simply its existing x-coordinate. Finally, all cells in the row, including feed-throughs, are packed left-to-right in the order of their target x-coordinates.

V.3.3.2. Single Cell Insertion

This algorithm performs intrarow cell insertion operations within a polycell row in order to reduce the total horizontal routing length of nets connected to cells in the row.

The algorithm is a simple one. Consider a row of N cells, with cell positions numbered from left to right. Let i be the cell position of the current candidate for insertion, and j the position of the insertion target. In an insertion operation, all cells in the interval i,j including j but excluding i are shifted (by the width of the candidate cell) one position in the direction of i. The cell originally in position i is transferred to position j. Starting with $2i=1$ as the insertion candidate position, insertions are attempted from left to right for all $N-1$ targets $j \neq i$. Then i is incremented and another $N-1$ trials are made, etc. On each trial insertion, the total horizontal routing length (Section V.3.2.2) is recalculated, and the operation is reversed only when it increases this routing length, i.e. neutral changes are retained. (A retained insertion results in a change in candidate cell, but not in candidate cell position.) A single insertion pass is complete when all N candidate positions have been tried. Thus, a pass contains $N(N-1)$ trial insertions, making running times considerably greater than those for an adjacent exchange pass, particularly with long rows.

V.3.3.3. Cell Reflection and Adjacent Exchange Algorithm – PRO(1D)

In the most general mode of operation, this placement procedure attempts to make the channels immediately above and below the selected row routable, and to reduce the sum of their channel densities. Let U be the sum of the unroutable net counts for both adjacent channels, D the sum of their channel densities, and S the sum of the spans. These criteria are handled hierarchically, with the most importance assigned to U and the least to S. Let Q represent a combined "figure of merit" for both channels. Then

Q is taken as improved if
    U is reduced
        or
    U is unchanged and D is reduced
        or
    U and D are unchanged and S is reduced.
Q is unchanged if U, D, and S are unchanged.
Q is regarded as deteriorated if
    U is increased
        or
    U is unchanged and D is increased
        or
    U and D are unchanged and S is increased.

The operation of these procedures is as follows:

(1) Starting at the left of the polycell row, reflect each successive cell about a y-axis passing through its center. After each reflection, reevaluate Q. If Q is improved retain the reflection; else reverse it. Continue to make left-to-right "reflection passes" until a complete pass yields no improvement in Q. Then go to (2).

(2) Starting at the left end, proceed along the row, exchanging each cell with its neighbor to the right. After each exchange, reevaluate Q and retain the operation only if Q is improved. Continue to make left-to-right "exchange passes" until a complete pass yields no further improvement in Q. Then go to (3).

(3) The current channel density, span, and unplaced net count for each channel are reported to the user. If successive reflection and exchange passes are profitless, terminate execution; else, go to (1).

If, at the beginning of the run, the user had requested to "keep neutral changes", the reflection or exchange operations will also be retained if they leave Q unchanged.

When the user chooses to only reduce the channel density, U is not considered in evaluating the figure of merit. If the user chooses to work on routability alone, D and S are ignored. A cell which has been designated as "HARD" placed is not reflected or exchanged.

Experience on many IC layouts has shown that in the great majority of instances routability is not difficult to achieve, even when there is an initially high unroutable net count U. A few cell exchanges and reflections are usually sufficient to unravel even fairly large constraint loops with little or no penalty in channel density. This fact, coupled with a 20-fold increase in execution time for calculating routability as opposed to calculating density, has resulted in the basic operating guideline: reduce density first; then make the channel routable.

VI. LTX ROUTING

In accordance with the modular approach of LTX, the highly complex LSI routing problem is decomposed into a set of simpler sub-problems. As stated in Section II, each net is assigned to one or more routing areas (domains and/or superdomains) depending upon the locations of its terminal connections and the overall topology. This assignment process is followed by applying a very efficient channel routing algorithm to each routing area. Routing results can be edited interactively while guaranteeing the integrity of the circuit connectivity.

The following detailed description of LTX routing is divided into three main sections: net decomposition/assignment, channel routing, and routing edits.

VI.1. Net Decomposition/Assignment

The function of this part of the routing package is to decompose the original netlist for the entire circuit into a set of netlists, one for each domain and superdomain. This decomposition/assignment process is accomplished in three discrete steps. Although the description is given for a single net, all nets are actually processed before going on to the next step.

VI.1.1. Net Decomposition

Each terminal of the net is tested to determine whether or not it lies within one of the user-defined domains. If so, the terminal is assigned to a netlist for that domain. Otherwise, it is assigned to the interdomain netlist. When all terminals for a particular net have been processed, a scan is made as to where they have been assigned. If the net has been found to span more than one domain or if the terminals have been assigned to the interdomain netlist and one or more domains, additional connection points called "domain pin-outs" must be created. Such a pin-out is created for each domain where the net has a component and is added to both that domain's netlist and the interdomain netlist. An intradomain net thus contains all of the original net's terminals within that domain, plus a domain pin-out if and only if the original net had at least one terminal outside that domain. The interdomain net consists of the original net's terminals that were not located inside a domain plus all of the domain pin-outs created for that net (if any). Figure 3 depicts the type of connectivity relationship established by the decomposer. Since the decomposition is based upon an inside/outside criterion, the pin-outs signify only topological connection, and not location.

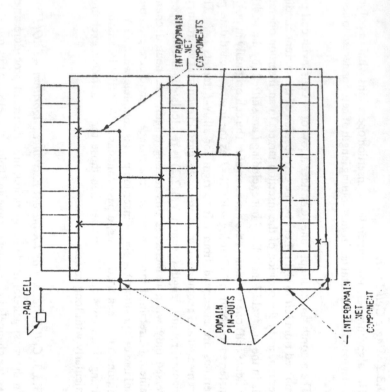

Figure 3. The decomposition of a net into intradomain and interdomain components

VI.1.2. *Interchannel Feed-Through*

With polycell rows of even moderate length, it is usually more efficient to feed an interdomain net from channel to channel vertically across the row, rather than out the end of one channel, up or down the side, and then into the other channel. Since for many of the cells in our polycell libraries, a terminal on the front has an electrically equivalent terminal on the back (see Section III.1.), such equivalent terminals can often be used to conveniently establish the necessary interchannel connection. If equivalent terminals are not available, specialized "feed-through" cells can also be inserted into the row for this purpose. In this phase of the net decomposition/assignment process, both these types of interchannel connections (that have not already been preconnected by the user) are made automatically. Restriction to polycell style layout is evident.

We assume that the domains are arranged in a "matrix" of rows and columns, although not every position need contain an actual domain, nor need the rows and columns be of uniform width. For nets which interconnect more than one domain in the same domain column, the program eliminates unnecessary pin-outs by using electrically equivalent terminals on the cells to establish the interchannel connections whenever possible, and by creating feed-through cells where equivalent terminals are not available. It will exchange logically equivalent connections in order to gain the benefit of an advantageous electrical equivalence. The user can specify the maximum number of polycell rows to be crossed, internally with equivalences and feed-throughs. This number can be changed for selected nets. Figure 4 illustrates the use of both types of interchannel connections.

The action taken by the equivalence and feed-through module can thus have a profound effect on the topology of certain nets. For example, it may cause some interdomain nets to be deleted in their entirety through the replacement of all their pin-outs by internal connections. On the other hand, an intradomain net component may be created for a domain in which the net was originally absent. Other interdomain nets may have been simplified (compare Figures 3 and 4) while the remainder have been left unchanged. The last module in the net decomposition/assignment process, described below, assigns these interdomain nets to the appropriate superdomains while maintaining net integrity.

VI.1.3. *Net Assignment*

All remaining interdomain nets must now be assigned to the appropriate vertical channel or channels. It is convenient to categorize these nets as follows:

Type (A): Nets containing only domain pin-outs where all of the domains lie in the same column (this type of net can occur if the number of cell rows that had to be crossed was greater than the feed-through range specified by the user).

Type (B): All others.

The algorithm that assigns these nets to superdomains analyzes the topology of each net (in general, a two-dimensional problem). There are two passes in the assignment algorithm. During the first pass, net type is determined, and nets of Type (B) are actually assigned. Processing for nets of Type (A) is deferred until the second pass.

In the assignment process, the interdomain nets are essentially decomposed into component nets in the superdomains, with a netlist created for each superdomain. Each domain pin-out is assigned to a superdomain, and has specified its row and column position, its x-coordinate (coinciding with the appropriate domain edge), and a preferred vertical location code — low, high, intermediate, or "don't care" with respect to its domain (channel). Actual y-coordinate specification takes place later, during channel routing. Occasionally, intradomain nets may be affected, as discussed below.

If only one specific vertical channel is involved (e.g., Type (B) nets for which all domains belong to two adjacent columns or one domain column plus cell terminals in one but not both of the adjacent vertical channels, etc.), the net is assigned to that channel. For nets which span two or more vertical channels, a more complicated procedure is necessary. Consider, for example, Figure 5. Since the net must interconnect two vertical channels, some domain must necessarily have two domain pin-outs for this net (one at the domain left edge, and the other at its right edge). If the (intradomain) net already exists in this domain, one additional pin-out must be created and added to the net. If not, a feed-through intradomain net consisting solely of two newly created domain pin-outs is constructed. A portion of the interdomain net is now assigned to each superdomain spanned by the net. The choice of which of several equivalent topologies to use is relatively arbitrary. The assignment program generally tries to minimize vertical wire length but avoids "staircasing" the net through the various vertical channels.

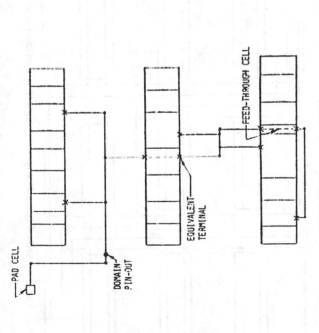

Figure 4. The net of Fig. 3 after the feed-through module has inserted an equivalent terminal and a feed-through cell

229

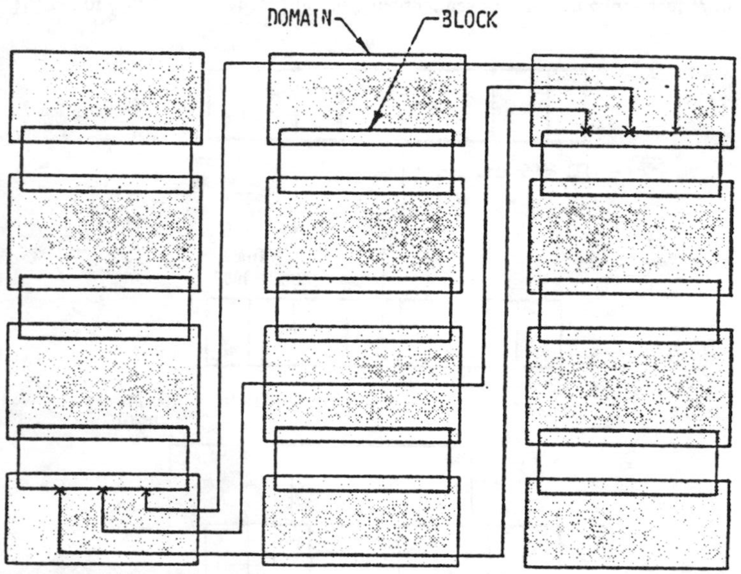

Figure 5. Alternative vertical channel assignments for nets in a multi-column layout

Type (A) nets are handled after all nets of Type (B) have been assigned. Type (A) nets can be assigned to either of the vertical channels adjacent to the domain column. Therefore, the assignment is made in a manner which minimizes the sum of the vertical channel densities while attempting to balance the number of tracks per channel.

At the conclusion of the net assignment phase, there exists a netlist for each routing area (i.e. each domain and superdomain), and all domain pin-outs have x-coordinates. No area has actually been routed, but the assignment process itself constitutes a type of crude topological routing. Subsequently, when a domain is actually routed, its pin-outs are assigned to specific routing tracks at the appropriate domain edge and hence get precise y-coordinates. Once all of the domains have been routed, all pin-outs have specific locations. The vertical channels (superdomains) can then be routed since the locations of the connection points for all nets are known. Because all routing areas are treated as channels, LTX uses a channel router rather than a more general two-dimensional approach. Details concerning channel routing and the specific algorithm used are given in the next section.

VI.2. Channel Routing

VI.2.1. Introduction

A basic problem in the design of large-scale integrated circuits is to complete the necessary interconnections in as small an area as possible. Many routing algorithms have been used with varying degrees of success. Some employ general purpose two-dimensional routing strategies such as those described by Lee [16], Hightower [17] and Mattison [18]. These methods, however, often yield poor results because a net routed early may be arbitrarily placed in a location that blocks some subsequent net. Various net ordering schemes have been proposed in an attempt to minimize this problem but unfortunately, the blockages persist.

In LTX, we assume that the original routing problem is amenable to a decomposition into a series of routing subproblems, each of which is a channel routing problem. Mattison [18] actually performs this decomposition even though he does not use a channel router. In descriptive terms, a channel typically consists of the space between two parallel rows of terminals. For this situation, the interconnection problem can be transformed into a one-dimensional routing problem with constraints. If the constraints are non-cyclic and the channel height is adjustable, 100% completion of the routing is always guaranteed.

Channel routers are usually easier to implement than the more general routers mentioned previously, and they execute much faster because the problem under consideration is inherently simpler. In addition, channel routers operate on one routing track at a time, thus minimizing the amount of data that must be core resident. Since the data storage requirements are small, channel routers can be implemented on minicomputers. Channel routing algorithms have been described by Hashimoto and Stevens [14] ("unconstrained left-edge") and by Kernighan, Schweikert and Persky [15] ("constrained left-edge with zoning" and "optimal" via branch and bound). As LSI circuit layouts become more modular, the channel routing approach tends to yield better results. The channel routing algorithm presented here generally outperforms those mentioned previously.

VI.2.2. Channel Routing Concepts

Before describing the new router in detail, we will set forth some of the terms and concepts that apply to channel routing in general. In order to clarify the definitions and descriptions, we will use simple examples, which are not necessarily realistic, in this part of the discussion.

Consider a rectangular channel with terminals along its top and bottom edges. The terminal abscissas usually lie on a grid that has a uniform spacing equal to that of adjacent terminal positions. A net is defined by a set of terminals that must be interconnected via some routing path. Some nets may also have a connection point (domain pin-out) at one or both ends of the channel. Such a point corresponds to a terminal that has a known abscissa (that of the specified channel end) but whose ordinate will be determined by the channel router. Let us assume that all of the horizontal routing occurs on one level while the vertical segments occur on another level. For integrated circuits, the horizontal segments are typically metal while the verticals are polysilicon and/or diffusion. In order to interconnect a horizontal and vertical segment, a contact must be placed at the intersection point. The task of a channel router is to route all of the nets successfully in the minimum possible area.

Since the channel length is fixed, the area goal is equivalent to minimizing the channel height; this height is equal to the spacing between adjacent horizontal tracks (a constant imposed by the technology) times the number of such tracks required to accomplish the necessary connections, plus appropriate clearances for the top and bottom tracks.

Let us now consider the constraints under which channel routers usually operate. The requirement that all horizontal segments occur on one level with all verticals on another may theoretically cost a track or more as is shown in Figures 6A and 6B. Such a saving can seldom be achieved in real problems. The overlap-

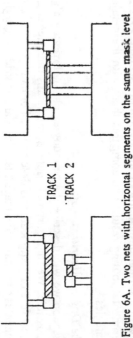

TRACK 1
TRACK 2

Figure 6A. Two nets with horizontal segments on the same mask level
Figure 6B. Nets of Fig. 6A with horizontal segments on distinct mask levels

ping of nets in this manner is generally undesirable because of increased capacitive coupling between the nets. The imposition of this constraint, however, has some benefits. In order to ensure that two distinct nets are not shorted, it is only necessary to ensure that no vertical segments overlap and no horizontal segments overlap. If a top terminal and a bottom terminal have the same abscissa and they are to be connected to distinct nets, the horizontal segment of the net connected to the top terminal must be placed above the horizontal segment of the net connected to the bottom terminal; otherwise, the vertical segments would partially overlap. Since the abscissas of these vertical segments are given (the terminal positions are known and fixed), many such vertical constraints must be satisfied. It is of course possible that constraint chains of arbitrary length (A above B, B above C, etc.) as well as constraint loops (A above B at one terminal position, B above A at another is a loop of size 2) can occur.

If there is no constraint loop, all channel routing algorithms will be able to complete the required routing. The "Left-Edge" algorithm [14] always yields optimum results if there are no vertical constraints (e.g., a channel which has terminals on one side only). As was shown in Kernighan, et al. [15], it often yields suboptimal results if vertical constraints are present. Figure 7A, an example of

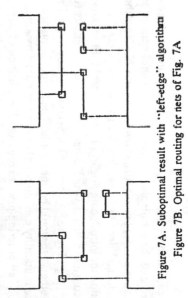

Figure 7A. Suboptimal result with "left-edge" algorithm
Figure 7B. Optimal routing for nets of Fig. 7A

such suboptimality, is taken from this reference. It is quite obvious that the routing in Figure 7B is optimal. In the "Optimal Channel Router" [15] a branch and bound algorithm is presented which guarantees an optimum result. Unfortunately, that algorithm contains the restriction that each net consists of a single horizontal segment; thus Figure 8A is considered optimal. However, by using a more versatile routing style, a better result can be obtained and is shown in Figure 8B; the number of routing tracks used is equal to the lower bound given by the "channel density" defined in Section V.3.2.3.

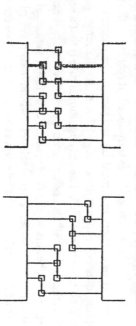

Figure 8A. Best obtainable routing with a single horizontal segment per net
Figure 8B. Best obtainable routing with multiple horizontal segments permitted

To illustrate the above, consider Figure 9A. The constraint in column 1 requires net A to be above net B while that in column 4 requires net B to be above net C. As shown, the routing is accomplished in three horizontal tracks. Note, however, that the channel density is only two. Figure 9B shows a possible procedure whereby the

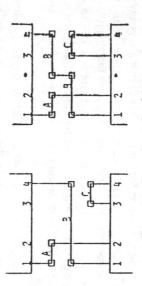

Figure 9A. Constraint chain of three nets
Figure 9B. Nets of Fig. 9A routed with a "dogleg" in net B

nets are routed in two tracks. The key feature is the "dogleg" in net B which transfers part of the horizontal section from track 2 to track 1. Whether or not such doglegs can be introduced beneficially is very much data-dependent. In this particular example, routing in two tracks is possible if and only if there is sufficient space between columns 2 and 3 (i.e., column * exists). Otherwise, there would be no available location for the dogleg.

Now consider the problem that was posed in Figure 2A. There are two distinct nets to be routed, but there is a constraint loop. If only one horizontal segment per net is permitted and the terminals are fixed, this channel cannot be routed. The obvious solution is a dogleg in net A as shown in Figure 2C. Note that in this case, however, the channel density is two but the routing requires three tracks.

The motivation for introducing doglegs is thus twofold. First, doglegs enable us to break long constraint chains which otherwise would prevent us from routing at or near density (e.g., Figure 8A versus 8B). Secondly, they may permit us to route a channel that otherwise would be unroutable. The former is much more important, however, since routability can almost always be obtained by some minor adjustment in the polycell ordering prior to routing. The addition of doglegs has certain disadvantages, however. From an electrical standpoint, each dogleg adds one or two additional contacts and this increases the capacitance. From a topological standpoint, if a dogleg occurs at a position where the local density is already equal to the channel density, the required number of horizontal segments increases by one and the subsequent routing cannot be completed in channel density (Figure 2C is an example of this situation). This is one reason why it is worthwhile to minimize the span (number of terminal columns with local density equal to the channel density) since the likelihood of this event is very dependent on the span. It is thus desirable to add as few doglegs as possible. Any routing style that extends a net beyond its original endpoints or permits its horizontal sections on different tracks to overlap (see [19] for such examples) can only increase the apparent local density. While sometimes beneficial, the incidence of such situations did not seem to justify the coding effort required to exclude detrimental cases.

In Section VI.2.3, we will describe the new dogleg routing algorithm for a rectangular channel with all terminals located on a grid (the typical case). Section VI.2.4 shows how this algorithm can be trivially modified to handle the cases where terminals are not on a grid, where there are tolerance problems for adjacent contacts on the same track, etc. These modifications are included in our implementation of the dogleg router. The situation where some tracks are partially blocked (e.g., some cells extend into the channel) is also easily handled. Section VI.2.4 concludes with a formal description of the algorithm. Section VI.2.5 presents some experimental routing results and compares them with the results of other routing strategies.

VI.2.3. *The Dogleg Algorithm – Simplified*

In this section we will describe the new dogleg channel routing algorithm as simply as possible. We shall purposely defer some details that, necessary as they are, might confuse or complicate this description. For example, we assume here that the channel is a rectangle and that all terminals are on grid, although neither of these assumptions is actually necessary. Consideration of these details, along with a formal algorithm description, are given in Section VI.2.4. We will also present the motivation for various decisions that were crucial during the development of this algorithm.

The primary goal is to be able to route the typical channel using as few tracks as possible, and hopefully equal to the channel density. Since doglegs increase the apparent local density, and since the corresponding added contacts increase the capacitance (generally an undesirable result), the number of doglegs should be kept as small as possible, subject to the goal of track minimization.

By studying the output of non-dogleged routers, it was found that when they required many tracks in excess of the channel density, the usual cause was quite evident. There was a long constraint chain with a few crucial nets (typically clock lines) that were heavily connected to both sides of the channel. If these multi-terminal nets could be dogleged efficiently, a far more compact routing would be achieved. Based upon this observation, it was decided that doglegs would only be introduced at a terminal position for the net. This decision severely constrains the potential number of doglegs in that no 2-terminal net is ever dogleged, 3-terminal nets have only one potential site for doglegs, etc. Note that this rule is independent of horizontal net length. The number of added contacts is also minimized by introducing doglegs only at a position where the net already has a terminal (one added contact per dogleg as shown in Figure 8B instead of two as shown in Figure 9B).

An obvious procedure, therefore, is to order the terminals within a net based upon their abscissas, and decompose the original net into a series of 2-terminal subnets such that the n'th subnet consists of terminals n and $n+1$. In this decomposition, each interior terminal becomes the end of two subnets. Apply this procedure to all nets (a 2-terminal net yields a single subnet) and then use the constrained left-edge algorithm on the subnets with the following modification. When a subnet ends, the next subnet of the same net (if any) can be placed in the same track (i.e., they can overlap at the common terminal). Subnets from different nets are not permitted to overlap. This procedure works reasonably well but it often adds many more doglegs than are necessary. In order to minimize this undesirable result, a parameter called the "range" was introduced. It represents the minimum number of consecutive subnets that must be assigned to the current track. The above description thus corresponds to a range of one. As the range gets larger, fewer doglegs tend to be introduced. The only additional rule is that a sequence of subnets less than the range will also be accepted if and only if there is no unplaced subnet for this net that begins where the sequence ends (e.g., the actual net end is reached). Without such a rule, 2-terminal nets would never get placed. Our implementation of this routing algorithm permits a range from 1 to 9 for doglegs as well as the value N to indicate no doglegs.

Another feature, not related to doglegs, was included in this router. Rather than route all tracks starting from the top-left, a more symmetric choice of tracks was made. This is possible because each track is equivalent to the first track for some reduced problem. Therefore, we alternate between top and bottom tracks except for the situation where all terminals are on one side of the channel. This alternation between top and bottom usually produces routing with a smaller total vertical length (summed over all nets) than the traditional one-sided approach, reducing the average capacitance per net. In addition, for two adjacent tracks (both on top or both on bottom), if one starts from the left end of the channel, the other starts from the right. If we consider the example of Figure 7, starting with top-left we arrive at 7A. Any other start (bottom-left, top-right, or bottom-right) yields 7B. Since the first track can be any one of four possibilities and the second any one of two (opposite side of channel but can start at either end), there are eight possible sequences. The routing sequence is thus another parameter available to the user.

Since we have a two-parameter procedure with ten ranges and eight routing sequences, the program can try various combinations automatically. If the user specifies a particular range and sequence, the algorithm is employed for those values only. The user has, however, the option to go automatically through all ten ranges for a particular routing sequence, all eight routing sequences for a particular range, or all eighty possible combinations of range and routing sequence. The program, of course, saves the routing results and the parameter values that were best, based upon the smallest number of tracks first and the smallest number of doglegs (added contacts) second. A more formal description of the dogleg algorithm is given at the end of the next section.

VI.2.4. The General Dogleg Algorithm

Sometimes the terminals in a channel may not all be on grid. This situation can occur, for example, if polycells with one terminal spacing share a channel with PLA's or pads with a different terminal spacing. For certain technologies, there is often the restriction that adjacent terminals cannot have contacts on the same routing track unless they are connected to the same net. A constraint similar to this must always be satisfied when terminals on opposite sides of the channel are not on grid. We will, therefore, introduce a few additional definitions and examples before concluding with a formal description of the general algorithm. Since all terminals and paths have nonzero widths, the following discussion assumes the use of centerlines to represent the terminal and path positions.

Let us first define the minimal terminal spacing to be the minimum distance between two adjacent terminals on the same side of the channel. Consider the two vertical paths emanating from such a pair of terminals. In general, the routing would be some variant of that shown in Figure 10. Obviously, there is some

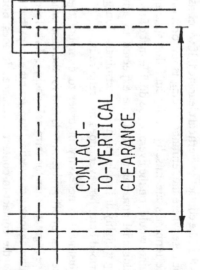

Figure 10. Definition of contact-to-vertical clearance

minimum required distance between the contact and the adjacent vertical segment based upon the technology being used. We call this distance the contact-to-vertical clearance. (If this minimum clearance was greater than the minimal terminal spacing, routing in the style considered here would be impossible.)

Another important clearance parameter is the minimum spacing between adjacent contacts on the same track as shown in Figure 11. If the contact width is

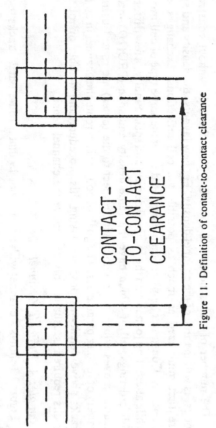

Figure 11. Definition of contact-to-contact clearance

greater than the vertical path width, the contact-to-contact clearance must be equal to or greater than the contact-to-vertical clearance. Note, however, that no relationship is expressed between the terminal spacing and the contact-to-contact clearance. The former can, in fact, be greater than, equal to, or less than the latter.

If the terminals are all on grid, the vertical constraint relationship as described previously is quite simple. At any terminal position, a vertical constraint exists if and only if both top and bottom terminals are connected to different nets. In the more general case where terminals are not on grid, a terminal on one side of the channel can generate a constraint relationship with two adjacent terminals on the opposite side. As illustrated in Figure 12, this situation arises whenever both XL

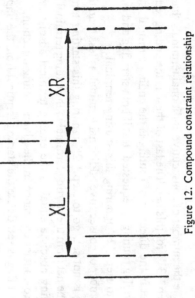

Figure 12. Compound constraint relationship

233

and XR are less than the contact-to-vertical clearance. If we consider the hypothetical but extreme case where the terminal spacing is uniform on both sides of the channel but the terminals are misaligned by half this spacing, the number of vertical constraints would, on the average, be double that of the normally aligned case. These additional constraints would be expected to adversely affect the routing, making it highly desirable to have the terminals "on grid." The routing algorithm described here, however, does not require it.

A minor adjustment is necessary in order to compute the channel density precisely for the case of off-grid terminals and for the case where the contact-to-contact clearance is greater than the terminal spacing. As before, simply apply the left-edge algorithm ignoring the vertical constraints except that each net must first be extended by the contact-to-contact clearance. It does not matter how this extension is accomplished so long as it is done consistently for all nets (e.g., add this amount to the right end of all nets). A vertical net (left end abscissa equal to right end abscissa) should be ignored entirely since it has no horizontal section and no contacts.

In the following algorithm description, only a single track is actually being routed. When that track is finished, the nets or net segments that are still unplaced constitute a reduced problem. The algorithm terminates when the track being routed remains empty. If there are no unplaced nets, the routing is complete. If, however, some unplaced nets remain, this is proof that one or more constraint loops exist. It is usually a simple matter to break such loops by means of cell reflections or exchanges.

We will assume that the netlist data has been formatted such that for each net the terminals are ordered by ascending abscissas with the channel side (top or bottom) indicated. There may be many nets whose left and/or right end is the respective channel edge. By sorting and cross-referencing various data lists before starting the routing calculations, many of the searches described below become unnecessary or trivial.

THE GENERAL DOGLEG ALGORITHM

1. Initialize the channel side (top or bottom), the starting abscissa (left or right edge of channel) and the routing d rection (right, if we start from the left edge, and vice versa) based upon the type of track being routed.

2. Search for a net terminal equal to the starting abscissa or closest to it when going in the routing direction. This is the starting terminal. If none, the track is done. If this terminal is the actual start of a net (not an interior terminal), set the start flag. Otherwise, clear the start flag.

3. If the terminal is on the opposite side of the channel from the routing track, a search must be made to see if it is constrained by some other as yet unplaced net (i.e., is there some terminal on the routing side within the contact-to-vertical clearance that is not yet routed?). If so, the net is blocked. Go to step "2" ignoring this starting terminal. If the terminal is not not blocked, clear the end flag.

4. Determine the number of consecutive additional terminals that are not blocked (as in step "3") for this net. The last of these is the ending terminal. If the end of the net in the direction of routing is the ending terminal, set the end flag.

5. If non-dogleg routing is requested, test if both start flag and end flag are set. If not, go to step "2" ignoring this starting terminal.

6. If the number of consecutive additional terminals is less than the range and the end flag is not set, go to step "2" ignoring this starting terminal.

7. Assign the net from the starting terminal to the ending terminal to this track. Set the starting abscissa equal to the ending terminal abscissa modified by the contact-to-contact clearance in the given routing direction.

8. If the start flag is set, delete the starting terminal from the unplaced net list. If the end flag is set, delete the ending terminal from the unplaced net list. Delete all interior terminals (if any) in this net between the starting and ending terminals. If some middle part of a net is thus deleted, the remaining end pieces constitute two distinct nets for the remainder of the algorithm. Go to step "2".

The case of a non-rectangular channel is relatively easy to handle. Typically, this situation occurs when some cell(s) protrudes into the routing channel. Instead of having a routing track which extends for the entire channel length, the track consists of two or more distinct segments. For each such segment, employ the above algorithm except that in step "1" the initialization should be to the appropriate segment end and in steps "2-4" the terminals must be tested to ensure that they lie within the length of the track segment. A terminal that does not lie within this range is automatically blocked.

VI.2.5. Specific Channel Routing Results

The number of tracks needed to route a typical channel, without constraint loops, is within three of the channel density. For new layouts, the density and span are first minimized as much as possible via cell reflections, exchanges and insertions. Constraint loops are then eliminated using the same type of cell moves while attempting to maintain the minimal density and span. Finally, a few different cell orderings, all of which have the same density and span (PRO(1D) makes neutral moves which are random with respect to the router) are routed using a variety of routing parameters. The goal, of course, is to route the channel in as few tracks above density as possible. Using this procedure, the number of routing tracks required is usually only 5% above the channel density.

Example: A 'Difficult' Channel

Figure 13 shows a channel with a density of 19 that is routed in 21 tracks using the dogleg router. An attempt was made to find the optimal routing without

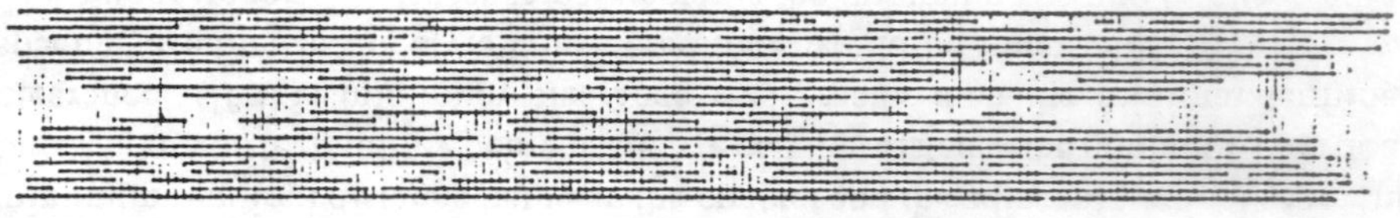

Figure 13. A "difficult" channel

doglegs using the program of Reference [15]. The program was aborted after four hours (the branch and bound algorithm did not finish) with the best routing obtained during these computations requiring 26 tracks. Use of the standard left-edge algorithm required 39 tracks to complete this routing problem.

After all of the routing is completed, the user is in a position to make modifications and edits while still maintaining the integrity of the connections. The next section describes the routing edit features available on LTX.

VI.3. Routing Edits

In LTX there are three distinct types of routing edits based upon the routing procedures described previously. The first of these refers to the terminal equivalences that are available on the polycells. The equivalence and feed-through module automatically utilizes the first useful equivalence pair that it finds for a net in order to cross a polycell row. The equivalence editor permits the user to preconnect certain electrically equivalent terminals to ensure their use. The Inter-channel Feed-Through program (see Section VI.1.2) recognizes these preconnections.

The equivalence editor can also be used to exchange nets between logically equivalent terminals. This may enable the user to uncross nets in a particularly critical area or perhaps decrease the length of some crucial net path. The equivalence editor rejects all operations that would destroy net integrity. After editing, the user must go back and rerun the entire routing package starting with the net decomposition phase.

The next routing edit is specifically for the Type (A) nets in Section VI.1.3 above, which could have been assigned to either of two vertical channels. The user can reverse the assignment of any such net simply by invoking the editor and interactively selecting the net. Only nets of Type (A) can be selected (all other requests will be ignored). It is subsequently necessary to run the channel router on all domains which have pin-outs connected to such edited nets (since the pin-out sides have changed), and to reroute the vertical channels.

The final type of routing edit applies to the actual routing within a channel (horizontal or vertical). The user can create doglegs anywhere along a net trunk, can move a trunk to another track which is vacant, can exchange trunks, etc. The motivation for such edits might be to save a routing track in a channel that was not routed in density, or perhaps to decrease the capacitance of some critical paths. All such edits are performed by the editor if and only if they do not violate any design rule constraint (see Section VI.2.4).

The LTX routing package can thus be used with or without interactive edits on the part of the user. Integrity of the net connectivity is guaranteed.

VII. EXAMPLES

systems allow some user interaction, or have steps which are bridged manually, the few published examples which are available usually include an unknown amount of user skill.

Any attempt to compare individual algorithms against published material faces the difficult task of finding examples which include practical constraints (e.g., keeping 24-pin DIPs out of 16-pin sockets). Usually, the relevant material is extracted from a larger problem, and a great deal of supporting explanatory material is necessary. Individual comparisons have been made for the router [7], and some of the one- and two-dimensional placement techniques [5,6]; these will not be repeated here. New strategies, not previously presented, involve the row-leveling portion of the 2-D placement algorithms and the 1-D iterative single cell insertion strategy.

The example given here will serve to: 1) present a reasonable sequence of use for the various algorithmic procedures presented in Sections V and VI; 2) offer an example of the relative performance of these different procedures in the chosen sequence; 3) indicate typical running times; and 4) indicate the added performance achieved by subsequent combined interactive and automatic procedures.

The circuit in the example has 264 cells, 337 nets, and a total of 1623 cell grids. The cells are arranged in 8 rows with the 6 center rows arranged in 3 pairs of facing ("front-to-front") rows; the remaining rows face towards the center. These

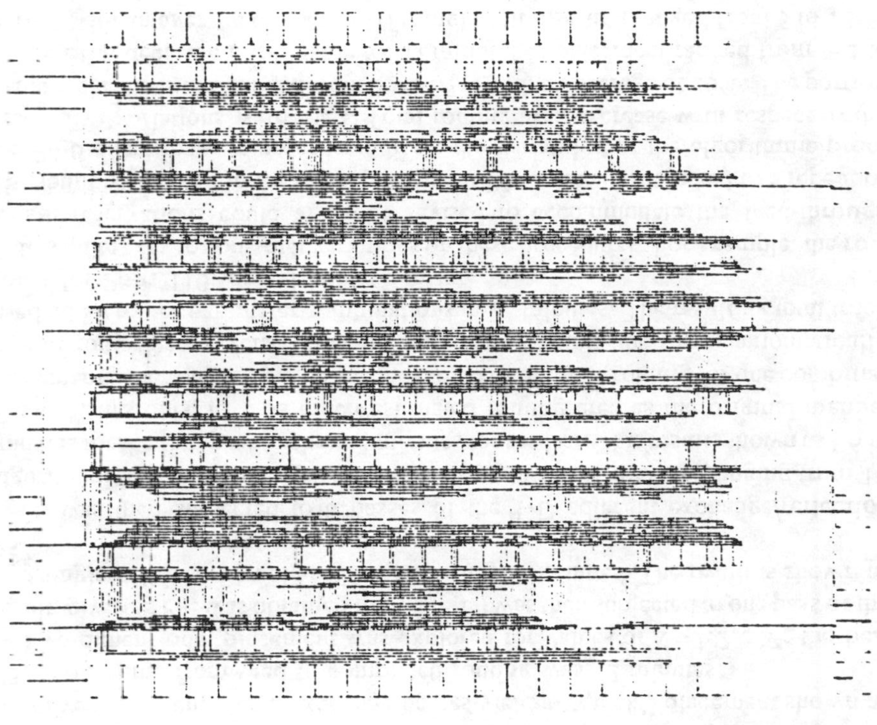

Figure 14. A chip layout with manual cell placement. Total channel density = 146. maximum row length = 233 grids.

VII.1. Discussion

Overall layout performance comparisons are quite difficult to make for a variety of reasons. First of all, relatively few systems like LTX exist. Since most of these

choices were made manually. (The "front" edge of a cell has, in this library, effectively all of a cell's terminals. An average of 2/3 of the terminals are also available through electrical equivalences on the "back" edge.)

The basis of our comparison is a completely manual cell placement (see Fig. 14). In the sequence presented below we will be concerned with estimated routing area (ERA — square grids), which we approximate as the maximum row length (MRL — grids), including feed-through cells, times the sum of the horizontal channel densities (D — grids). Channel density is used instead of the actual routing track count for convenience and consistency. Since the router routinely achieves routing within about 5% of density, density is quite a good approximation.

Statistics for the manual placement are shown in Table 1 as placement P0.

For the 2-D placement procedures, each row was temporarily divided into 5 blocks spaced 800 units apart, center-to-center, horizontally. The "front-to-front" row pairs were also spaced 800 units vertically. "Back-to-back" row pairs were spaced 1600 units vertically with 1200 units for "front-to-back" pairs. Pad blocks were all spaced 1200 units horizontally or vertically from the nearest cell block. These spacings were chosen manually to represent the relative costs of routing the various cell placement combinations. Block lengths were approximately equal. Total block length for each row was initially uniform at 4906 units (223 grids). The grand total of block lengths was 10% more than the total cell grids, to avoid "box-filling" problems. The block leveling factor, beta (see Section V.2.3), was taken as $H/h = 10400/22 = 473$.

VII.2. *Procedural Sequence*

In the attempt to provide an example relatively free of the designer's skill, a completely mechanical sequence is presented first. Generally the designer reviews the design at each stage and, if the results are poor, he will recycle back to some previous stage, interactively making crucial changes. While this manual interaction has been omitted from the mechanical sequence, we wish to re-emphasize its importance in achieving highest quality results.

A 40-cell seed placement was chosen manually. Constructive placement of the remaining cells took 40 seconds. Each pairwise exchange pass took 3.5 minutes. Constructive placement followed by two exchange passes (shown as placement P1 in Table 1) quickly achieved a result comparable to the original manual result P0.

Two-dimensional placement was explored for increasing construct/exchange interlace frequencies, K. Values of $K=1,2,...,7$. each followed by exhaustive pairwise exchange, were calculated in 2.3 hours of automatic, unattended operation. The best results (see P2, Table 1) were obtained for $K=2$, after only 22 minutes.

During the 2-D placement phase, feed-through cells do not exist. It is necessary to "reserve" some of the row length for the expected distribution of feed-through cells. Since feed-throughs tend to occur more heavily in the middle rows, placement P2, which made no reservations, had a longer middle row than the original layout P0.

The block lengths were revised to accommodate the feed-through distribution encountered in P2. Starting over with the original 40-cell seed placement, constructive placement plus 2 exchange passes yield a "quick" placement shown as P3. (Total time from seed to completed routing was 12 minutes.)

Two-dimensional placement was explored for values of $K=1,2,...7$: the best result, given as P4, was obtained for $K=5$. P4 was then subjected to one pass of the 1-D single-cell insertion algorithm, requiring 22 minutes. The result is shown as P5.

P5 was then subjected to 8 passes of the 1-D adjacent exchange/reflection algorithm PRO(1D). The 8 passes required a total of 40 minutes, ranging from 11 minutes for the first pass to about 2 minutes for the last. This result, shown as P6 in Table 1, has only 73% as much estimated routing area as the original manual placement. The router takes 21 seconds to do a complete routing for one combination of router parameters. To achieve the best routing, the router automatically tried all 80 router parameter combinations in 22 minutes. The resulting total track count for P6 is 110, only 1 above density.

It is quite easy to construct variations on these procedures. For example, the row length reservations could again be revised to accommodate the feed-through distribution obtained in the quick placement P3, and then subjected to a repeat of the P4-6 operations. In order to indicate the variability of the algorithmic processes, five variations were tried. Total routing area decrease with respect to the manual placement P0 ranged from 25 to 27% with an average of 26%. The portion of the ERA decrease attributable to: 1) initial 2-D placement ranged from -2 to $+16\%$ with an average of 6%; 2) 1-D single cell insertion ranged from 5 to 12%, with an average of 9%; 3) 1-D adjacent exchange/reflection ranged from 7 to 14% with an average of 12%.

As noted above, manual review is normally interjected at each stage. The following work involving a combination of interactive and automatic operations on the results of the mechanical sequence indicates the kind of additional performance which is obtained by permitting manual intervention.

Starting from P6, interactive interrow cell moves (as described in Section IV.3) were used to decrease the longest row from 227 to 214 grids. Then a combination of single cell insertion passes, adjacent exchange/reflection passes, feed-through re-dispersals, interactive moves, and equivalent terminal editing was used by a skilled operator during sessions totaling five hours at the machine. The final result is given in Table 1 as P7, and corresponds to an estimated routing area reduction of 38% with respect to the original manual placement. Result P7 with a density of 98 was then routed. A layout requiring 108 routing tracks was produced using a default combination of router parameters. The router was then run using all 80 possible parameter combinations for each channel. This run resulted in the 102 track layout (4 above density) shown in Figure 15. Additional interactive and algorithmic work would undoubtedly further decrease the layout size.

For the convenience of persons wishing to compare the results of other placement and routing procedures with those presented here, the input data defining this example are contained in an Appendix available from the authors.

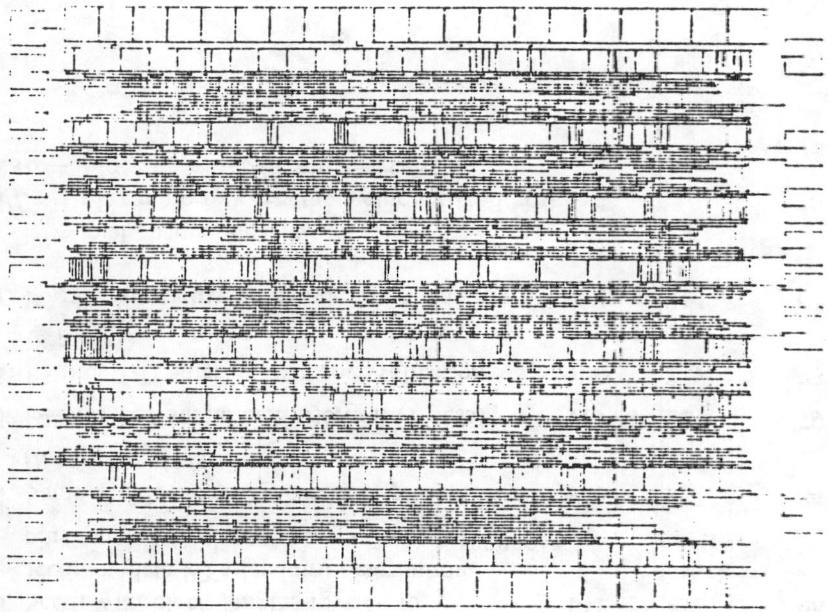

Figure 16 shows a somewhat larger chip designed with LTX. It contains 511 cells and a total of 2589 transistors, with the routing occupying 61% of the active area. From initial data input it took eight working days to achieve this layout, including time lost when a logic change had to be accommodated during the design cycle.

VIII. CONCLUSION

In assessing a CAD system, it must be measured against the available alternatives. For LTX these alternatives have been its predecessor, GRAFOS [3], and interactive drafting programs. Typically, LSI "hand layouts" using only interactive drafting aids have required close to a man-year of effort, while polycell layouts done with GRAFOS, which does offer some polycell-oriented features, have taken about three or four man-months. LTX has shortened LSI layout times to a week or two at most, including ample allowance to achieve compact layouts, while its automated handling of connectivity has greatly reduced the chances of error. Packing densities generally greater than those of non-LTX polycell layouts have been obtained. The reentrant features of the system enhance the short layout time in permitting circuit logic changes to be accommodated rapidly and gracefully.

Acknowledgments

LTX is the work of many hands. Output programs have been contributed by L. M. Finne, H. K. Gummel, and B. R. Chawla. S. J. Kent has contributed to the programming effort and W. E. Carter has supplied technical support. Through their trials and tribulations, the initial users of LTX, in particular E. J. Fulcomer, J. Sosniak, and R. L. Mortenson, have furthered the improvement of the system. The manuscript has benefited from the helpful comments of H. K. Gummel.

REFERENCES

1. H. Y. Chang, G. W. Smith, Jr. and R. B. Walford. "LAMP: System Description." B.S.T.J. 53. (1974). pp. 1431-1449.
2. B. R. Fowler. "XYMASK." Bell Laboratories Record 47. (July 1969), pp. 204.
3. G. Persky and H. K. Gummel. "GRAFOS — A Symbolic Routing Language." Proc. of the 10th Design Automation Workshop. (1973). pp. 173-181.
4. G. Persky, D. N. Deutsch. and D. G. Schweikert, "LTX — A System for the Directed Automatic Design of LSI Circuits". Proc. 13th Design Automation Conf. San Francisco (1976). 399-407.
5. D. G. Schweikert. "PLAC — A Partioning-Placement Program for Modular IC Layout." Proc. 13th Design Automation Conf.. San Francisco (1976). 408-416.
6. G. Persky. "PRO — An Automatic String Placement Program for Polycell Layout." Proc. 13th Design Automation Conf., San Francisco (1976). 417-424.
7. D. N. Deutsch. "A Dogleg Channel Router." Proc. 13th Design Automation Conf.. San Francisco (1976). 425-433.
8. P. Kozak, H. K. Gummel, and B. R. Chawla. "Operational Features of an MOS Timing Simulator." Proceedings of the 12th Design Automation Conference. (1975). pp. 95-101.
9. G. W. Smith, Jr., "Net-Span Minimization: An N-Dimensional Placement Optimization Criteria". internal BTL memorandum. (1 Nov 1972).
10. M. Hanan. "On Steiner's Problem with Rectilinear Distance". SIAM J.Appl.Math. vol. 14 (1966). 255-265.
11. D. G. Schweikert and B. W. Kernighan. "A Proper Model for the Partitioning of Electrical Circuits". Proc. 9th Design Automation Workshop. Dallas (1972). 57-62.

Figure 15. Circuit of Fig. 14 after automatic and interactive LTX placement procedures. Total channel density = 98. maximum row length = 214 grids. This figure is shown at an expanded scale relative to Fig. 14.

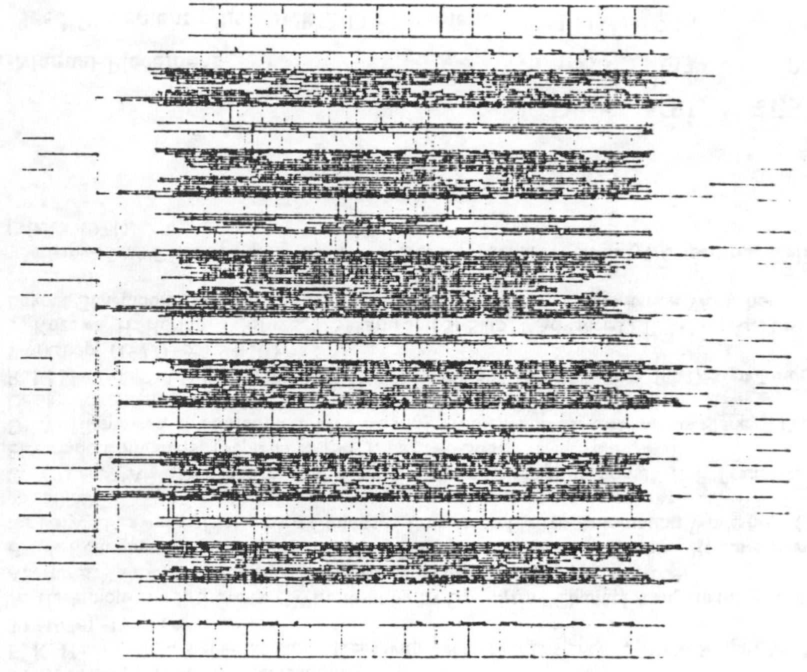

Figure 16. LTX layout of a chip having 511 cells.

12. A. G. Fraser, private communication.
13. F. K. Hwang, "On Steiner Minimal Trees with Rectilinear Distance", SIAM J. Appl. Math, vol. 30 (1976), 104-114.
14. A. Hashimoto and J. Stevens, "Wire Routing by Optimizing Channel Assignment within Large Apertures," Proc. 8th Design Automation Workshop. (1971), pp. 155-169.
15. B. W. Kernighan, D. G. Schweikert and G. Persky, "An Optimum Channel-Routing Algorithm for Polycell Layouts of Integrated Circuits," Proc. 10th Design Automation Workshop. (1973), pp. 50-59.
16. C. Y. Lee. "An Algorithm for Path Connections and Its Applications," IRE Transactions on Electronic Computers. (September 1961), pp. 346-365.
17. D. W. Hightower, "A Solution to Line-Routing Problems on the Continuous Plane." Proc. 6th Design Automation Workshop. (1969), pp. 1-24.
18. R. L. Mattison, "A High Quality, Low Cost Router for MOS/LSI." Proc. 9th Design Automation Workshop. (1972), pp. 94-103.
19. T. Kozawa, H. Horino, T. Ishiga, J. Sakemi and S. Sato. "Advanced LILAC — An Automated Layout Generation System for MOS/LSIs." Proc. 11th Design Automation Workshop. (1974), pp. 26-46.
20. "Moving-Head Disc Operating System." Hewlett-Packard Company, Cupertino, California (March 1971).

TABLE 1

	D→	MRL	%ERAo
P0. Manual Placement	146	233	100
P1. Seed Placement plus quick 2-D placement (K=1, Lp=497200)	144	235	99
P2. Seed plus best 2-D placement (K=2, Lp=472400)	141	236	98
P3. Seed, with block lengths revised to accommodate P2's feedthru distribution, plus quick 2-D placement (K=2, Lp=488800)	141	232	96
P4. Same as P3, but using best 2-D placement (K=5, Lp=444400)	126	227	84
P5. P4 plus 1-D single cell insertion (1 pass)	119	227	79
P6. P5 plus 1-D adjacent exchange/reflection (8 passes)	109	227	73
P7. P6 plus interactive/automatic optimization	98	214	62

D = total channel density

MRL = Maximum row length, including feedthrus

$ERAo$ = $146*233$ = 34018 square grids

THE BERKELEY BUILDING-BLOCK(BBL) LAYOUT SYSTEM FOR VLSI DESIGN

N. P. Chen, C. P. Hsu and E. S. Kuh

Department of Electrical Engineering and Computer Sciences
and the Electronics Research Laboratory
University of California, Berkeley, California 94720, USA

Automatic layout design of custom VLSI circuits depends on the building-block hierarchical approach in which macrocells of arbitrary shapes and sizes are given together with the net list. Our aim is to design an intelligent and practical automatic layout system which will interface with other design aids at Berkeley. The BBL system has a general purpose database, a smart global router which can dynamically adjust placement and efficient detailed routers, namely: the channel router and the switch-box router. The System incorporates several novel ideas and is based on a number of graph-theoretical algorithms. Experimental results indicate that the System is extremely effective.

1. INTRODUCTION

The present status of automatic layout design for integrated circuits is limited to the gate array and the polycell approach. Although papers have been written describing automatic hierarchical, building-block layout [1-10], to the best knowledge of the authors no such system has been adopted for production use. One main reason is that in comparison with manual design the chip size is anywhere between 30% and 50% larger. In our view the inefficiency of each existing method is due to one or more of the following reasons: (1) poor placement and no provision for placement adjustment, (2) crude routing algorithms, and (3) in the case where channel routing is used, it relies on channel routing exclusively and thus faces the difficult problem of cyclic constraints. Our present approach is sufficiently different in philosophy from others, and we believe that our System will compete favorably with manual design in terms of area usage.

The BBL System is a hierarchical, automatic layout system for IC design. It allows component modules with rectilinear boundary of arbitrary size as building blocks. A software configuration of the BBL System is shown in Fig. 1. The programs of the System are classified into the following categories: input/output processors, placement programs, global routing programs, detailed routing programs, interactive editors, and display utilities. The database is constructed in such a way that different placements, feedthrough assignments and routing algorithms can be implemented and tested by using the BBL System. The users can input the initial configuration either by a text file or by using the interactive graphics editors developed at the University of California, Berkeley.

A constructive placement program is under development and is not reported in this paper. The automatic routing system takes relative placement and the signal netlist as input. The signal netlist specifies all terminals to be connected. The present paper focuses on the global routing strategy and its interface with two detailed routers, namely: the channel router and the switch-box router

[11,12]. A key feature of our approach is the built-in automatic placement adjustment scheme which moves modules (blocks) whenever necessary to facilitate 100% interconnection.

In Section 2 we will briefly discuss the database. Section 3 presents the BBL routing system which includes prerouting analysis, global routing and detailed routing. Section 4 gives a brief discussion of power and ground routing. We conclude the paper with some experimental results and suggestions for future work.

2. THE DATABASE

A general purpose database is used in the BBL System. It is based on a hierarchical structure and serves as both the input and the output for any placement and routing programs. Logical signal information and geometrical layout descriptions of the entire chip are represented hierarchically according to the user specified hierarchy.

The geometrical information is represented by modules. Each module may be a functional block or a circuit component. The boundary of a module can be any rectilinear polygon, and a module may have any number of different boundaries on different layers (sometimes called protection frames [13]). Different modules may have the same geometry which is represented by the same geometric structure. A cell library may be created for different geometric structures. Two lists of terminals are associated with a module, one for logical terminals and the other for physical terminals. Electrical equivalence and logical equivalence information of terminals are properly represented. Every net is hierarchically decomposed into a set of signals, where every signal represents the net information on one level of the hierarchy. A sophisticated structure is used for routing. It has the ability to represent any routing results based on straight line connections.

The communication between the BBL system and other software tools developed at the University of California, Berkeley is established through this database.

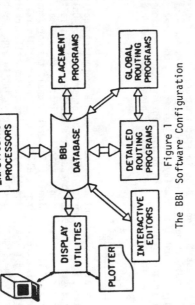

Figure 1
The BBL Software Configuration

Reprinted with permission from *Proc. Int. Conf. on Very Large Scale Integration*, Edited by F. Anceau and E. J. Aas, 1983, pp. 37-44, North-Holland, Amsterdam.

3. THE BBL ROUTING SYSTEM

The flow chart of the routing system is shown in Fig. 2. The main strategy is to assign a minimal amount of space for routing but to provide means for block shifting when additional space is called for. This entails a critical look at the cause of wiring conjestion, and has led to the principal concept of "Bottleneck." Figure 3 illustrates the meaning of the bottleneck. There are two horizontal bottlenecks, bottleneck I between block A and block C and bottleneck II between block C and block B. We call a routing region a horizontal bottleneck if (1) there exists a horizontal line in the region, which intersects the vertical edges of the neighboring blocks, and (2) none of such horizontal lines between the two blocks intersects with any other vertical edges. We define a vertical bottleneck similarly.

The complete set of bottlenecks gives the critical regions of the routing plane where congestion of routing is most likely to occur and hence needs to be carefully controlled. It also serves as a link between blocks whereby all information is easily updated when some blocks must move to a new position. It should be pointed out bottlenecks could disappear or could be generated as placement changes. Bottlenecks thus become a data structure which store current information on routing and geometry of blocks. In Fig 4 we give an illustration

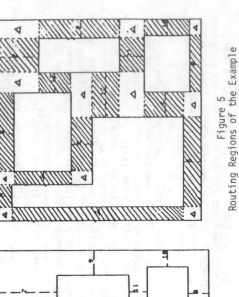
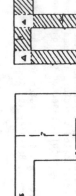

Figure 5
Routing Regions of the Example
W: Channel Router Routing Region,
△: Switch-box Router Routing Region.

Figure 4
Four-Module Example with Seven
Horizontal Bottlenecks and Seven
Vertical Bottlenecks

of a chip which contains four blocks. Horizontal and vertical bottlenecks are marked by dash lines. In the figure aa'-b'b depicts the active region of bottleneck 3.

Routing is divided into two steps, global routing and detailed routing. In global routing we assign signal nets to specific regions. No over-the-block routing is allowed. In detailed routing we use two layers and assign nets to tracks, usually horizontal on one layer and vertical on the other. Our strategy is to use the channel router for the active regions of the bottlenecks and the switch-box router for the remaining regions. This is illustrated in Fig. 5.

3.1. Prerouting Analysis

From a given relative placement we first perform a prerouting analysis which consists of two passes, one horizontal and the other vertical. Only the horizontal analysis is discussed because the vertical analysis is the same except for the direction. First, a signal span is defined as the minimal rectangular region containing all its terminals. For each signal, the horizontal prerouting analysis starts with the finding of all the bottlenecks within its signal span. The probability of a signal going through a particular bottleneck is calculated by assuming that the signal has equal chance of going through any bottleneck within the signal span. Finally, we sum up the probability over all signal nets to obtain the expected number of tracks for each bottleneck. Based on this information the preliminary placement adjustment is made, which allocates routing space between blocks. This new placement is then used as the initial placement for global routing. The System will work on absolute coordinates from this point on. A complete set of bottlenecks is created for this placement and maintained throughout the global routing and the detailed routing processes to follow.

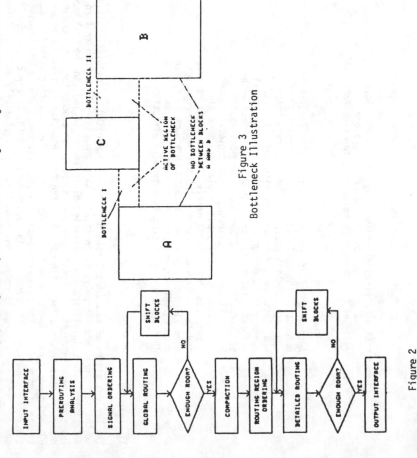

Figure 3
Bottleneck Illustration

Figure 2
BBL Routing System Flowchart

3.2. Global Routing

The routing order is determined according to the routing space available for each signal span. The smallest one is assigned first. The idea behind this is that it is more difficult to route signals with smaller available routing space because they will more likely cause congestion. The routing space pertaining to a given signal is approximated by the sum of available tracks of all bottlenecks within the signal span.

A "global routing graph" is defined as follows: A node corresponds to a routing region between two or more neighboring bottlenecks. An edge represents a bottleneck which connects two nodes if and only if the two nodes represent two neighboring regions of the bottleneck. The global routing graph for the example is shown in Fig. 6. Each edge is given a weight which reflects the congestion of the associated bottleneck and the length of its active region. The formula used for the edge weight is

$$W_i = aL_i + \frac{b}{N_i{}^c}$$

where N_i is the number of the available tracks of the bottleneck i and L_i denotes the length of its active region; a, b and c are parameters to be chosen by the designers. The edge weight L_i is updated each time a signal is routed. At the beginning, the length factor L_i is dominant because there are many available tracks in each bottleneck. As global routing proceeds, the available number of tracks becomes less and the congestion factor N_i will be more important. Thus our aim is to find the minimum weight Steiner tree for each signal net. A newly developed "Steiner Tree on Graphs" (STOG) algorithm is used [14]. Because the problem is NP-complete, STOG is a heuristic algorithm based on a 3-point "Steiner Tree on Graphs" algorithm. This algorithm has a reasonable computational complexity and yields very good results from our experimental tests

Near the end of global routing, i.e., as bottlenecks begin to fill up, we need to check whether there is enough room for each bottleneck. Blocks must be shifted to allow extra room when the number of tracks needed exceeds that allocated by the prerouting analysis. In shifting blocks, an attempt is made to minimize the increase of total chip area. At the conclusion of global routing, a compaction is performed both horizontally and vertically to get rid of excess space not used. We are now ready for detailed routing.

3.3. Detailed Routing

As mentioned earlier we divide up the routing regions into two types: the channel routing regions and the switch-box routing regions. (See Fig. 5.) We have two highly efficient routers: the channel router [11] and the switch-box router [12]. The channel router almost always completes the routing with the tracks allocated from global routing. The channel router can handle a rectangular region with fixed terminals on the two edges of the channel and floating terminals on the remaining two sides. After channel routing, floating terminals become fixed in position. The switch-box router can handle an arbitrary rectilinear region with fixed or floating terminals on all sides. However, we cannot predict ahead of time whether the space allocated is always enough to complete the switch-box routing. Therefore it is necessary to have the flexibility of shifting blocks whenever needed. Based on these considerations, we follow the following rules in deciding the routing order of regions: (1) To route the channels before the switch boxes for those bottleneck regions which border the switch-box region in question. This is needed to fix the terminals on the boundary of the switch-box region. (2) To start with regions in the lower left corner of the chip and to proceed to the upper right. (3) To always shift blocks upward and to the right in order to guarantee that regions in the lower left direction already routed will not be disturbed.

Further improvements have been made to interface the two modes of detailed routing. The most significant one is to enlarge the routing space assigned to the switch box wherever possible. When the active region of a bottleneck contains no terminals near its two ends, the routing space assigned to the channel can be pushed toward the center until a terminal is met. This has no effect on channel routing but results in noticeable relief of those congested switch boxes.

4. POWER AND GROUND ROUTING

The power and ground (P&G) nets will be allowed different wire width according to power consumption information provided by the user. P&G nets have priority to use the metal layer which can provide higher current densities and lower resistance. BBL handles P&G routing in a similar way as signal routing but with some modifications in both the global and the detailed routing.

In global routing, P&G nets will be given higher priorities than signal nets so that they lead to minimum length global routes. Also the edge weight for the bottleneck which contains a P&G net is assigned a larger number so that other P&G nets will not go through the same bottleneck thus minimizing the crossings of P&G nets.

The calculation of wire width is done by means of the tree representation of a P&G net. A bottom-up procedure has been implemented with the width information stored in the bottlenecks and the P&G terminals.

In channel routing, a preprocessor is used to route P&G nets on metal near the edges of the channel. Channel routers are then modified to exclude the P&G routing regions. A channel router which can handle irregular boundaries is used for signal net routing. A postprocessor will put every signal net which crosses a P&G net in a tunnel.

In the switch-box routing, P&G nets are routed first on metal. Protection frames are put around them according to their needed widths. Signal nets are then routed to avoid the protected area. A postprocessor is used to resolve crossings by tunnelling.

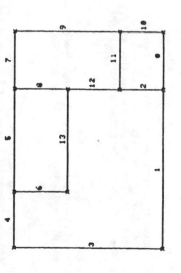

Figure 6

Global Routing Graph of the Example. Edge Number Corresponds to Bottleneck number.

5. CONCLUSION AND FUTURE WORK

The System is implemented in C language on a VAX 11/780 under the Berkeley UNIX operating system. An example from AMI with 33 functional blocks, 132 nets and 440 pins has been tested. It takes 5.5 minutes cpu time and 2.5 megabytes memory. The chip size is about 8.5% less than the original AMI design. Our result is shown in Fig. 7.

Future work to be completed includes a constructive placement algorithm, further refinements on detailed routing, interactive routing, feedthrough assignment, electrical equivalent pins and automatic hierarchy generation.

6. ACKNOWLEDGEMENT

Research sponsored by the National Science Foundation Grant ECS-8201580 and the Air Force Office Office of Scientific Research (AFSC) United States Air Force Contract F49620-79-C-0178.

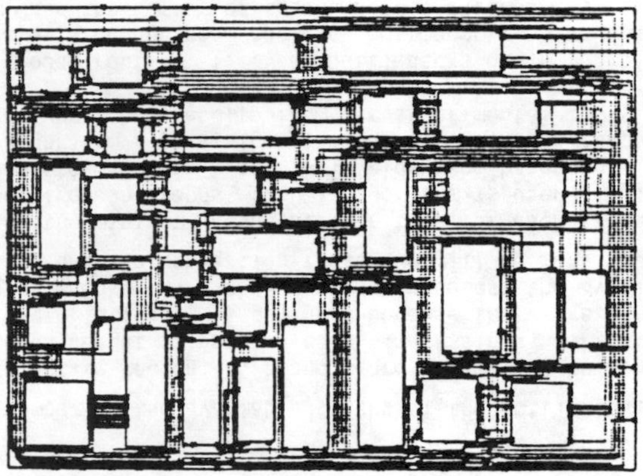

Figure 7
AMI Example with 33 Blocks, 132 Nets and 440 Pins.

REFERENCES

[1] Kani, K.,Kawanishi, H. and Kishimoto, A., ROBIN: A building block LSI routing program, Proc. IEEE Int. Symp. on Circuits and Systems, (1976) 658-661.

[2] Preas, B. T. and Gwyn, C. W., Methods for hierarchical automatic layout of custom LSI circuit masks (1978) 206-212.

[3] Preas, B. T., Placement and routing algorithms for hierarchical integrated circuit layout, Technical Report 180, Computer System Laboratory, Stanford University (1979).

[4] Lauther, U., A min-cut placement algorithm for general cell assemblies based on graph representation, Proc. 16th Design Automation Conference, (1979) 1-10.

[5] Soukup, J. and Royle, J., Cellmap representation for hierarchical layout, Proc. 17th Design Automation Conference, (1980) 591-594.

[6] Soukup, J., Circuit layout, Proc. of the IEEE, 69 (1981) 1281-1304.

[7] Wiesel, M. and Mylnski, D. A., An efficient channel model for building block LSI, Proc. IEEE Int. Symp. on Circuits and Systems, (1981) 118-121.

[8] Horng, C. S. and Lie, M., An automatic/interative layout planning system for arbitrarily-sized rectangular building blocks, Proc. 18th Design Automation Conference (1981) 293-300.

[9] Colbry, B. W. and Soukup, J., Layout aspect of the VLSI microprocessor design, Proc. IEEE Int. Symp. on Circuits and Systems (1982) 1214-1228.

[10] Riverst, Ronald L., The 'PI' (Placement and Interconnect) system, Proc. 19th Design Automation Conference (1982) 475-481.

[11] Yoshimura, Y. and Kuh, E. S., Efficient algorithms for channel routing, IEEE Trans. on Computer-Aided Design of Integrated Circuits and Systems, CAD-1, (1982) 25-35.

[12] Hsu, C. P., A new two-dimensional routing algorithm, Proc. 19th Design Automation Conference (1982) 46-50.

[13] Keller, K. Y., Newton, A. R., and Ellis, S., A symbolic design system for integrated circuits, Proc. 19th Design Automation Conference (1982) 460-466.

[14] Chen, N. P., New algorithms for Steiner tree on graphs, Proc. IEEE Int. Symp. on Circuits and Systems (1983) 1217-1219.

Part VI
Module Generation

Computer-Aided Layout
of LSI Circuit Building-Blocks†

Min-Yu Hsueh and Donald O. Pederson

Dept. of Electrical Engineering and Computer Sciences
University of California, Berkeley, CA 94720

Abstract

This paper describes the organization of and the compaction algorithm used in the program CABBAGE. CABBAGE can generate a compact actual layout from layout schematics, which are similar to circuit schematic diagrams, but with improved topology for layout purposes. The compacted layout can be used in the hierarchical build-up of a complete LSI circuit layout. The compaction operation is based on finding the longest path through a graph which represents the size and spacing requirements of the features in the layout schematic diagram. The program is written in FORTRAN and is implemented on a minicomputer. Circuits of the complexity of a T flip-flop require approximately 6 seconds to compact. The maximum circuit size that can be handled by the program is limited only by the data storage available on the computer.

1. Introduction

Building-blocks of an LSI digital circuit are logical partitions of an overall chip function; they can vary in size and complexity from that of a major function, such as a complete control section, down to a logic element level, such as gates. At present most of the building blocks implemented in random logic form are laid-out manually or with interactive graphics tools, such as commercially-available graphics processing systems. With either of these layout methods, most layout designers find it advantageous to convert the circuit schematic diagram into a layout schematic diagram which has an improved placement of elements and interconnection lines and a reduced number of crossovers before generating a compact actual layout meeting all design-rule specifications. An example of a layout schematic diagram is shown in Fig. 1. In other words, the layout operation is carried out in two steps; the *topology* of the layout is optimized before the exact *geometrical* requirements are imposed.

While most designers can generate with ease efficient layout schematics, the actual layout process is a tedious and error-prone task that few enjoy. This paper describes the organization and the compaction algorithm of the computer program CABBAGE (Computer-Aided Building-Block Artwork GEnerator), which generates correct and compact layout drawings from layout schematic diagrams input by the user.

2. An Overview of CABBAGE

CABBAGE consists of an interactive graphics input processor and a compactor. The graphics input processor allows the user to draw layout schematics interactively on a refresh CRT terminal. It supports a set of basic elements, such as different line types, transistors, contacts, etc., as well as the capability to merge into the existing drawing predefined "library building-block" files. Since the user

must either specify the size (width and/or length) of basic elements or use their default values, design-rules specifying minimum size requirements are enforced at this input phase.

The compactor operates on the user input to determine the exact location of each of the elements, based on the elements' size and spacing requirements. Since the particular compaction algorithm used here detects the most congested areas in the layout schematic diagram before calculating element locations, further improvement on the resulting layout is carried out only in these congestion areas to achieve computation efficiency. The compacted layout also can be converted back to the schematic form so that the user can make manual improvements on the intermediate result.

At the present time, CABBAGE is written specifically for the silicon gate N-MOS logic family. However, the compaction algorithm used in it is quite general, as the algorithm is concerned with only the compaction of rectangles.

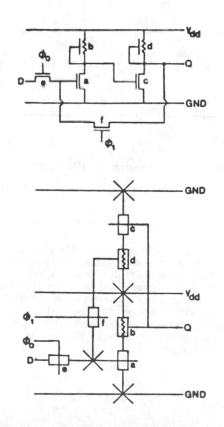

Fig. 1   Layout schematic diagram and its corresponding circuit schematic diagram (top drawing) of a D-type latch.

† This work is sponsored in part by Hewlett-Packard Co., Palo Alto, CA.

Reprinted from *IEEE Int. Symp. Circuits and Systems Proc.*, 1979, pp. 474–477.

3. The Compaction Algorithm

A number of layout schematic diagram compaction methods have been developed and implemented recently [1,2,3]. The detail of the compaction algorithm used in the program described in [1] has not been published. The other programs use either an exhaustive search method or a localized trial-and-error approach by confining the compaction process to a small portion of the overall layout. Such methods are inefficient and require approximately n^2 operations, where n is the number of features (elements and line segments) to be compacted [3].

In contrast, the compaction method described in this paper is based on the most-constraining structure principle. The most-constraining structure in a layout is a group of features which ultimately limits the size of the layout in a given dimension. Thus, modification (reducing the length) of such a structure is most likely to improve the compactness of a layout rapidly. The most-constraining structure can be found if the user's layout schematic diagram is converted into a directed graph to determine its longest path in a given dimension. The use of the longest path principle in layout automation is not new; among other applications it has been used to determine the placement and routing of building-block-based integrated circuits [4, 5] and the placement of hybrid circuit elements [6].

In the following the computer-based compaction process is described with the aid of the T flip-flop example shown in Fig. 2a. One of the N-MOS circuit implementations of this T flip-flop is shown in the layout schematic diagram form in Fig. 2b. This layout schematic describes the desired relative placement and the actual interconnection of transistors and line segments.

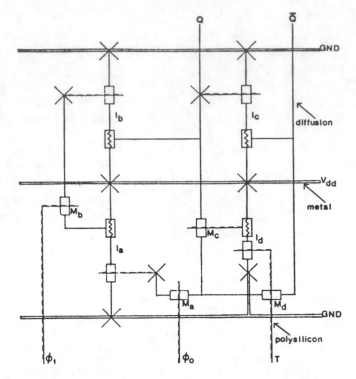

Fig. 2b Layout schematic diagram of the circuit shown in Fig. 2a.

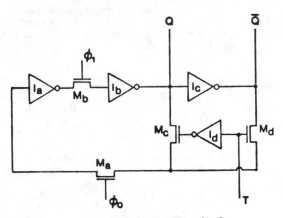

·Fig. 2a Circuit schematic diagram of a T-type flip-flop.

Compaction Procedure

Step 0: Layout features in the user input are sorted according to their minimum X and Y values. Node numbers (or path names) are assigned to features to indicate their electrical connections. These operations are done primarily in preparation for an efficient design-rule checking in the following steps. Note, however, a side benefit of being able to generate node numbers is that the same schematic input can be used to describe circuit connectivity to existing circuit and timing simulation programs.

Step 1: The layout schematic diagram is separated into horizontal and vertical portions. The resulting horizontal portion contains all vertically-connected features since these features are to be compacted horizontally. Similarly, the vertical portion contains all horizontally-connected features. This operation decouples movements of the features due to the horizontal compaction from those due to the vertical compaction, and vice versa. It significantly simplifies the required data structure and computational effort at the expense of the ability to compact in a general direction, such as diagonal compaction.

Step 2: For each portion obtained in Step 1, features connected on the same center line are grouped together. Such a grouping operation prevents connected features from drifting apart during the compaction. For example, in Fig. 2b, inverters I_a and I_b and their associated vertical line segments and contacts are all considered to be in the same group.

Step 3: Spacing design-rule requirements are now added to the portion under consideration. This is done using techniques similar to those employed in traditional design-rule checking programs [7]. Consider, for example, the ϕ_1 polysilicon line (the leftmost feature) in the horizontal portion of Fig. 2b. Spacing design-rules are applied to calculate the required separation between group center-lines of the ϕ_1 line and the groups bordering it on the right; these bordering groups being a) the ϕ_0 line, b) the transistors forming the inverter I_a, and c) the transfer transistor M_b and its connection line. Since spacing is measured from center line to center line, line widths and element sizes are automatically included in the above calculation. Line lengths are assumed to be variable, unless specified otherwise. At the end of the calculation, only the largest required separation is retained for each pair of groups.

246

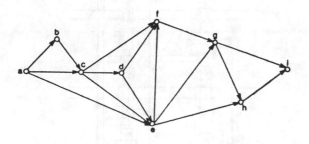

Fig. 3 Graph representation of the relation among the feature groups of the horizontal portion of the layout schematic diagram in Fig. 2b.

Step 4: After the design-rule requirements are added, a graph can be generated to represent the relations among the individual groups. Fig. 3 shows such a graph for the horizontal portion of the example in Fig. 2b. The nodes of the graph represent the individual groups and the branches the separation requirements between groups. The maximum separation requirements calculated in Step 3 are used as branch weights.

Step 5: The longest path is determined through the graph obtained in the last step. Fig. 4 shows the data structure used to calculate the longest path through the horizontal graph of the present example. The path length to a node is the maximum of

a) the existing path length and

b) the path length from its predecessor currently under consideration.

After each path length calculation the number of unprocessed predecessors of a given node is reduced by one. The longest path search operation completes when all nodes have zero unprocessed predecessor.

node	number of unprocessed predecessors	link	list of followers
a	0	→	b, c, e
b	1	→	c
c	2	→	d, e, f
d	1	→	e, f
e	3	→	f, g, h
f	3	→	g
g	2	→	h, i
h	2	→	i
i	2	→	$$

Fig. 4 Data structure used for calculating the longest path through the graph in Fig. 3.

Step 6: The path length to each node indicates the exact location of the associated features. The resulting movement in the compaction direction necessitates a new spacing design-rule calculation in the perpendicular direction. Compaction is carried out in this perpendicular direction next. Thus the compaction process iterates until no new design-rule requirement calculation is necessary.

Fig. 5a shows the result obtained after the user input is compacted in the horizontal direction. Fig. 5b shows the result of a vertical compaction performed on Fig. 5a.

The layout generated with the above procedure is correct but not necessarily the most compact. The size of the layout can be reduced further by modifying the composition of the longest path. In fact, break points in the longest path are good places to generate jogs. Since a break point occurs in a group when the feature limited by a previous group differs from the feature limiting a following group, the latter feature can be moved closer to the previous group in order to move in the following group. Fig. 5c shows the result after jog generation has been performed on Fig. 5b. Other layout improvement methods such as rotating and mirroring features are being developed. All such methods use information derived from the longest path to achieve computation efficiency.

4. Program Performance

CABBAGE is implemented on an HP-1000 Series E 16-bit word minicomputer with an HP-2648A graphics terminal as the primary user interface. The code sections for the interactive input processor and the compactor are both written in FORTRAN, and each of them is approximately 24K words long when compiled. The data area required to store the T flip-flop (Fig. 2) is approximately 2K words long. (The maximum main memory available for data is 4K words. In addition, there are a 32K-word block-transfer type fast memory and a 50-megabyte disk available for data storage.) The computation time for the same T flip-flop is approximately 1.5 seconds per compaction or jog generation. The major portion of the computation time is spent in design-rule checking to determine separation between groups. Since modern design-rule checking techniques typically have a complexity of $n^{1.5}$, where n is the number of features in the layout, the performance of the compactor will degrade slightly for larger circuits.

5. Summary

An algorithm is described in this paper which can be used to generate compact LSI circuit building-block layout from the user's layout schematic diagrams. The use of the layout schematic input allows the user to provide the computer program with a good head-start. The tedious layout work is performed by the program which also ensures the correctness of the resulting layout. Further improvements of the layout are carried out automatically or through manual modifications. In both cases, the most-constraining structure derived by the program are used as a guideline.

With the approach used in the program CABBAGE, the traditional, "after-the-fact" design rule checking, circuit connectivity extraction, and capacitance calculation are all integrated in a forward path. This path starts with the layout or circuit schematic diagram input, and terminates with a complete layout and a set of input ready for use in circuit and timing simulation programs. High-density LSI circuits thus can be built up efficiently from building-blocks of increasing complexity.

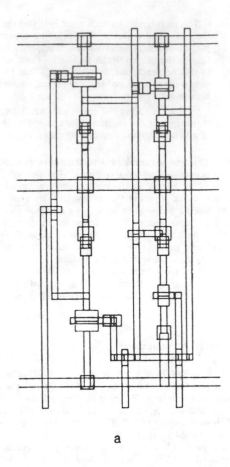

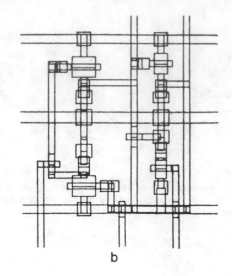

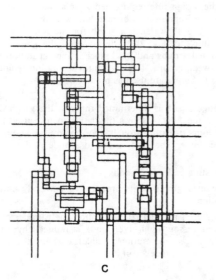

a

b

c

Fig. 5   a) T flip-flop (Fig. 2b) after compaction in the horizontal
direction, b) after compaction in the vertical direction, c)
after jog generation.

References:

[1]   Y. E. Cho, A. J. Korenjak, and D. E. Stockton, "FLOSS: An
      Approach to Automated Layout for High-Volume Designs,"
      Proceedings of the 14th Design Automation Conferences,
      June 1977, pp. 138-141.

[2]   J. D. Williams, "STICKS - A Graphical Compiler for High-
      Level LSI Design," AFIPS Conference Proceedings, Vol. 47,
      June 1978, pp. 289-295.

[3]   A. E. Dunlop, "SLIP: Symbolic Layout of Integrated Circuits
      with Compaction," Computer-Aided Design, vol. 10, no. 6,
      Nov. 1978, pp. 387-391

[4]   K. Kani, H. Kawanishi, and A. Kishimoto, "ROBIN; A
      Building-Block LSI Routing Program," Proceedings of the 1976
      International Symposium on Circuits and Systems, pp. 658-
      661.

[5]   B. T. Preas and C. W. Gwyn, "Methods for Hierarchical
      Automatic Layout of Custom LSI Circuit Masks," Proceedings
      of the 15th Design Automation Conferences, June 1978, pp.
      206-212.

[6]   K. Zibert, and R. Saal, "On Computer-Aided Hybrid Circuit
      Layout," Proceedings of the 1974 International Symposium on
      Circuits and Systems, pp. 314-318.

[7]   H. S. Baird, "Fast Algorithms for LSI Artwork Analysis,"
      Proceedings of the 14th Design Automation Conferences,
      June 1977, pp. 303-311.

One-Dimensional Logic Gate Assignment and Interval Graphs

TATSUO OHTSUKI, SENIOR MEMBER, IEEE, HAJIMU MORI, ERNEST S. KUH, FELLOW, IEEE,
TOSHINOBU KASHIWABARA, AND TOSHIO FUJISAWA, FELLOW, IEEE

Abstract—This paper gives a graph-theoretic approach to the design of one-dimensional logic gate arrays using MOS or I^2L units. The incidence relation between gates and nets is represented by a graph $H=(V,E)$, and a possible layout of gates and nets is characterized by an interval graph $\hat{H}=(V, E \cup F)$, where F is called an *augmentation*. It is shown that the number of tracks required for between-gate wiring is equal to the clique number (chromatic number) of $\hat{H}$, and hence the optimum placement problem is converted to that of *minimum clique number augmentation*. This turns out to be an NP-complete problem. Instead a polynomial-time algorithm for finding a minimal augmentation is presented, where an augmentation is *minimal* if no proper subset of it is an augmentation. An algorithm for gate sequencing with respect to a given augmentation is also presented.

I. INTRODUCTION

RECENT advances in microelectronics have brought renewed interests in layout design, a field which has so far depended primarily on manual effort and experiences of design engineers. There exist very few fundamental results and theoretical methods. One main difficulty seems to lie in obtaining analytic characterizations of problems which are practically meaningful. The problem of linear placement is a notable exception. The problem is to determine the optimum positions of modules which lie in a single row. It is important in the design of one-dimensional logic gate arrays using MOS or I^2L units, and the main purpose is to minimize the chip areas [1]–[3]. Recently, linear placement has also been found useful in solving the general placement problem [4]. Although there exist various algorithms for obtaining suboptimum linear placement, most depend on heuristics and the method of branch and bound [5], [6]. In this paper we introduce a new formulation of the problem using graph theory. It converts the practical problem of finding a minimum-track placement into that of generating an optimum interval graph with minimum chromatic number. Both theoretical results and efficient computational algorithms are obtained.

Manuscript received January 17, 1979; revised April 24, 1979.
T. Ohtsuki and H. Mori are with the Central Research Laboratories, Nippon Electric Company, Ltd., Kawasaki, Kanagawa 213, Japan.
E. S. Kuh is with the Department of Electrical Engineering and Computer Sciences and the Electronics Research Laboratory at the University of California, Berkeley, CA 94720.
T. Kashiwabara and T. Fujisawa are with the Department of Information and Computer Sciences, Osaka University, Toyonaka, Osaka 560, Japan.

In Section II, we give the problem formulation with a brief sketch of the elementary properties of interval graphs. In Section III, we present a method of obtaining a proper gate sequence and a track assignment. In Section IV, a method is presented to find a minimal augmentation which plays a key role in our approach. Examples and comments next follow.

II. PROBLEM FORMULATION

In Fig. 1, a NAND gate is shown together with its circuit schematic and layout schematic. A complex logic function can be realized with gates interconnected as a one-dimensional array. In the detailed layout, in-gate wiring is done along the vertical columns and between-gates connections are made horizontally. At the intersection of vertical gates and horizontal wires, there is either a terminal of a transistor as A, C, D, and P or a straight connector as B shown in Fig. 1(c). Also, the width of each gate and the separation distance between two neighboring gates are predetermined according to physical constraints. Therefore, the total horizontal dimension of an array is considered fixed. The vertical spacing determines the total chip area and is to be minimized. This can be achieved by finding an optimum sequence of the gates. For example, an optimum realization of logic array is shown in Fig. 2. Each gate is represented by a vertical line. Horizontal wires are assigned to tracks.

The gate sequencing problem is a linear placement problem. In general, there are given a set of modules, numbered as $1, 2, \cdots, m$ and a set of nets, numbered as $1, 2, \cdots, n$ together with a net list which specifies the connection pattern. A net corresponds to a horizontal wire which interconnects modules assigned to the net. Thus a placement may be considered to be a permutation $\pi = (\pi_1, \pi_2, \cdots, \pi_m)$ of the first m integers such that module π_i is placed in the ith position on a row. The most commonly used objective function for minimization is the total wire length. However, in the gate sequencing problem, as we have discussed, a different criterion must be used. We want to minimize the necessary number of tracks.

Before going further we introduce a few definitions related to interval graphs. A graph $G=(V,E)$ is an *interval graph* if there exists a set of finite closed intervals

Reprinted from *IEEE Trans. Circuits and Systems*, vol. CAS-26, pp. 675–684, Sept. 1979.

249

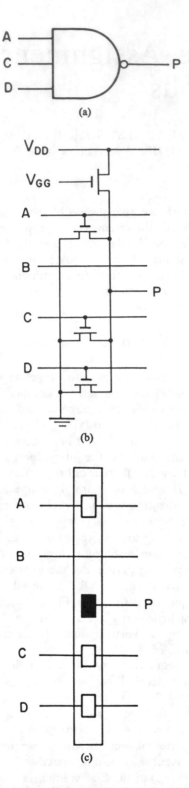

Fig. 1. (a) A NAND gate. (b) Its circuit schematic. (c) Its layout schematic.

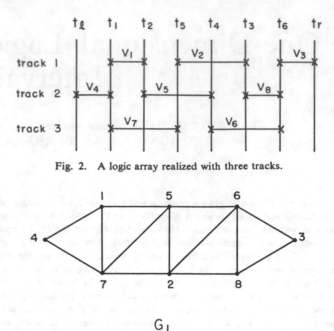

Fig. 2. A logic array realized with three tracks.

Fig. 3. Interval graph G_1 corresponding to the placement in Fig. 2.

intervals intersect when they are placed on the same track. It should be noted that the intervals are closed, and hence two intervals intersect each other even if their common point represents an end point of the respective intervals. The interval graph which corresponds to the placement in Fig. 2 is shown in Fig. 3.

For a given permutation of modules, the intervals should be placed in tracks so that the required number of tracks is kept as small as possible. This is similar to the coloring problem in graph theory. If we designate a specific color for each track, then the restriction that no two overlapping intervals can be placed in the same track is exactly the same as the restriction on graph coloring that no two adjacent vertices are colored by the same color. Therefore, the necessary number of tracks is the chromatic number of the corresponding interval graph. It is well known that the chromatic number of an interval graph is equal to the clique number or the maximum cardinality of all cliques [7].

Let us first introduce the necessary terminology. Let

$$T = \{ t_l, t_1, t_2, \cdots, t_m, t_r \} \tag{1}$$

be the set of modules or gates, where t_l and t_r are boundary gates representing the nets to be extracted to the left and to the right, respectively. Let

$$V = \{ v_1, v_2, \cdots, v_n \} \tag{2}$$

be the set of nets. We assume that there are m internal gates and n nets. The set of nets connected to gate t will be denoted by $V(t)$, and the set of gates to which net v is connected will be denoted by $T(v)$. Without loss of generality, we assume that

$$|V(t_i)| > 1, \qquad \text{for all } t_i \tag{3}$$

and

$$|T(v_j)| > 2, \qquad \text{for all } v_j. \tag{4}$$

$\{ I(v) \}_{v \in V}$ such that $(u,v) \in E$ implies and is implied by $I(u) \cap I(v) \neq \varnothing$. The set of intervals is called a *realization* of the graph G.

For each permutation of modules we can draw horizontal intervals corresponding to nets, from which we obtain an interval graph corresponding to the gate sequence. The set of vertices of the interval graph is the set of nets. The vertices are adjacent if and only if the corresponding

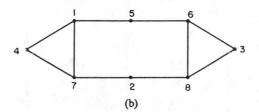

gates	nets
t_ℓ	4
t_1	1 , 4 , 7
t_2	1 , 5
t_3	2 , 8
t_4	5 , 6
t_5	2 , 7
t_6	3 , 6 , 8
t_r	3

(a)

(b)

Fig. 4.   Net list specification for the logic array in Fig. 2 in terms of a table and its connection graph H.

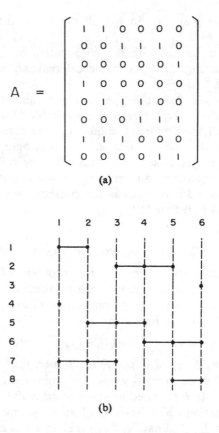

$$A = \begin{bmatrix} 1 & 1 & 0 & 0 & 0 & 0 \\ 0 & 0 & 1 & 1 & 1 & 0 \\ 0 & 0 & 0 & 0 & 0 & 1 \\ 1 & 0 & 0 & 0 & 0 & 0 \\ 0 & 1 & 1 & 1 & 0 & 0 \\ 0 & 0 & 0 & 1 & 1 & 1 \\ 1 & 1 & 1 & 0 & 0 & 0 \\ 0 & 0 & 0 & 0 & 1 & 1 \end{bmatrix}$$

(a)

(b)

Fig. 5.   (a) v.d.c. matrix of the interval graph G_1 in Fig. 3. (b) The canonical realization of the v.d.c. matrix A.

The net list specification can be given by a table listing all the gates and the nets which are associated to each gate. Based on this, we construct a graph, called a *connection graph*,

$$H = (V, E) \tag{5}$$

where V is the set of n vertices representing the n nets. The edges are defined by

$$E = \{(x,y)|\exists t \text{ s.t. } x,y \in V(t)\} \tag{6}$$

that is, two vertices are adjacent if and only if there exists a common gate of the two corresponding nets. For the example of Fig. 2 the table specification and its connection graph H are given in Fig. 4. It is obvious that H is a subgraph of the interval graph in Fig. 3.

From the above, it is clear that the linear placement problem to minimize the number of tracks can be stated in terms of graph theory as follows: given a graph H, find a supergraph by adding a set of edges, which is an interval graph and has the least clique number.

For the time being we ignore the boundary gates t_l and t_r because their fixed locations introduce further constraint on the problem. This will be taken care of in Section IV. In the following we will briefly present some fundamental properties of interval graphs. Special attention will be placed to the matrix representation of interval graphs.

It is obvious that every clique of an interval graph has its vertices corresponding to mutually overlapping intervals. A clique is called *dominant* if it is maximal, i.e., if it is not a proper subset of another clique. The total number of dominant cliques does not exceed $|V|$ [8].

Let $G = (V, E)$ be an interval graph with dominant cliques $C_1, C_2, \cdots, C_p$. Then a v.d.c. matrix (vertex versus dominant clique matrix) $A = [a_{ij}]$ is an $n \times p$ 0, 1 matrix defined by

$$a_{ij} = \begin{cases} 1, & \text{if vertex } i \in C_j \\ 0, & \text{otherwise} \end{cases}$$

where $|V| = n$. The matrix is said to have the *consecutive ones property* if the ones in each row occur in consecutive positions. Fulkerson and Gross give the following theorem [9].

Theorem 1

A graph is an interval graph if and only if there exists an ordering of dominant cliques such that the v.d.c. matrix has the consecutive ones property.

The consecutive ones property of a v.d.c. matrix defines a *canonical realization*. In the ith row, let $l(i)$ be the leftmost column having one and $r(i)$ be the right most column having one. Then $l(i) \leqslant r(i)$. Let the interval $[l(i), r(i)]$ correspond to the ith vertex. In this way we obtain a set of intervals called the canonical realization of matrix A. The following theorem due to Fulkerson and Gross [9] is almost self-evident.

Theorem 2

Let A be a v.d.c. matrix, having the consecutive ones property, of an interval graph $G = (V, E)$. Then the cannonical realization of A is a realization of G.

For the graph G_1 in Fig. 3, the dominant cliques are $\{1,4,7\}$, $\{1,5,7\}$, $\{2,5,7\}$, $\{2,5,6\}$, $\{2,6,8\}$, and $\{3,6,8\}$. The v.d.c. matrix with the consecutive ones property is shown in Fig. 5(a). Its canonical realization is given in Fig. 5(b).

Finally, it should be pointed out that efficient algorithms on interval graphs are available: for an interval graph, all the dominant cliques can be listed in $0(|V|+|E|)$ time by the application of a lexicographic breadth first search [8]. Based on a data structure called PQ-trees we can determine an ordering of dominant cliques for which the v.d.c. matrix has the consecutive ones property in $0(|V|+|E|)$ time [10].

III. Gate Sequencing and Track Assignment

As stated in Section II, the problem of finding an optimum linear placement with minimum tracks can be formulated in terms of graph theory. Given a connection graph $H=(V,E)$, find a supergraph

$$\hat{H}=(V,E\cup F) \tag{7}$$

by adding a set of edges F to H such that the resulting graph $\hat{H}$ is an interval graph with minimum clique number. The set F is called an *augmentation* if $E\cap F=\varnothing$. An augmentation which leads to a least clique number is what we want. Unfortunately, this problem turns out to be NP complete [11]. A related problem of finding an augmentation F with a minimum $|F|$ is also NP complete [11]. Thus, like many combinatorial problems, we will aim at a *minimal augmentation*. A minimal augmentation is one such that no proper subset is an augmentation. In the example of Fig. 3, $G_1=(V,E\cup F_1)$ is an interval graph and may be viewed as that generated from the connection graph $H=(V,E)$ in Fig. 4. It is easy to see that F_1 in G_1 is a minimal augmentation.

In the present section we assume that the supergraph H has been obtained from H with a minimal augmentation F. The dominant cliques of H are next determined and a v.d.c. matrix with the consecutive ones property is obtained as described in the preceding section. The first problem we consider here is to find a proper gate sequence from the $n\times p$ v.d.c. matrix A, where p represents the number of dominant cliques. Let

$$C=\{C_1,C_2,\cdots,C_p\} \tag{8}$$

be the complete set of dominant cliques of $\hat{H}$, where C_j corresponds to the jth column of A. Each C_j includes a set of vertices represented by the nonzero positions in the jth column of A. From the matrix A, we also know that, for each vertex v_i, there associates a closed interval $[l(i),r(i)]$ in the ith row as discussed before.

To discuss the gate sequencing problem, it should be noted that, for any gate t, there exists a dominant clique C_j in $\hat{H}$ such that $V(t)\subseteq C_j$, where $V(t)$ is the set of vertices pertaining to gate t. This statement is obvious since $V(t)$ is a clique in H, and so in $\hat{H}$.

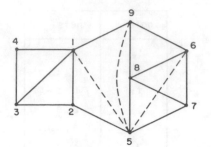

Fig. 6. Connection graph H and a minimal augmentation $F=\{(1,5),(5,9),(5,6)\}$ for Example 1.

TABLE I
Net Specification for Example 1

Gates	V(t)
t_ℓ	1,4
t_1	3,4
t_2	1,2,3
t_3	2,5
t_4	1,9
t_5	9
t_6	6,8,9
t_7	5,7,8
t_r	6,7

To determine C_j for each $V(t)$, we first form an interval

$$[L(t),R(t)]=\bigcap_{v_i\in V(t)}[l(i),r(i)]. \tag{9}$$

Then $L(t)<j<R(t)$ implies and is implied by $V(t)\subseteq C_j$.

Now we pick up arbitrarily for each gate t a dominant clique $C_{d(T)}\supseteq V(t)$, where

$$L(t)<d(t)<R(t). \tag{10}$$

Then we can obtain a gate sequence, or in other words an ordering of gates

$$t_{\pi(0)},t_{\pi(1)},\cdots,t_{\pi(m)},t_{\pi(m+1)}$$

such that

$$d(t_{\pi(0)})<d(t_{\pi(1)})<\cdots<d(t_{\pi(m)})<d(t_{\pi(m+1)}) \tag{11}$$

where $(\pi(0),\pi(1),\cdots,\pi(m),\pi(m+1))$ has to be considered to be a permutation of integers $1,\cdots,m$ and letters l and r.[1] Now we can state the following assertion.

Theorem 3

Let $t_{\pi(0)},\cdots,t_{\pi(m+1)}$ be a gate sequence satisfying (10) and (11), and let $H^*=(V,E^*)$ be the interval graph corresponding to the gate sequence. Then H^* coincides with $\hat{H}=(V,\hat{E})=(V,E\cup F)$ provided that F is a minimal augmentation.

---

[1] Two virtual gates t_l and t_r should be placed at the left-most and right-most ends, respectively, as pointed out in the preceding section. This constraint will be taken into consideration in the following section.

Proof: We first show $E^* \subseteq \hat{E}$. To prove this let $(u,v) \in E^*$. For the given gate sequence, let $I(u)$ and $I(v)$ be intervals corresponding to nets u and v, respectively. Then $(u,v) \in E^*$ implies that $I(u)$ and $I(v)$ overlap. Therefore, there exist gates t_1 and t_2 such that $u \in V(t_1)$, $v \in V(t_2)$ and $d(t_1) \leqslant d(t_2)$. Due to (9) and (10) we have $l(u) \leqslant d(t_1) \leqslant d(t_2) \leqslant r(v)$. Similarly we obtain $l(v) \leqslant r(u)$ and hence

$$[l(u), r(u)] \cap [l(v), r(v)] \neq \varnothing.$$

Since intervals $[l(u), r(u)]$ and $[l(v), r(v)]$ correspond to u and v, respectively, in the canonical realization of A, there follows that $(u,v) \in \hat{E}$.

If E^* is a proper subset of $\hat{E}$, then $E^* - E$ is an augmentation which is a proper subset of a minimal augmentation $F = \hat{E} - E$. This is a contradiction. Therefore $E^* = \hat{E}$ and $H^* = \hat{H}$.   Q.E.D.

A gate sequence satisfying (11) can be obtained in $0(am)$ time, where a is the average value of $|V(t)|'s$.

Example 1: Consider a net specification given in Table I, where $m = 7$ and $n = 9$. The corresponding connection graph H is given in Fig. 6. By adding edges of a minimal augmentation $F = \{(1,5), (5,9), (5,6)\}$, $\hat{H} = (V, E \cup F)$ becomes an interval graph, since a v.d.c. matrix A of $\hat{H}$ has the consecutive ones property as shown below.

$$A = \begin{bmatrix} 1 & 1 & 1 & 1 & 0 & 0 \\ 0 & 1 & 1 & 0 & 0 & 0 \\ 1 & 1 & 0 & 0 & 0 & 0 \\ 1 & 0 & 0 & 0 & 0 & 0 \\ 0 & 0 & 1 & 1 & 1 & 1 \\ 0 & 0 & 0 & 0 & 1 & 1 \\ 0 & 0 & 0 & 0 & 0 & 1 \\ 0 & 0 & 0 & 0 & 1 & 1 \\ 0 & 0 & 0 & 1 & 1 & 0 \end{bmatrix} \quad (12)$$

For each gate t the interval $[L(t), R(t)]$ is computed and is summarized in Table II, with a possible choice for $d(t)$. Therefore, the gate sequence

$$t_l, t_1, t_2, t_3, t_4, t_6, t_5, t_7, t_r$$

is legitimate for the corresponding sequence $d(t)$'s is non-decreasing. The realization for this gate sequence is shown in Fig. 7.

The problem we consider next is to place the set of nets (vertices) on proper tracks so that the number of used tracks is equal to the clique number of $\hat{H} = (V, E \cup F)$. Once a gate sequence has been determined, an efficient algorithm, so-called "left edge" algorithm, is available for obtaining such a track assignment [12]. However, it should be stressed here that a track assignment can be obtained directly from the interval graph $\hat{H}$ without fixing a gate sequence. To be more precise, a v.d.c. matrix with the consecutive ones property derived from $\hat{H}$ associates the order of dominant cliques of $\hat{H}$. Now the "left edge" algorithm applied to the dominant clique ordering, rather than gate ordering, leads to an optimum track assignment.

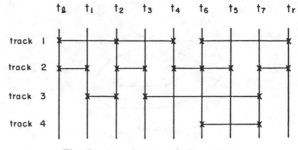

Fig. 7.   A realization for Example 1.

TABLE II
INTERVALS $[L(t), R(t)]$ for EXAMPLE 1

Gates	[L(t), R(t)]	d(t)
t_l	[1, 1]	1
t_1	[1, 1]	1
t_2	[2, 2]	2
t_3	[3, 3]	3
t_4	[4, 4]	4
t_5	[4, 5]	5
t_6	[5, 5]	5
t_7	[6, 6]	6
t_r	[6, 6]	6

It is not difficult to prove that this process can be done in $0(|V|)$ time.

IV.   MINIMAL AUGMENTATION

In this section we first explain how to obtain a minimal augmentation. Then we present a method to take care of boundary gates.

Given a graph $H = (V, E)$ and an ordering $v_1, v_2, \cdots, v_n$ of vertices, define for $1 \leqslant i \leqslant n$

$$V_i = \{v_1, v_2, \cdots, v_{i-1}, v_i\}.$$

Let the section graph determined by V_3 be

$$H_3 = (V_3, E_3)$$

where

$$E_3 = \{(v_i, v_j) | v_i, v_j \in V_3, (v_i, v_j) \in E\}.$$

This is necessarily an interval graph for a graph having fewer than four vertices is an interval graph.

The section graph determined by V_4

$$H_4 = (V_4, E_4)$$

is not an interval graph if H_3 is a path and v_4 is connected to its two end vertices as is illustrated in Fig. 8(a). Then by adding an edge connecting v_4 and the intermediate vertex, the graph becomes an interval graph as shown in Fig. 8(b). Let F_4 be the set consisting of this single edge in this case and be the empty set otherwise. Thus $\hat{H}_4 =$

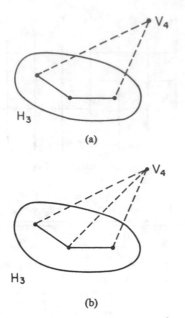

(a)

(b)

Fig. 8.   (a) H_4 which is not an interval graph. (b) $\hat{H}_4$, an interval graph.

$(V_4, \hat{E}_4) = (V_4, E_4 \cup F_4)$ is an interval graph. It should be clear that F_4 is a minimal augmentation with respect to H_4, which consists of edges incident with v_4.

The above idea can be extended to the general case $4 \leqslant i \leqslant n$. We can assume

$$\hat{H}_{i-1} = (V_{i-1}, \hat{E}_{i-1})$$

is given as an interval graph. Then we add vertex v_i and edges in E connecting v_i and vertices of V_{i-1}, that is,

$$H_i = (V_i, E_i)$$

where

$$E_i = \hat{E}_{i-1} \cup \{(v_i, v) | v \in V_{i-1}, (v_i, v) \in E\}.$$

Let F_i be a minimal augmentation with respect to H_i, which consists of edges incident with v_i. Such a set exists because the graph becomes an interval graph if edges are added such that v_i is adjacent to all the vertices of V_{i-1}. This is seen as follows. Let $\{I(v)\}_{v \in V_{i-1}}$ be a realization of interval graph $\hat{H}_{i-1}$. Let $I(v_i)$ be an interval covering all the intervals $\{I(v)\}_{v \in V_{i-1}}$. Then this is a realization of the graph obtained by adding edges so that v_i is adjacent to all the vertices of V_{i-1}.

Thus by adding edges of F_i we obtain an interval graph

$$\hat{H}_i = (V_i, \hat{E}_i) = (V_i, E_i \cup F_i).$$

Repeating this process we finally obtain an interval graph

$$\hat{H} = (V, E \cup F)$$

where

$$F = F_4 \cup F_5 \cup \cdots \cup F_n.$$

Theorem 4

The set F thus obtained is a minimal augmentation.

Proof: To prove the assertion by contradiction, assume that a proper subset F' of F is an augmentation. Then there exists a unique partition of F' such that $F' = F_4'$

$\cup \cdots \cup F_n'$ and $F_i' \subseteq F_i$ for $4 \leqslant i \leqslant n$. Let k be the first index such that $F_k' \subsetneq F_k$. Note that $\hat{H}_k' = (V_k', E_k \cup F_k')$ is an interval graph for it is a section graph of interval graph $(V, F \cup F')$. Therefore, F_k' is an augmentation with respect to $H_k = (V_k, E_k)$. Since F_k' is a proper subset of a minimal augmentation F_k with respect to H_k, this is the desired contradiction.   Q.E.D.

Based on this theorem an algorithm of finding a minimal augmentation has been developed, which runs in $0(|V| \cdot (|E| + |F|))$ time. This is presented in a separate paper [11].

So far we have not paid any attention to the constraint that two distinguished gates t_l and t_r are to be placed at the end-most positions. According to the method of finding a gate sequence as described in the preceding section, we can always obtain a gate sequence subject to the constraint if a v.d.c. matrix, having the consecutive ones property, of interval graph $\hat{H} = (V, E \cup F)$ is given such that the first column includes $V(t_l)$ and the last column includes $V(t_r)$. An augmentation is said to satisfy the boundary condition or, in short, to be a *B augmentation* if there exists such a v.d.c. matrix. A minimal B augmentation is one such that no proper subset is a B augmentation.

Remark: It is easy to obtain such a v.d.c. matrix if it exists [10].

Thus we have shown above that if F is a minimal B augmentation then there is a gate sequence placing t_l and t_r as the left-most and the right-most gates, respectively, for which the interval graph corresponding to the gate sequence is exactly $\hat{H} = (V, E \cup F)$.

Conversely, assume that there is a gate sequence which places t_l and t_r at the correct positions. Let $H^* = (V, E^*) = (V, E \cup F^*)$ be the corresponding interval graph. For notational simplicity, let $t_l, t_1, \cdots, t_m, t_r$ be the gate sequence. For each gate t, let $C(t)$ be the clique consisting of nets, each of which pertains to t or passes through t. It should be obvious that all dominant cliques are found among $C(t_l), C(t_1), C(t_2), \cdots, C(t_m), C(t_r)$. Reading this from left to right, let $C_1, C_2, \cdots, C_p$ be dominant cliques obtained in this order. Then a v.d.c. matrix A having $C_1, \cdots, C_p$ as the 1st, $\cdots$, the pth columns has the consecutive ones property, and

$$V(t_l) \subseteq C(t_l) \subseteq C_1 \qquad V(t_r) \subseteq C(t_r) \subseteq C_p. \qquad (13)$$

Hence F^* is a B augmentation. This justifies our policy of finding a minimal B augmentation and then determining a proper gate sequence subject to the constraint on the two distinguished gates t_l and t_r.

In what follows our main concern is how to obtain a minimal B augmentation. Before going further we assume hereafter that

$$V(t_l) \cap V(t_r) = \varnothing. \qquad (14)$$

If there exists a net $v \in V(t_l) \cap V(t_r)$, net v connects the left most gate t_l and the right-most gate t_r; it occupies one full track. Such a net v can be excluded at the outset, and

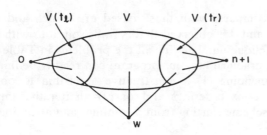

Fig. 9. Graph $H'=(V',E')$ formed from $H=(V,E)$.

only afterwards it needs to be added to occupy a new track.

Instead of applying the algorithm of finding a minimal augmentation to H, we form a new graph H' from H and then apply the algorithm to H'. We add three more vertices 0, $n+1$, and w to the connection graph $H=(V,E)$. The new graph $H'=(V',E')$ is defined as follows and is shown in Fig. 9:

$$V'=V\cup\{0,n+1,w\}$$

and

$$E'=E\cup\{(0,v)|v\in V(t_l)\}\cup\{(n+1,v)|v\in V(t_r)\}$$
$$\cup\{(w,v)|v\in V\}$$

thus

$$|V'|=|V|+3 \quad \text{and} \quad |E'|<|E|+2|V|. \tag{15}$$

Let $v_1,\cdots,v_k$ be an arbitrary ordering of vertices of $V-V(t_l)-V(t_r)$, and let α' be an ordering of V' such that vertices of $V(t_l)$, vertices of $V(t_r)$, 0, $n+1$, w, $v_1,\cdots,v_k$ are ordered in this way. We apply the algorithm to H' with ordering α' and obtain a minimal augmentation with respect to H'. We can prove the following theorem.

Theorem 5

Set F thus obtained is a minimal B augmentation with respect to connection graph $H=(V,E)$.

A proof is given in the appendix. From (15) this can be done in $O(|V|\cdot(|E|+|F|))$ time provided that H is connected and hence $|V|<|E|+1$.

Example 2: Consider a net specification as shown in Table III, where $m=3$ and $n=5$. The connection graph H is shown in Fig. 10(a), which is an interval graph as seen in Fig. 10(b). The extended graph H' is shown in Fig. 10(c). Since $V(t_l)=\{1,3,4\}$ and $V(t_r)=5$, we take an ordering α'

$$1,3,4,5,0,6,w,2.$$

Then $F=\{(2,4)\}$ is a minimal B augmentation. Dominant cliques of $\hat{H}=(V,E\cup F)$ are $C_1=\{1,2,3,4\}$ and $C_2=\{3,4,5\}$. A v.d.c. matrix A is the following:

$$A=\begin{bmatrix} 1 & 0 \\ 1 & 0 \\ 1 & 1 \\ 1 & 1 \\ 0 & 1 \end{bmatrix}.$$

It is easily seen that $V(t_l)\subseteq C_1$ and $V(t_r)\subseteq C_2$.

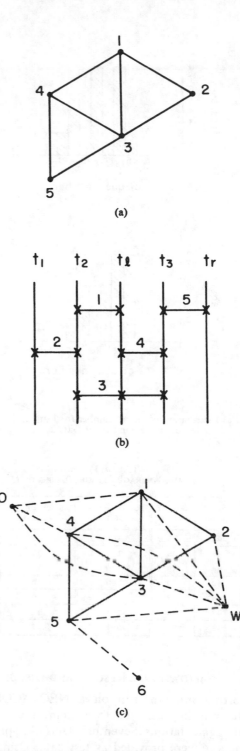

Fig. 10. (a) Connection graph H. (b) A realization of H. (c) Extended graph H'.

TABLE III
NET SPECIFICATION FOR EXAMPLE 2

Gates	V(t)
t_l	1,3,4
t_1	2
t_2	1,2,3
t_3	3,4,5
t_r	5

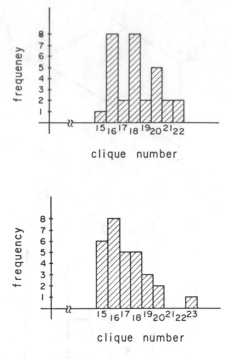

Fig. 11. (a) Distribution of clique number for Data 1. (b) Distribution of clique number for Data 2.

TABLE IV
LEAST AND AVERAGE CLIQUE NUMBER OF H

Data No.	No. of gates	No. of nets	No. of edges of H	Least clique number	Average clique number
1	48	48	161	15	18.2
2	48	49	173	15	17.1
3	48	48	176	13	15.4
4	48	50	288	18	21.2
5	48	48	222	9	13.2
6	48	48	294	12	16.3
7	48	49	359	17	20.3

V. COMPUTATIONAL RESULTS AND DISCUSSIONS

A Fortran program run on a NEC ACOS 77/700 computer was implemented to examine the property of minimal augmentations. Seven problems of approximately the same size were provided as test data. Using the program, 30 minimal augmentations for each problem were obtained based on randomly generated vertex orderings, and then the least and average clique numbers of the resulting interval graphs were calculated. Table IV summarizes the computational results with the problem sizes. Fig. 11 illustrates the distribution of clique numbers of the resulting interval graphs for Data 1 and 2.

The results suggest us a good heuristic method to obtain nearly optimum solution; that is to obtain several minimal augmentations based on randomly generated vertex orderings and then to choose the best solution. In these problems, the computer time for obtaining a minimal augmentation was about 4 s. The heuristic algorithm has

been compared with those based on branch and bound [5], [6] and it is observed that is finds solutions with less or equal clique numbers for all the problems in Table IV.

For practical use, our program can readily be improved by appending (1) a constructive algorithm to obtain a good vertex orderings and/or (2) an iterative improvement scheme starting from a minimal augmentation.

VI. APPENDIX
PROOF OF THEOREM 5

We prove the theorem step by step. Let $H_0 = (V_0, E_0)$ be the section graph of the connection graph $H = (V, E)$ determined by $V_0 = V(t_l) \cup V(t_r)$. Our algorithm of finding a minimal augmentation with respect to H' starts with finding a minimal augmentation F_0 with respect to H_0 to construct an interval graph $\hat{H}_0 = (V_0, E_0 \cup F_0)$. We first show a key property of $\hat{H}_0$. Since every vertex of H_0 belongs to either $V(t_l)$ or $V(t_r)$ and both $V(t_l)$ and $V(t_r)$ are cliques, it is clear that two distinct dominant cliques C_l and C_r including $V(t_l)$ and $V(t_r)$, respectively, are uniquely determined, unless V_0 itself is the only one dominant clique. Based on this we give the following lemma.

Lemma 1: C_l and C_r are placed at the endmost positions for any v.d.c. matrix having the consecutive ones property of $\hat{H}_0$.

Proof: Suppose C_l lies between C_r and another dominant clique, say C_j. Then, from the consecutive ones property, $v \in C_j \cap C_r$ implies $v \in C_l$. Since every vertex belongs to C_l or C_r, it follows that $C_j \subseteq C_l$. This contradicts the maximality of C_j. Hence C_l must be placed at one of the end-most positions. The assertion for C_r can also be proved in the same way. Q.E.D.

Our algorithm is next followed by the stage of adding vertices $\{0, n+1, w\}$ and the edges incident with them.

Let $H'_0 = (V'_0, E'_0)$ be the section graph of $H' = (V', E')$ determined by $V'_0 = \{0, n+1, w\} \cup V_0$. We next prove that $\hat{H}'_0 = (V'_0, E'_0 \cup F_0)$ is an interval graph, where F_0 is the augmentation of H_0 given above. For this purpose we give a realization of $\hat{H}_0$ as follows. Consider a v.d.c matrix with the consecutive ones property of $\hat{H}_0$, and let C_1 and C_q be the first and last columns (dominant cliques), respectively. Then, from Lemma 1, either $C_l = C_1$ and $C_r = C_q$ or $C_l = C_q$ and $C_r = C_1$. Without loss of generality we assume the former. In the canonical realization, the intervals for a vertex $v_i \in C_l$ and a vertex $v_j \in C_r$ are given by $I(v_i) = [1, r(i)]$ and $I(v_j) = [l(j), q]$, respectively. Now, for each vertex $v_i \in V(t_l)$ and each vertex $v_j \in V(t_r)$, we modify the corresponding intervals from $[1, r(i)]$ to $[0, r(i)]$ and from $[l(j), q]$ to $[l(j), q+1]$, respectively. Clearly, the set of intervals after the modification is still a realization of $\hat{H}_0$. Let $[0, 0]$, $[q+1, q+1]$, and $[1, q]$ be intervals for vertices $0, n+1$ and w, respectively. We add these intervals to the above set. Then it is easy to see that the resulting set of intervals is a realization of $\hat{H}'_0 = (V'_0, E'_0 \cup F_0)$.

Since $\hat{H}'_0$ is already an interval graph, there is no need to add edges, which leads to the following lemma.

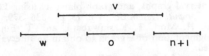

Fig. 12.   Interval for v overlaps with that for 0.

Lemma 2: If $x \in \{0, n+1, w\}$ and $y \in V'_0$, then $(x, y) \notin F$.

Next we show the following lemma.

Lemma 3: In any realization of $\hat{H}' = (V', E' \cup F)$, the interval for w lies between those for 0 and $n+1$.

Proof: Suppose on the contrary that the interval for 0 lies between those for $n+1$ and w as shown in Fig. 12. Let $v \in V(t_r)$, then the interval for v overlaps with those for $n+1$ and w, and hence with that for 0. Therefore $(0, v) \in E' \cup F$, which contradicts the assumption (14) or the assertion of Lemma 2. In the same way we can prove that the interval for $n+1$ cannot lie between those for 0 and w.
Q.E.D.

Based on the preceding two lemmas we prove the following.

Lemma 4: If $v \in V - V_0$, then $(0, v) \notin F$ and $(n+1, v) \notin F$.

Proof: For definiteness, assume $(0, v) \in F$ and lead to a contradiction. Let $\{I(v)\}_{v \in V'}$ be a realization of interval graph $\hat{H}' = (V', E' \cup F)$. Without loss of generality we can assume from Lemma 3 that intervals $I(0)$, $I(w)$, and $I(n+1)$ lie in this order from left to right. Let $I(v) = [l_v, r_v]$, $I(0) = [l_0, r_0]$, and $I(w) = [l_w, r_w]$. Since $I(v)$ overlaps with $I(0)$ and $I(w)$, there follows

$$l_v \leqslant r_0 < l_w \leqslant r_v.$$

Now modify $I(v)$ from $[l_v, r_v]$ to $[r_0 + \epsilon, r_v]$, where ϵ is a sufficiently small positive number. Then the existing overlap of $I(0)$ with $I(v)$ disappears. On the other hand, all other existing nonempty overlaps remain as they stand. We can show this in the following way. Let $u \neq 0$ be a vertex such that $(u, v) \in E' \cup F$, and assume that the existing overlap of $I(u)$ with $I(v)$ disappears after the modification.
Then,

$$r_u < r_0 + \epsilon$$

for an arbitrary small positive number ϵ, where $I(u) = [l_u, r_u]$. This means

$$r_u \leqslant r_0$$

and then $I(u)$ cannot overlap with $I(w)$. Hence $u = 0$, which is a contradiction.

The argument made above clearly indicates that $F - \{(0, v)\}$ is an augmentation, which contradicts the minimality of F.
Q.E.D.

The assertions made in Lemmas 2 and 4 are combined together to give the following key lemma.

Lemma 5: $(x, y) \in F$ implies $x, y \in V$.

Due to the lemma, $\hat{H} = (V, E \cup F)$ is the section graph of interval graph $\hat{H}' = (V', E' \cup F)$ determined by V. Thus

$\hat{H}$ is an interval graph and hence F is an augmentation with respect to $H = (V, E)$. Next we show that F satisfies the boundary condition.

Let A' be an $(n+3) \times p'$ v.d.c. matrix of $\hat{H}'$ having the consecutive ones property. Due to the assumption (14) and Lemma 5, $V'(t_l) = \{0\} \cup V(t_l)$ is the only one dominant clique, of H', which contains vertex 0. Similarly, $V'(t_r) = \{n+1\} \cup V(t_r)$ is the unique dominant clique, of $\hat{H}'$, which contains vertex $n+1$. It is also clear that all the other dominant cliques contain vertex w. Let $\{I(v)\}_{v \in V'}$ be the canonical realization of A' and let $C'_1, \cdots, C'_{p'}$ be dominant cliques corresponding to the 1st, $\cdots$, p'th columns of A', respectively. Then due to Lemma 3, $I(0) = [1, 1]$, $I(w) = [2, p'-1]$, and $I(n+1) = [p', p']$. Therefore $C'_1 = V'(t_l)$ and $C'_{p'} = V'(t_r)$.

Now we discard the rows corresponding to vertices $0, n+1$, and w. Then there remains an $n \times p'$ matrix A'' having the consecutive ones property. Note that the first column corresponds to $V(t_l)$ and the last to $V(t_r)$. It should also be noted that all the dominant cliques in $\hat{H}$ have corresponding columns in A''. Therefore, by further discarding nondominant cliques and some redundant dominant cliques, we obtain a v.d.c. matrix A of H having the consecutive ones property. Thus the first column of A includes $V(t_l)$ and the last column $V(t_r)$. This implies that F satisfies the boundary condition.

What remains to be shown is the minimality of F. To prove this by contradiction, assume that a proper subset F^* of F is a B augmentation. Let A^* be a $n \times q$ v.d.c. matrix of $H^* = (V, E \cup F^*)$ having the consecutive ones property. Let $\{I(v)\}_{v \in V}$ be its canonical realization. Since $V(t_l)$ is included in the first column, $v \in V(t_l)$ implies $l_v = 1$ where $I(v) = [l_v, r_v]$. Similarly $v \in V(t_r)$ implies $r_v = q$. Then, for every vertex $v \in V(t_l)$, modify $I(v)$ from $[1, r_v]$ to $[0, r_v]$ and, for every vertex $v \in V(t_r)$, from $[l_v, q]$ to $[l_v, q+1]$. This is still a realization of H^*. Then add three intervals $[0, 0]$, $[1, q]$, and $[q+1, q+1]$ for vertices 0, w, and $n+1$. It is easily seen that this is a realization of graph $(V', E' \cup F^*)$ and hence F^* is an augmentation with respect to $H' = (V', E')$. As F^* is a proper subset of a minimal augmentation F with respect to $H' = (V', E')$, this is the desired contradiction. This completes the proof of Theorem 5.

REFERENCES

[1]   A. Weinberger, "Large scale integration of MOS complex logic: A layout method," *IEEE Solid-State Circuits*, vol. SC-2, pp. 182–190, Dec. 1967.

[2]   R. P. Larsen, "Computer-aided preliminary layout design of customized MOS arrays," *IEEE Trans Computer*, vol. C-20, pp. 512–123, May 1971.

[3]   E. Wittengeller, "Computer-aided design of large-scale integrated I^2L logic circuits," *IEEE Solid-State Circuits*, vol. SC-12, pp. 199–204, Apr. 1977.

[4]   S. Goto and E. S. Kuh, "An approach to the two-dimensional placement problem in circuit layout," *IEEE Trans. Circuits Syst.*, vol. CAS-25, pp. 208–214, Apr. 1978.

[5]   H. Yoshizawa, H. Kawanishi, and K. Kani, "A heuristic procedure for ordering MOS arrays," in *Proc. 12th Design Automation Conf.*, 1975, pp. 384–389.

[6]   T. Asano and K. Tanaka, "A gate placement algorithm for one-dimensional arrays," *J. Information Processing*, vol. 1, no. 1, pp. 47–52, 1978.

[7] C. Berge, *Graphs and Hypergraphs*. London: North-Holland, 1970.

[8] D. J. Rose, R. E. Tarjan, and G. S. Lueker, "Algorithmic aspects of vertex elimination on graphs," *SIAM J. Comput.*, vol. 5, no. 2, pp. 266–283, June 1976.

[9] D. R. Fulkerson and O. A. Gross, "Incidence matrices and interval graphs," *Pacific J. Math.*, vol. 15, no. 3, pp. 835–855, 1965.

[10] K. S. Booth and G. S. Lueker, "Testing for the consecutive ones property, interval graphs, and graph planarity using PQ-tree algorithms," *J. Comput. System Sci.*, vol. 13, pp. 335–379, 1976.

[11] T. Ohtsuki, H. Mori, T. Kashiwabara, and T. Fujisawa, "On minimal augmentation of a graph to obtain an interval graph," to be published.

[12] A. Hashimoto and J. Stevens, "Wire routing by optimizing channel assignment within large apertures," *Proc. 8th Design Automation Workshop*, pp. 155–169, 1971.

A Dense Gate Matrix Layout Method for MOS VLSI

ALEXANDER D. LOPEZ, MEMBER, IEEE, AND HUNG-FAI S. LAW, MEMBER, IEEE

Abstract—A rapid and systematic method for performing chip layout of VLSI circuits is described. This method utilizes the configuration of a matrix composed of intersecting rows and columns to provide transistor placement and interconnections. This structure, which is orderly and regular, gives high device-packing density and allows ease of checking for layout errors. Resulting layouts may be updated to new design rules automatically. This method has been used in the layout of a 20 000-transistor section of a VLSI circuit.

I. INTRODUCTION

A VERY DIFFICULT, time consuming, and, therefore, costly task in the design of custom VLSI chips is the transformation of a logic design to final mask artwork. With hundreds of thousands of device features forming transistors and circuit interconnections to be arranged on a chip, there is an ever increasing need to stylize the topology in order to allow a systematic approach to chip layout.

There are a number of considerations which need to be taken into account in the development of a topological style which will best serve the requirements for VLSI circuit layouts. The style should have an ordered structure. This is important from the standpoint of computer-aided design (CAD) which benefits from a simplified arrangement of geometries. The topological style should not sacrifice very much silicon area at the expense of layout ease, since area also enters into the cost of a chip. The layout method should be adaptable to a team effort. Frequently, to expedite the chip layout process, the layout task for a VLSI circuit is divided among several chip designers, each operating on different portions of the chip. Problems often arise when the different portions of the chip are finally combined. Thus the time savings derived from dividing the layout task may be offset by increased chip-area usage caused by complexities of interconnection between the different portions of the chip. Yet another consideration is the ability to update a given VLSI layout to new design rules without having to redo the layout.

This paper introduces a new layout style called Gate Matrix, for CMOS VLSI circuits in the polysilicon-gate technology. The Gate Matrix simplifies and unifies the layout procedure by using an orderly structure, a matrix composed of intersecting rows and columns. The columns are implemented in the polysilicon level and serve the dual role of transistor gates and interconnection. The rows are diffusion and at the intersection with a column form transistors. Extension of this style to single-metal polysilicon-gate NMOS technologies is believed to be straightforward. Gate Matrix can be viewed

Manuscript received March 10, 1980; revised March 26, 1980.
The authors are with Bell Laboratories, Murray Hill, NJ 07974.

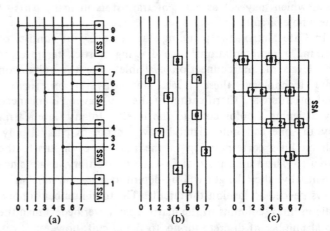

Fig. 1. (a) Illustration of a standard cell configuration. (b) A series of polysilicon lines (matrix columns) which correspond to the inputs and outputs of the circuit with the transistors placed on their gating line. (c) The transistors have been grouped (matrix rows) and the interconnections are made with METAL or diffusion.

as something of a generalization of the Weinberger layout style [1] which was developed for PMOS NAND gates in the metal-gate technology.

The concept of this style will be presented along with a detailed procedure for producing a layout in this style. Next, the implementation of this style through computer aids will be described and finally, results of the layout of a 20 000-transistor section of a VLSI circuit will be discussed.

II. CONCEPT BEHIND THE METHOD

In the well-known standard cell method [2], [3], the layout is composed of two parts, i.e., the logic which is contained within cells that are grouped together and an adjacent wiring channel comprised of two types of isolated conductors, usually polysilicon and metal. The wiring channel serves the purpose of interconnection amongst the logic cells. The concept of Gate Matrix is to transpose the transistors in the logic cells onto the wiring channel. This is done in such a fashion that all transistors having a common input are placed on a common polysilicon line, which is one of the conductors of the wiring channel. This common line now serves a dual purpose, i.e., it is the gate of many transistors which lie on the line and it serves as the common connect among the transistors which have a common input. The polysilicon lines, which are equally spaced and parallel, become the columns of the Gate Matrix. The rows are formed by grouping those transistor diffusions which associate with each other either in serial or parallel fashion. Fig. 1 illustrates the concept graphically.

Reprinted from *IEEE Trans. Electron Devices*, vol. ED-27, pp. 1671–1675, Aug. 1980.

III. THE LAYOUT

The procedure for producing a Gate Matrix layout will be described in two parts. First, we will show how a small circuit (a subcircuit of the whole) is laid out. This will be followed by the approach taken to lay out a larger circuit. As will be seen, for larger circuits it is helpful to plan ahead especially with regard to the column position of polysilicon lines which may act as gates of transistors in many parts of the whole circuit.

In Gate Matrix, the planning stage of a layout consists of making a representational line drawing or stick figure using the levels of interconnection available. For the polysilicon gate technology, these levels are: POLYSILICON, METAL, and DIFFUSION. Clarification of the line drawing is enhanced by assigning a color code to the levels. In any Gate Matrix layout, one can immediately draw a series of parallel polysilicon lines corresponding to the number of inputs to the circuit. These lines may, at first, be of arbitrary length since eventually, their length will be determined by the number of rows needed to fashion the circuit. The total number of polysilicon lines (columns of the matrix) can never be less than the total number of discrete inputs to the circuit; however, it can be more if an output from the circuit is chosen to be polysilicon. Furthermore, the position of a polysilicon output in the matrix is arbitrary and may be chosen for convenience. In fact, it is possible (and often done) that a polysilicon output of a part of the circuit (a column) then becomes the input to a second stage of the circuit (also a column). We should remember here that the polysilicon columns were designed to serve as the residences of transistor sites irrespective of whether the column acts as a connect, as a transistor gate, or both. In the line drawing, a transistor is represented by a rectangle which is drawn on the gating polysilicon column. Subsequent transistor placements will be determined by two factors, i.e., the input column and the serial or parallel association among the transistors. Thus when placing two or more associated transistors on their respective columns, they are drawn on the same row. Once the rows have been defined, further interconnections may be done with METAL or diffusions which run parallel to, and in between, the polysilicon columns. These diffusions are drawn as single straight lines. Contact to either the diffusion or polysilicon is represented by a dot. The METALS which run in the same direction as the rows will have the same pitch as the rows. Fig. 2 shows the representational line drawing for a 1-bit half-adder circuit which illustrates graphically the layout procedure discussed.

As an example of the approach to take for a larger circuit, consider the implementation of a 4-bit half-adder by cascading four of the 1-bit half-adders. The interconnections required for this circuit are shown in Fig. 3. As can be seen from this figure, it would be desirable to have the carry output of the half-adder [column 3 in Fig. 2(b)] as the outermost polysilicon column on the right side of each half-adder. This would eliminate the need of an extra METAL to connect the carry out of the lower significant bit by allowing the sharing, between two bits, of the same polysilicon line. In the first bit, the polysilicon line is the carry and in the second bit the same line is the input A. This also applies to the third and fourth bits. Fig. 4 shows the layout of the 4-bit half-adder which has

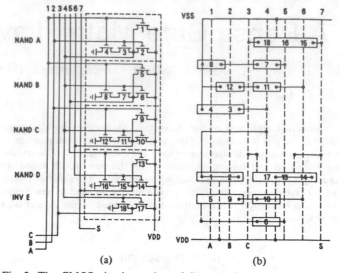

(a)   (b)

Fig. 2. The CMOS circuit consists of four two-input NAND gates and one INVERTER. Transistors of NAND gate A are now placed on columns 1 and 2 with the n channel in series and the p channel in parallel. The output from this NAND gate goes to column 4 which also serves as the gate of INVERTER E (transistors 17 and 18). The series connection of NAND gate D n channel is via the diffusion row because they exist in adjacent columns, but in NAND gate B the series connector is through METAL in order to bridge columns 2 and 3.

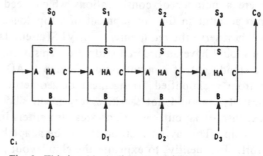

Fig. 3. This is a block diagram of a 4-bit HALF-ADDER.

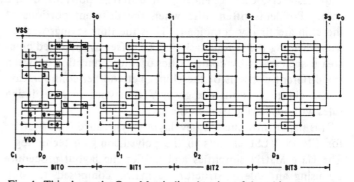
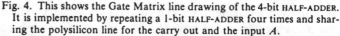

Fig. 4. This shows the Gate Matrix line drawing of the 4-bit HALF-ADDER. It is implemented by repeating a 1-bit HALF-ADDER four times and sharing the polysilicon line for the carry out and the input A.

been modified to take advantage of sharing the polysilicon column. The point to be made here is that in the layout of a large circuit comprised of a number of matrices, one can choose the column position of a particular polysilicon line which best serves the overall layout.

The layout of a matrix contains a rectangle which defines the boundary of the matrix, but is not called for when generating the masks. All inputs and outputs in a matrix must cross the boundary, thus guaranteeing connectivity between the

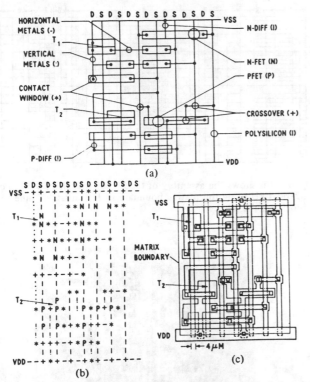

Fig. 5. (a) The representational line drawing. The encircled regions correspond to each of the symbols used to decode this line drawing. Notice the resemblance of the feature and its symbol. The heading labels D or S serve to define polysilicon or diffusion. (b) and (c) are the symbolic description and final artwork of (a), respectively. The matrix boundary is also illustrated in (c) along with a dimensional scale.

matrices. This is illustrated in Fig. 5(c). The boundaries are used in placing matrices next to each other during the assembly of a large circuit.

IV. PHYSICAL CHARACTERISTICS OF THE MATRIX

The Gate Matrix topology is made up entirely of "Manhattan" geometries. This is true for all of the mask levels used in processing. This characteristic not only benefits the fabrication of the device, but also makes it easy to determine circuit parasitics from the mask artwork. The pitch of the rows and columns is a function of the process design rules which are being used. The pitch of the rows is determined by the minimum allowed separation between two discrete transistors (diffusion) without cross coupling. The pitch of the columns is similarly determined by the technology design rules and, specifically, by the room required to accomodate a diffused region with a contact window in it between two polysilicon columns. Thus the matrix pitch, for both its rows and columns, is set up to allow the use of minimum size transistors. In this style, POLYSILICON runners are always the same width, which is equal to the transistor gate length, WINDOWS are all the same size throughout the matrix, and METAL, with the exception of power buses, is also always the same width.

V. MASK ARTWORK GENERATION

A second main aspect of Gate Matrix is the method used to generate the actual mask artwork. We have seen so far that the complete circuit can be described by representational line drawings of the various subcircuits which make up the whole. The task now is to somehow translate the line drawing to final

artwork. The basis for the solution to this problem lies in the use of a symbolic representation that takes full advantage of the regularity in the structure. While we will propose a symbolic method, it should be noted that it differs from the classical symbolic method [4], [5]. Now we will briefly review previously used symbolic layout methods.

Symbolic layouts are constructed by the placement of symbols on a grid which serve to create the topology for a given circuit. The number of symbols used varies according to the technology. Each symbol represents geometries which may include any number of mask levels. The number of symbols is a function of the degree of freedom in the layout. Naturally, the higher the degree of freedom, the larger the number of symbols that is required. In the usual symbolic layout method there is no particular structure or style associated with the layout so that, in essence, it can be viewed as a hand placement of symbols on a coarse grid with the constant aim of conserving silicon area. Of course, the designer is alleviated of the task of having to hand draw the actual mask geometries.

The nature of the Gate Matrix, composed of intersecting columns of polysilicon and rows of diffusion, allows a high degree of simplicity both within the composites of the symbols and also in terms of the number of symbols required to describe the layout. We refer now to Fig. 5 which shows a representational line drawing [see Fig. 5(a)] and its corresponding symbolic description [see Fig. 5(b)] and final artwork [see Fig. 5(c)]. The choice of the symbols used to encode the line drawing is based on a similarity of the symbol to the shape of the geometry it represents. Because of this, unless there is an ambiguity, a single symbol may represent more than one geometry on the line drawing. This is an advantage to the designer since it reduces the number of symbols needed to describe the layout.

Before defining the set of symbols it is first best to summarize the unique characteristic of the gate matrix.

1) Polysilicon runs only in one direction and is constant in width and pitch.

2) Diffusion runners exist between polysilicon columns.

3) Metal runs in both directions and is constant in width except for power buses.

4) Transistors can exist only on the polysilicon columns.

Notice that in Fig. 5 polysilicon columns are labeled S and that in between them the columns are labeled D. These labels will allow a program to distinguish automatically different kinds of contact or crossovers. The following is a description of the set of symbols that is used in Gate Matrix. These symbols are identified also in Fig. 5(a).

N n-channel IGFET,
P p-channel IGFET,
$+$ crossover (metal over diffusion; metal over polysilicon; intersecting vertical and horizontal metals),
$*$ contact (to polysilicon or diffusion),
1 polysilicon or n-diffusion runner,
$!$ p-diffusion runner,
$:$ METAL in vertical direction,
$-$ METAL in horizontal direction.

Each IGFET symbol produces a minimum size transistor. However, the channel width can be increased by including multiples of the symbol. This is illustrated in Fig. 5 transistors $T1$ and $T2$. Only one symbol (+) is required to specify the crossing of all three levels of interconnection; namely, metal over polysilicon, metal over diffusion, and horizontal metal crossing or touching vertical metal. The contact symbol (∗) is used to denote metal contact to polysilicon or to diffusion. The symbol "1" is used to represent either polysilicon or n-diffusion runners. This is automatically distinguished by a computer program which translates the symbolic code to mask artwork. The symbol for a p-type diffusion (!) is required to distinguish it from the n-type diffusion which can exist in the same column. The symbols for metal either (:) or (-) denote vertical or horizontal metal, respectively. These eight symbols are the only ones required to encode any line drawing for a CMOS circuit. The task of encoding the line drawing to its symbolic description is trivial. The computer program which is used to translate the symbolic code to mask (x, y) data looks at the overall symbolic description and recognizes all symbols which should be combined. For example, the continuous set of (OR) symbols, [see Fig. 5(b)] will become a single geometry. In like fashion, transistor diffusions which are symbolically adjacent will be merged into one geometry. Fig. 5(c) shows the final mask artwork generated by the translation program. The input to this program is shown in Fig. 5(b).

Finally, the layout must be checked for logic integrity. The mask artwork is input to a proprietary layout verification program [6] that reconstructs the circuit topology and characterizes active and passive circuit elements. These characterizations are then used by the simulation programs MOTIS [7] and SPICE [8]. As a byproduct, the layout verification program generates a transistor-level-circuit diagram which greatly aids in the confirmation that the circuit produced from the mask artwork is indeed the intended one.

VI. UPDATING DESIGNS TO NEW DESIGN RULES

As technological capabilities evolve, finer design rules become possible. It is desirable to exploit these new design rules in many existing complex designs for lower cost and better performance. It may be desired to reuse existing complex designs either as complete entities on smaller chips, or as modules in newer, more complex designs. The Gate Matrix style lends itself well to accomplishing this in a highly automatic way. This is because mask (x, y) data are generated from the symbolic description directly through the use of a design specification file which contains the process design rules. When the design rules change, new smaller designs can be generated using any previously existing symbolic description automatically after the design specification file has been updated.

VII. APPLICATION TO A VLSI CIRCUIT

So far, the method of Gate Matrix has been used successfully to layout a 20 000-transistor section of a VLSI circuit. This section is part of a 32-bit CPU. Fig. 6 is the assembly of this section showing only the matrix boundaries. The section is composed of nine major blocks. The block corresponding to an ALU (block 8) is shown in the photomicrograph of Fig. 7. Fig. 8 shows a photomicrograph of the entire section. The inclusion of a boundary for each matrix has made it very

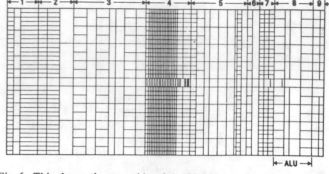

Fig. 6. This shows the assembly of the 20 000-transistor section of the 32-bit CPU using the matrix boundaries. It consists of nine blocks. Block 8 is the ALU.

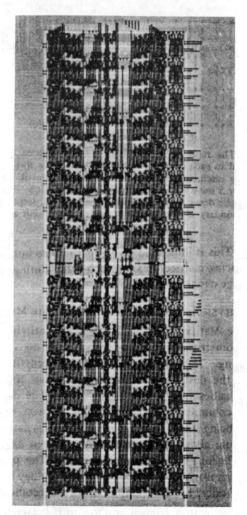

Fig. 7. A photomicrograph of the ALU block.

simple to assemble the block and, finally, to assemble the section by again placing the blocks next to each other. As can be seen in Fig. 6, the boundaries of the matrices on the edges of the block then delineate the block boundaries, which enables us to assemble the section with guaranteed meshing amongst the blocks.

Naturally, the transistor packing density is not the same for each matrix in this 20 000-transistor section. Some circuits tend to pack better than others and, for this reason, it is best to report on the overall packing density for this VLSI circuit. Fig. 6 shows the dimensions in mils for this section. The over-

Fig. 8. A photomicrograph of the 20 000-transistor section of the CPU.

all transistor packing density is 2.0 square mils/transistor. This number is competitive with hand-packed densities. A section of this same VLSI circuit, namely the ALU, had been previously laid out by hand packing before it was decided to pursue a stylized method and, in fact, the hand-packed version was slightly less dense; furthermore, the Gate Matrix version was done in significantly less time.

VIII. Discussion

The experience which we have had with the layout of a 20 000-transistor section of the CPU has shown that the use of a highly stylized topology has a number of advantages that will prove to be valuable in large chip designs. We found, for example, that since the representational line drawing was so easy and fast to do, it allowed a high degree of freedom to optimize the layout for area reduction. The Gate Matrix layout of the VLSI circuit went through two distinct types of modification. The first involved the optimization of the layouts for area reduction by manipulating the line drawings. Based on the efforts of four designers assigned to different parts of the VLSI section, this optimization resulted in a 10-15-percent area reduction. The second type of layout modification involved optimizing for circuit performance from the standpoint of transistor sizes based on results of circuit simulation. This type of modification could be handled easily by operating on the symbolic description of the layout. It was simply a matter of adding more N or P to the already encoded line drawing. On occasion, if an encroachment was encountered, a transistor could be relocated again, using the symbolic description.

IX. Summary

In conclusion, we would like to summarize some of the more important aspects in the paper.

1) The method of Gate Matrix is an approach to stylize the topology of circuit layout, useful in VLSI.

2) The method lends itself to the division of effort within a design team, a trend in VLSI.

3) Transistor-packing density based on a 20 000-transistor layout was 2.0 square mils/transistor and is competitive with the hand-packed layouts.

4) The resulting layout using Gate Matrix is updatable to new design rules with a minimum of effort to update the design file and with no disturbance to the layout.

Acknowledgment

The authors would like to thank the members of their group for exercising and verifying this method. They also wish to thank D. G. Schweikert and H. K. Gummel for their advice and criticism of the manuscript and especially H. K. Gummel and P. A. Swartz for providing them with their circuit recognition and layout verification program. Finally, they wish to express their gratitude to R. H. Krambeck and B. T. Murphy for their encouragement and support.

References

[1] A. Weinberger, "Large-scale integration of MOS complex logic: A layout method," *IEEE J. Solid-State Circuits*, vol. SC-2, pp. 182–190, Dec. 1967.

[2] T. Kozawa, H. Horino, T. Ishiga, J. Sakemi, and S. Sato, "Block and track method for automated layout generation of MOS/LSI arrays," in *Proc. Int. Solid State Circuits Conf.* (Philadelphia, PA, 1972), pp. 62–63.

[3] G. Persky *et al.* "LTX-A system for the directed automatic design of LSI circuits," in *Proc. 13th Design Automation Workshop* (San Francisco, CA, 1976), pp. 399–407.

[4] R. P. Larsen, "Versatile mask generation techniques for custom microelectric devices," in *Proc. 15th Annu. Design Automation Conf.* (Las Vegas, NV, 1978), pp. 193–198.

[5] D. Gibson, and S. Nance, "Symbolic system for circuit layout and checking," in *Proc IEEE ISCAS Conf.* (Phoenix, AZ, 1977).

[6] H. K. Gummel and P. A. Swartz, private communication.

[7] B. R. Chawla, H. K. Gummel, and P. Kozak, "MOTIS–An MOS timing simulator," *IEEE Trans. Circuits Syst.*, vol. CAS-22, pp. 901–910, Dec. 1975.

[8] L. W. Nagel and D. O. Pederson, "Simulation program with integrated circuit emphasis," in *Proc. 16th Midwest Symp. Circuit Theory* (Waterloo, Ont., Canada, Apr. 12, 1973).

Author Index

Subject Index

T. C. Hu received the B.S. degree in engineering from the National Taiwan University in 1953, the M.S. degree from the University of Illinois, Urbana, in 1956, and the Ph.D. degree in applied mathematics from Brown University, Providence, RI, in 1960.

From 1960 to 1966, he was a member of the Mathematical Science Department at T. J. Watson Research Center, Yorktown Heights, NY. From 1966 to 1974, he was a permanent member of the Mathematics Research Center and a Professor of Computer Science Department at the University of Wisconsin, Madison, WI. He has been a Professor of Computer Sciences at the University of California, San Diego, La Jolla, CA, since 1974. He was also a consultant to the Office of Emergency Preparedness, Executive Office of the President 1968–1972. He was Associate Editor of the *Journal of SIAM* and the *Journal of ORSA*.

Dr. Hu is the author of two books: *Integer Programming and Network Flows* in 1969 (translated into German, Russian, and Japanese) and *Combinatorial Algorithms* in 1982, both published by Addison-Wesley. He is the coeditor of *Mathematical Programming* with S. M. Robinson. He has written over 60 technical papers including the paper "Multi-terminal network flows" with R. E. Gomory (awarded the outstanding paper in 1961 by ORSA). His fields of interest include network and graphs, mathematical programming, combinatorial algorithms, and CAD in VLSI. Dr. Hu is a member of ACM, Society of Industrial and Applied Mathematics, and Operations Research Society of America.

Ernest S. Kuh (S'49–A'52–M'57–F'65) was born in Beijing, China, in 1928. He attended the Shanghai Jiao-tong University from 1945 to 1947. He received the B.S. degree from the University of Michigan, Ann Arbor, in 1949, the S.M. degree from the Massachusetts Institute of Technology, Cambridge, in 1950, and the Ph.D. degree from Stanford University, Stanford, CA, in 1952.

From 1952 to 1956, he was a member of the Technical Staff at the Bell Telephone Laboratories, Murray Hill, NJ. He joined the Department of Electrical Engineering at the University of California, Berkeley, in 1956. From 1968 to 1972 he served as Chairman of the Department of Electrical Engineering and Computer Sciences. From 1973 to 1980 he served as Dean of the College of Engineering, University of California, Berkeley. He is author of over 70 technical papers in circuits, systems, electronics and computer-aided design, and is co-author of three books.

Dr. Kuh served as a consultant to the IBM Research Lab, San Jose, CA. He was a member of the Advisory Board of the General Motors Institute, the Scientific Advisory Committee of Mills College, and the Engineering Advisory Board of the National Science Foundation. At present, he serves on the NAS-NRC Committee on Education and Utilization of the Engineer, 1980–2000.

Dr. Kuh is a Fellow of the IEEE and AAAS, a member of the National Academy of Engineering and the Academia Sinica, and Honorary Professor of the Shanghai Jiao-tong University. He has received a number of awards and honors: NSF Senior Postdoctoral Fellow (1962), Miller Research Professor (1965–1966), University of Michigan Distinguished Alumnus Award (1970), IEEE Guillemin–Cauer Award (1973), Alexander von Humboldt Senior Scientist Award (1977), IEEE Education Medal (1981), Lamme Award of the American Society for Engineering Education (1981), British Science and Engineering Research Fellowship (1982), and IEEE Centennial Medal (1984).